WETLANDS OF THE INTERIOR SOUTHEASTERN UNITED STATES

Wetlands of the Interior Southeastern United States

Edited by

C. C. TRETTIN

Oak Ridge National Laboratory, Oak Ridge, TN, USA

W. M. AUST

Virginia Polytechnic Institute, Blacksburg, VA, USA

and

J. WISNIEWSKI

Wisniewski & Associates, Inc., Falls Church, VA, USA

Reprinted from *Water, Air and Soil Pollution* 77(3–4), 1994

Kluwer Academic Publishers

Dordrecht / Boston / London

A C.I.P. Catalogue record for this book is available from the Library of Congress.

ISBN-13: 978-94-011-6581-5 e-ISBN-13: 978-94-011-6579-2
DOI: 10.1007/978-94-011-6579-2

Published by Kluwer Academic Publishers,
P.O. Box 17, 3300 AA Dordrecht, The Netherlands.

Kluwer Academic Publishers incorporates
the publishing programmes of
D. Reidel, Martinus Nijhoff, Dr W. Junk and MTP Press.

Sold and distributed in the U.S.A. and Canada
by Kluwer Academic Publishers,
101 Philip Drive, Norwell, MA 02061, U.S.A.

In all other countries, sold and distributed
by Kluwer Academic Publishers Group,
P.O. Box 322, 3300 AH Dordrecht, The Netherlands.

Printed on acid-free paper

TABLE OF CONTENTS

PART VI
WETLAND RESTORATION AND CREATION

FOREWORD

The early 1990's marked an environmental watershed for our country. Under two federal administrations significant environmental legislative, regulatory and institutional changes took place which affected our Nation's wetland resources. In just a few years, we have seen rapid evolution in the way in which we view wetlands with more emphasis on specific wetland types and the geographic provinces in which they occur. This Southern Appalachian Man and the Biosphere (SAMAB) conference on "Wetland Ecology, Management and Conservation" represents just one example of our desire to understand wetlands in non-coastal regions of the southern United States. The backdrop to this conference was one where the government, universities, and private sector have come together to create a broader and more sophisticated understanding of environmental stewardship for our water resources, especially wetlands.

Although enforcement of environmental legislation by federal and state government agencies - limited by manpower shortages, budgetary constraints and undermined by weak enforcement - remains strong as measured by world standards; the realization that environmental degradation of wetlands is likely to get much worse necessitates a greater commitment and increased resource allocation for wetland protection and management. These continued pressures on the wetland resource will create substantial opportunities for the application of environmentally-sound technologies and interdisciplinary modeling teams to keep abreast of the factors influencing wetland integrity and function in the last half of the 1990's. This conference clearly demonstrates the teaming and modeling opportunities in action. From hydrogeomorphic classification and functional assessment systems to characterization of wetland types and conservation and restoration technologies, SAMAB has initiated this dialog and should continue its efforts to promote such informative discussions in the future.

There is a plethora of "hands on" wetlands meetings covering topics of classification, management, scientific functional studies, replacement, etc. No one conference, workshop, or certification process can cope with the galloping evolution of knowledge relating to our understanding of wetlands. To ensure that our wetlands research and management programs are of the highest quality, it will be up to every stakeholder from this conference to continue the process by providing:

- Research focus and leadership;
- Cross-disciplinary partnerships; and
- Peer review.

The proceedings of this conference should be kept handy on your bookshelf and reviewed often. When sufficient numbers of informational beauty marks become warts, let's vow to get together again. Happy reading!

Russ Lea
Associate Vice Chancellor for Research and
Professor, College of Forest Resources
North Carolina State University
Box 7003
Raleigh, North Carolina 27695-7003 U.S.A.

PREFACE

Wetlands are widely recognized for the their important role in the sustainability of landscape functions and for societal values derived from wetland-dependent processes. The wetland resource in the southern United States is particularly important because it comprises approximately 50% of the total wetland area in the nation. Eight southern states (Alabama, Georgia, Kentucky, North Carolina, South Carolina, Tennessee, Virginia, and West Virginia) contain approximately 7.8 million hectares of wetlands, approximately 21% of the national total (Dahl 1990). Wetlands in those states include many different types ranging from coastal marshes, bottomland swamps, pocosins, riparian zones, and mountain bogs. Most wetland research in the southern United States has focused on the coastal plain region or in the Mississippi delta, encompassing wetland types such as bottomland swamps, pocosins, and flatwoods.

The objective of the Southern Appalachian Man and the Biosphere (SAMAB) conference on "Wetland Ecology, Management, and Conservation," held during 28-30 September 1993 in Knoxville, Tennessee, was to convene a forum that focused on the diverse wetland types that are characteristic of the piedmont plateau, ridge and valley, and mountain regions of the southern United States. These wetland types include bottomland swamps, riparian zones, and mountain bogs and fens. The conference was organized to provide a forum for presentation and discussion of current state-of-the-art research, and for developing recommendations for future research and management programs. Scientists were invited to present papers on selected subjects that were broadly grouped into the following thematic areas: Wetland Ecology, Functions and Values, and Wetland Conservation and Management. A total of 23 invited papers were presented. Working group sessions focused on: Research Needs, Functional Assessment and Restoration, and Wetland Protection and Conservation. This volume consists of 20 peer-reviewed papers that were presented at the Conference, including a summary paper of the major findings reported at the conference and working group sessions.

The SAMAB program is a component of the United Nations Education, Scientific, and Cultural Organization (UNESCO) - Man and the Biosphere Program. SAMAB was created in 1988 to provide solutions to resource management and economic development problems. That function is accomplished through participation and cooperation of eight federal agencies, two state agencies, and non-governmental organizations. The members of SAMAB promote efficient use of the region's resources, increased environmental awareness, environmentally safe economic development, and sharing of scientific research. SAMAB members include: US Forest Service, US Fish and Wildlife Service, US Department of Energy - Oak Ridge National Laboratory, Economic Development Agency, Tennessee Valley Authority, National Park Service, US Environmental Protection Agency, US Geological Survey, the states of Georgia and North Carolina, and the SAMAB Foundation.

Dr. Carl C. Trettin
Co-editor

Dr. W. Michael Aust
Co-editor

Dr. Joe Wisniewski
Co-editor

Reference

Dahl, T.E. 1990. Wetland losses in the United States 1780's to 1980's. US Department of Interior, Fish and Wildlife Service, Washington, DC. 13 p.

ACKNOWLEDGMENTS

We acknowledge the contributions from all authors who prepared papers for the conference and this book, and the reviewers who provided valuable comments and suggestions for improving the manuscripts. We acknowledge the Environmental Sciences Division - Oak Ridge National Laboratory (ORNL), Tennessee Valley Authority (TVA), the US Forest Service (USFS), and the Southern Appalachian Man and the Biosphere program (SAMAB) for their support of this conference. We especially thank the following individuals from those organizations, Drs. Robert I. Van Hook and Stephen G. Hildebrand (ORNL) for approving and facilitating development of the conference, Mr. George Martin (USFS), for his encouragement and support, Mr. Hubert Hinote (SAMAB) for his enthusiastic support, and Ms. Janet Herrin (TVA) for supporting the conference and the excellent banquet speech. Special gratitude is extended to Ms. Pat Presley (ORNL), who served as conference manager, and provided impeccable management and organization which was instrumental to achieving a successful conference. The moderators of the conference sessions, Dr. Kevin Moorhead, University of North Carolina - Asheville, Dr. William McKee, Jr., US Forest Service, Dr. Gerry Edwards, TVA, and Dr. Ted Shear, North Carolina State University, did an excellent job keeping the program on schedule and facilitating discussion. Support for the Conference banquet was provided by the Reservoir Operations, Planning, and Development Division, Tennessee Valley Authority. Members of the Conference Planning Committee included Ralph Jordon (TVA), Wes James (TVA), Burline Pullin (TVA), Allen Ratzlaff (US Fish and Wildlife Service), Babmi Teague (National Park Service), Gerry Edwards (TVA) and Linda Mann (ORNL); their contributions to the program are gratefully acknowledged. We acknowledge Dr. Billy McCormac, Editor-in-Chief of Water, Air and Soil Pollution, Dee McCormac, the copy editor, for the technical guidance necessary to produce the book, and Kathy Barnes (ORNL) for the cover graphics.

Dr. Carl C. Trettin
Co-editor

Dr. W. Michael Aust
Co-editor

Dr. Joe Wisniewski
Co-editor

ACKNOWLEDGMENTS

PART I

CONFERENCE SUMMARY STATEMENT

WETLANDS OF THE INTERIOR SOUTHEASTERN UNITED STATES: CONFERENCE SUMMARY STATEMENT

C. C. TRETTIN[1], W. M. AUST[2], M. M. DAVIS[3], A. S. WEAKLEY[4], and J. WISNIEWSKI[5]

[1]Environmental Sciences Division, Oak Ridge National Laboratory, P.O. Box 2008, Oak Ridge, TN 37831; Present Address: Center for Forested Wetlands Research, 2730 Savannah Highway, Charleston, SC 29414

[2]Department of Forestry, Virginia Tech. University, Blacksburg, VA 24061

[3]Ecological Research Division, U.S. Army Corps of Engineers Waterways Experiment Station, 3909 Halls Ferry Rd, Vicksburg, MS 39180

[4]North Carolina Natural Heritage Program, P.O. Box 27687, Raleigh, NC 27611

[5]Wisniewski and Assoc. Inc., 6862 McLean Province Circle, Falls Church, VA 22043

Abstract. The wetland resources in the southern United States are diverse, being characterized by bottomland hardwoods, forested riparian zones, swamps, marshes, bogs, and fens. Recognizing the importance of the wetland resource, the need to develop information on the diversity of wetland types, and the evolving public debate regarding the protection and management of wetlands, this conference was organized to (1) provide a forum for the presentation and discussion of current research and information on wetland ecosystems, (2) to develop a basis on which to improve wetland conservation and management, (3) to provide a forum to encourage collaboration in the study and management of wetland resources, and (4) to suggest actions that would benefit wetland conservation and management. Twenty-three invited technical papers and three working group sessions addressed topics encompassing the full spectrum of wetland issues, including resource status, ecological and hydrological processes, management and conservation, and restoration and creation. Major findings discussed at the conference included the need to distinguish between functions and values, the development of a hydrogeomorphic classification system for assessing wetland functions, and assessment methodologies for planning and implementing effective wetland restoration projects. Papers summarizing the current understanding of wetland soil and vegetation processes in the region highlighted the important role wetlands play in landscape function, yet the understanding of those processes is incomplete. Insights developed from the study of wetlands in the interior southeastern United States have widespread applicability to other regions because of similarities in hydrogeomorphic setting and vegetation communities, and the management and conservation issues.

1. Motivation for the Conference

The wetland resources in the southern United States are diverse, being characterized by bottomland hardwoods, forested riparian zones, swamps, marshes, bogs, and fens. There is no reliable estimate of the area of these individual wetland types; however, the current estimate of the total wetland area demonstrates that the resource is large and that the cumulative wetland loss has been significant (Table I). Wetlands in the these southern states comprise approximately 24% of the wetland resource in the U.S (Cubbage and Flather, 1993). Although most wetlands in the southern states are located in the coastal plain, wetlands in the piedmont, ridge and valley, and mountain physiographic provinces are an important component of the landscape. Recognizing the importance of the wetland resource, the need to develop information on the diversity of wetland types, and the evolving public debate regarding the protection and management of wetlands, this conference was organized to (1) provide a forum for the presentation and discussion of current research and information on wetland ecosystems, (2) to develop a basis on which to improve wetland conservation and management, (3) to provide a forum to encourage collaboration in the study and management of wetland resources, and (4) to suggest actions that would benefit wetland conservation and management.

Water, Air and Soil Pollution 77: 199–205, 1994.

Table I. Estimates of the area of wetland loss during the last 200 years in the south-central U.S. (from Dahl, 1990).

State	Wetland Area -1780's X 10^6 ha	Wetland Area -1980's X 10^6 ha	Proportion of Wetlands Destroyed (%)
Alabama	3.0	1.5	50
Georgia	2.7	2.1	23
Mississippi	4.0	1.7	59
North Carolina	4.5	2.3	49
South Carolina	2.6	1.8	27
Tennessee	0.8	0.3	59
Virginia	0.7	0.4	42
West Virginia	0.04	0.04	24

This conference was organized by the Southern Appalachian Man and the Biosphere Program (SAMAB) in two parts. The first was the presentation of invited technical papers. These papers were selected to provide a current assessment of wetland functions, wetland regulation and assessment, management effects, and wetland restoration and creation. The second part of the conference consisted of three working group sessions (1-Research and Information Needs; 2 -Wetland Functional Assessment and Restoration; and 3-Wetland Protection and Conservation) which were designed to summarize the topic and provide recommendations to scientists, resource managers, and regulators.

This paper provides a summary of the major findings and research recommendations presented at the technical session of the conference and the working group sessions.

2. Conference Summary

The following discussion summarizes important findings reported at the technical sessions of the conference.

2.1 WETLAND FUNCTIONS AND VALUES

• Distinctions must be made between wetland functions and values, they are not synonymous terms. Wetland functions are derived from inherent ecosystem processes. Five basic functions can be recognized: hydrology, productivity, biogeochemistry, decomposition, and community dynamics. Each comprise biotic and abiotic processes that affect the structure, composition and dynamics of the wetland ecosystem. While these basic functions are common to all wetlands, they will be expressed differently among wetland types. Value is an anthropocentric interpretation of the quality or importance of an ecosystem function or process. Examples of values ascribed to wetlands include: hunting, fishing, timber production, assimilation of nutrients in waste water or runoff, and flood control, to name a few. (see Richardson, 1994).

• A hydrogeomorphic classification system provides the basis for functional classification of wetlands. Because wetlands require saturated soil to sustain anaerobic conditions, a hydrogeomorphic system provides the means to incorporate geomorphic, hydrologic, and edaphic information into a classification system which reflects properties or processes that affect wetland functions and values (see Brinson, 1993).

• Wetlands are disproportionally important in providing landscape diversity, maintaining biological diversity, and providing refugia for threatened and endangered species. Management and conservation of these wetland ecosystems should involve landscape-level analyses that consider interactions between the upland and aquatic communities at different spatial and temporal scales. Conservation of individual wetlands is largely ineffective without an understanding of the landscape functions of the ecosystem. (see Pearson, and Weakley and Schafale, this volume).

• Much more research is needed to understand the relationships between biogeochemical and hydrological processes, and how those processes affect wetland vegetation dynamics. (see Walbridge, and Weakley and Schafale this volume).

• Wetlands are important habitat for endangered and threatened species, and maintenance of wetland quality is necessary to sustain the refugia. Habitat for some species requires continual maintenance, which is probably impractical in perpetuity. Improved understanding of hydrologic and biogeochemical affects on habitat are needed. (see Murdock, this volume).

2.2 WETLAND MANAGEMENT

• A functional assessment system is needed to assess impacts from management and conservation efforts, or consequences of wetland loss. An assessment approach using a hydrogeomorphic classification and regional reference data on wetland functions offers the opportunity to assess functions before and after planned activities in the wetland. (see Brinson, 1993).

• Disturbance either within the wetland or in adjoining uplands may adversely affect wildlife habitat quality, vegetation structure, and vegetation composition. (see Wigley and Roberts, and Weakley and Schafale, this volume).

• Riparian zones are important areas for improving runoff water quality. (see Hubbard and Lowrance, this volume).

• Best Management Practices (BMPs) have been developed for all states in the Southeastern United States. However, the specific content of the BMPs and their implementation varies by state. No data is available for comparing the effectiveness of different systems, although evidence suggests that there is no difference between mandated and voluntary BMPs. (see Aust, this volume).

• Regulatory contexts are evolving to effect wetland conservation through training and application of statutes. (see Ainslie, and Wakeley this volume).

2.3 WETLAND ASSESSMENT AND RESTORATION

• Functional mitigation of unavoidable loss of wetlands is possible. Newly established assessment methodologies provide an effective basis for planning, designing, and implementing wetland restoration and creation projects. The three critical components to successful wetland restoration and creation are siting, design criteria, plan development and implementation. (see Davis, Bartoldus, and McCuskey et al., this volume).

3. WORKING GROUP SUMMARIES AND RECOMMENDATIONS

The working groups were organized to summarize the current status of the topic and to develop recommendations to further scientific advancement, technology transfer, and policy and program development. Panelists delivered their perspectives on the topic, and that was followed by a facilitated discussion session among all working group participants. The following discussion provides a summary of the discussions and recommendations developed by each working group.

3.1 RESEARCH AND INFORMATION NEEDS (Working group panelists: Mike Aust, Bill McKee, Jr., Dan Smith).

Studies on the ecology and management of wetland ecosystems have occurred predominately in the last 10 years. Despite that relatively short period, considerable information has been developed documenting hydrological processes, biogeochemical cycles, and vegetation dynamics. Because of the diversity of wetland types, geomorphic setting, and management regimes, available information is not adequate to understand interrelated biotic and abiotic wetland ecosystem processes. The following were research and information needs identified by the working group participants:

• Characterization (hydrology, soils, vegetation communities) of wetland types, particularly forested riparian zones and mountain bogs.

• Processes affecting wetland functions are poorly understood and require further research. This is particularly necessary to support the functional assessment of wetlands.

• Establish common research sites throughout the south-central US in which hydrology, soil and vegetation processes could be studied using a common protocol.

• Landscape functions of wetlands have not been adequately considered, particularly with respect to landscape diversity, habitat, biological diversity.

These research needs could be facilitated by (a) increased emphasis on cooperative, inter-disciplinary research, (b) overcoming disciplinary boundaries or jurisdictions among agencies to enhance opportunities for cooperation in research and technology transfer, and (c) utilization of applied journals and publications to disseminate information.

The working group participants further recommended that:

• SAMAB serve as a clearing house of information on wetland resources, research, and funding opportunities. Such a function may include an Information Center, a Data Management Center, or electronic bulletin board.

• SAMAB serve as a catalyst for jointly funded, cooperative research among states, federal agencies, and universities. SAMAB is uniquely positioned to provide a synergetic basis for helping to secure research funds and to encourage cooperation among participants.

• SAMAB and its members work to encourage the expansion of the mission of the National Wetland Inventory to include the study of functional processes, and to explore opportunities for the National Biological Survey to serve a clearing house of wetland information and data that would be available for regional assessments or study.

3.2 WETLAND FUNCTIONAL ASSESSMENT AND RESTORATION (Working group panelists: Mary Davis, Clif Amundesen, Edward Houser, Sue McCuskey, and Tom Roberts).

The integrity, functions, and values of wetlands of the southern Appalachian region are impacted directly and indirectly by housing development, golf courses, roads, agriculture, mining, and other activities. Although the ecology of many of these wetlands is poorly understood, it is evident that their capacities to provide wildlife habitat, water quality improvement, and other valuable functions are being severely impacted. As the remaining wetland areas continue to be impacted, it is increasingly important that we are able to 1) evaluate what functions and values are being lost, 2) assess whether there has been functional replacement of wetlands in restoration/creation projects in the area, and if not, 3) develop an approach to achieve functional mitigation of the losses. The findings of the working group session were the following:

• Functional assessment methods have been developed for wetlands and modified for regional uses. Although North Carolina and Tennessee have assessment methods, no consistent method exist for the south-central U.S.

• Functional assessment methods often lack sensitivity to degree of pristine condition (i.e., a highly impacted wetland can be rated similarly to an intact, mature system).

• Actual functional assessments are commonly based on "Best Professional Judgment." There is little quality control on or consistent quantification of the functional capacities of wetlands.

• There is inadequate understanding of the roles of decomposer-producer-consumer processes in wetland functions, and simple assessment methods for estimating levels those processes are needed.

• Available information indicates that bottomland hardwood forests are the most commonly mitigated wetland type in the south-central United States.

• Information about the distribution and characterization of natural and restored/created wetlands in the Southern Appalachians is necessary to evaluate impacts on wetlands.

• Little information is available to determine whether functional wetland replacement is being achieved. Hydrology of a restored wetland is the most critical factor to establish and the most difficult to obtain.

• Technical information is needed on methods, equipment, and materials to restore and create the hydrology, sediments, and vegetation in wetland projects.

• Information about wetland restoration is difficult to access. Exchange of such information usually occurs at meetings and through personal contacts.

• Limited data bases exist on wetland mitigation, some of which probably pertain to wetlands in the south central U.S. A repository of wetland restoration/creation information would be a valuable aid for future wetland projects in the SAMAB and other regions.

The working group participants recommended the following:

• Hydrogeomorphologic-based functional assessment models need to be developed for wetlands that are sensitive to degree of impact and that are relatively easy to apply.

• The functional assessment models should be sensitive to wetland producer-decomposer-consumer processes.

• A systematic method to access data about wetland restoration/creation projects and natural wetlands is essential.

• Wetland protection varies within the SAMAB region due to differences in mitigation requirements and it should be made more consistent.

• Establish a data base for regional exchange of wetland information. The format should be a computerized data base that is easily accessible to a wide range of users. Information should be acquired from existing data bases, agencies, consultants, engineers, and others.

3.3 WETLAND PROTECTION AND CONSERVATION (Working group panelists: Alan Weakley, Dan Pittillo, Bob Johnson).

Major concerns expressed about the protection and conservation of wetland resources included adequacy of wetland inventory data, especially those with special or sensitive resources, land ownership, and capabilities for protection. To further develop a regional perspective on this subject, because not all states were represented at the workshop, a post-conference survey of six state agencies was conducted. Results from that survey indicate that while some wetland inventory information is available, the adequacy of data on sensitive wetland resources is incomplete (Table II). Regarding wetland protection, the survey response indicated the capability of protection is greatest if the land is held in public ownership as compared to private. However, some states (South Carolina, Tennessee, and Virginia) indicated that funds could be made available if additional matching moneys were organized.

Another major problem that was addressed was the historic loss and degradation of mountain fens and bogs. These are unique wetland communities in mountain regions that exhibit some similarities to northern wetlands. These wetlands provide important habitat for many sensitive plants and animals. Conservation of these wetlands is imperative, yet a vexing problem due to management and land ownership constraints. Other wetland types are also important and have not been adequately protected. Headwater riparian systems are common, however their importance to the maintenance of the hydrologic regime and water quality has not been recognized.

Recommendations from the working group were:

• Develop a detailed inventory and qualitative assessment of mountain bogs and fens as a basis for evolving conservation priorities.

• The current inventory of wetlands is inadequate for assessing protection strategies or assessing the distribution of wetland types. SAMAB is encouraged to take leadership the role in developing a regional wetland classification system, and states should take leadership in wetlands inventory.

• SAMAB should continue efforts to develop educational materials and programs for the public and policy-makers about the importance of wetland resources to ecosystem health and landscape processes.

• SAMAB should facilitate research focused on understanding ecosystem processes as a basis for improved wetland conservation and management.

Table II. Results from a survey of state agencies responsible for wetland protection in Alabama (AL), Georgia (GA), North Carolina (NC), South Carolina (SC), Tennessee (TN), and Virginia (VA) to determine status of wetlands inventories, unique ecological areas, and wetland conservation and protection programs. (Survey conducted by J. D. Pittillo and A. S. Weakley, unpublished data 1994).

	AL	GA	NC	SC	TN	VA
To what degree have wetland communities been inventoried?	very limited	very limited	extensive	limited	limited	moderate
Probability of locating new, ecologically significant wetlands?	very high	very high	moderate	high	low for large sites; high for sites < 1 ha	very high
What is the degree of protection for public lands?	moderate	moderate	excellent	very good	very good	moderate
What is the degree of protection for private lands?	low	low	very low	low	low	very low

ACKNOWLEDGMENTS

We would like to thank all the participants of the conference and working group sessions. Their participation was crucial to the engaging and informative discussions that made for a successful conference. Special thanks is extended to Dan Pittillo for his work on the survey of state agencies. Oak Ridge National Laboratory is managed by Martin Marietta Energy Systems, Inc., under contract DE-AC05-LR21400 with the U.S. Department of Energy.

REFERENCES

Ainslie, W. B., this volume.

Aust, W. M., this volume.

Bartoldus, C. C., this volume.

Brinson, M. M. 1993. A hydrogeomorphic classification for wetlands. Tech. Rep. WRP-DE-4, U.S. Army Corps. of Engineers, Washington, D.C. 101 p.

Cubbage, F. W., and C. H. Flather. 1993. Forested wetland area and distribution. *J. For.* 91:35-40.

Dahl, T. E. 1990. Wetland losses in the United States 1780's to 1980's. U.S. Dept. of Interior, Fish and Wildlife Service, Washington, D.C. 13 p.

Davis, M. M., this volume.

Hubbard, R. K. and R. R. Lowrance, this volume.

McCuskey, S.A., A.W. Conger, and H.O. Hillestad, and this volume.

Murdock, N. A., this volume.

Pearson, S. M., this volume.

Richardson, C. J. 1994. Ecological functions and human values in wetlands: a framework for assessing forestry impacts. *Wetlands* 14:1-9.

Wakeley, J. S., this volume.

Walbridge, M. R., this volume

Weakley, A. S. and M. P. Schafale, this volume.

Wigley, T. B. and T. H. Roberts, this volume.

PART II

WETLAND RESOURCES

CLASSIFICATION AND INVENTORY OF WETLANDS IN THE SOUTHERN APPALACHIAN REGION

John M. Hefner and Charles G. Storrs
Fish and Wildlife Service
U.S. Department of the Interior
Atlanta, Georgia 30345

ABSTRACT

The National Wetlands Inventory of the U.S. Fish and Wildlife Service has prepared large scale (1:24,000) wetland maps for nearly all of the Southern Appalachian Region. Traditional and digital cartographic products are available from the Earth Science Information Centers of the United States Geological Survey and from State-run distribution outlets. Most of the materials prepared by the NWI within the region were cooperatively funded by the States and other Federal Agencies.

NWI maps describe wetlands in terms of the life form of the dominant vegetation, substrata where vegetation is sparse or lacking, water chemistry, relative duration of inundation or saturation, and special modifiers. The maps display wetland polygons as small as 0.5 hectares in size and linear wetlands as narrow as 8 meters, showing the size, type of wetland, and relative position of the wetland on the landscape. The wetland inventory process is principally a remote sensing task, relying on the interpretation of high altitude color infrared aerial photography, supported with ground truth data and collateral information. The procedure has limitations related to scale, quality, and timing of the aerial photography; experience and training of the photo interpreters; and the wetland types which are to be classified and delineated. Since wetland maps provide a static depiction of a dynamic resource, the NWI conducts periodic wetland status and trends studies to evaluate wetland change in areal extent and the reasons for the change. Although trend surveys are routinely conducted nationally and selectively for regional and local areas, no study to specifically address the wetlands of the Southern Appalachian Region has been developed.

1. INTRODUCTION

The National Wetlands Inventory (NWI) of the U.S. Fish and Wildlife Service (Service) has been mapping and classifying wetlands, and analyzing wetland trends since the late 1970's. The information collected and disseminated by the NWI is intended as a tool to foster wise management of wetland resources.

The NWI is the fourth wetland inventory carried out by the Federal Government. The first two inventories, conducted in 1906 and 1922 by the Department of Agriculture, were intended to identify lands that could be improved by drainage and converted to productive croplands. The Service's previous wetland inventory was conducted in 1954 to identify important wetland habitat for wildlife, especially waterfowl. In the southern United States, the Service concentrated inventory efforts in the most important waterfowl wintering habitats of the Gulf of Mexico and Atlantic Coastal Plains and the Lower Mississippi Alluvial Valley. Consequently the wetlands of the Southern Appalachians were virtually ignored. Nevertheless, the release of the findings in Wetlands of the United States, usually referred to as Circular 39 (Shaw and Fredine, 1956), marked a major turning point in wetland conservation.

Since that survey, wetlands have undergone many alterations, both natural and man-induced. The recognition of these changes, coupled with our increased understanding of wetland values, led the Service to establish the NWI. During its 17 year history, the NWI has developed a variety of cartographic and narrative products. The project's principal focus has been the preparation of detailed large-scale wetland maps and periodic reports of the status and trends of the nation's wetlands. Wetland maps are in wide use for impact assessment of site-specific projects including facility and corridor siting, oil spill contingency plans, natural resource inventories, habitat surveys and other studies. National estimates of the current status and trends (i.e., losses and gains) of wetlands have been used to evaluate the effectiveness of existing Federal wetland programs and policies, and to identify national or regional problem areas. The first status and trends study by the NWI (Frayer et al. 1983) increased public awareness of wetlands and was instrumental in stimulating important wetland legislation, including the Emergency Wetlands Resources Act of 1986 (P.L. 99-645).

Water, Air and Soil Pollution 77: 209–216, 1994.

2. WETLAND CLASSIFICATION

At the inception of the NWI, a variety of regional wetlands classification schemes were in use. However, no single classification fully met the needs of a nationwide project. Therefore, a new classification system (Cowardin et al. 1979) was developed by a team of wetland ecologists, with the assistance of local, State, and Federal agencies, as well as many private groups and individuals. After extensive field testing and four major revisions, the classification was officially adopted by the Service in 1980.

The Service's wetland classification defines wetlands in the following manner: "Wetlands are lands transitional between terrestrial and aquatic systems where the water table is usually at or near the surface or the land is covered by shallow water. For purposes of this classification, wetlands must have one or more of the following three attributes: (1) at least periodically, the land supports predominantly hydrophytes, (2) the substrate is predominantly undrained hydric soil, and (3) the substrate is nonsoil and is saturated with water or covered by shallow water at some time during the growing season of each year" (Cowardin et al. 1979). This definition predates recent Federal efforts to define jurisdictional limits of wetlands under the purview of Section 404 of the Clean Water Act and is more comprehensive in scope. Lists of wetland plants (Reed, 1988), and hydric soils (U.S.D.A. Soil Conservation Service, 1991), have been developed in support of the Service's definition and have become integral to the Federal methodology for identifying jurisdictional wetlands.

The classification is hierarchical. At the most general level, wetlands and deepwater habitats are separated into five systems - Marine, Estuarine, Riverine, Lacustrine, and Palustrine. Each system groups wetlands and deepwater habitats according to hydrologic, geomorphologic, chemical and biological similarities. In the southern Appalachians, most wetlands are associated with the Palustrine system. Small acreages of wetlands are associated with the Riverine and Lacustrine systems, although these systems principally include deepwater habitats.

At the next level of the hierarchy, subsystems subdivide the systems on the basis of water depth and other hydrologic characteristics. At the taxonomic level below the subsystems are the classes, followed by subclasses. The 11 classes are based on either vegetative life form or substrate and flooding regime. Classes describing vegetated wetlands include Aquatic Bed, Moss-Lichen Wetland, Emergent Wetland, Scrub-Shrub Wetland, and Forested Wetland. Classes describing nonvegetated wetlands include Rock Bottom, Unconsolidated Bottom, Unconsolidated Shore, Rocky Shore, Streambed, and Reef. Subclasses provide additional life form detail (e.g. needle-leaved evergreen), or substrate information (e.g. sand). The classes and subclasses are easily recognized and can normally be identified by using remote sensing techniques.

At the most precise and detailed level of the classification are dominance types. These are named for the dominant plant species in vegetated wetlands or the predominant sedentary or sessile macroinvertebrate species in nonvegetated wetlands. At this point, the classification is open-ended and dominance types can be identified and named as required. For example, a western North Carolina bog vegetated primarily by rhododendron would be classified: SYSTEM: Palustrine; SUBSYSTEM: none; CLASS: Scrub-Shrub; SUBCLASS: Broad-leaved Evergreen; DOMINANCE TYPE: Rhododendron maximum.

To complete the wetland description, the classification includes modifiers which describe hydrology, water chemistry, soil type, and the impact of beavers or man. Modifiers can be applied at the class, subclass, and dominance type levels. The "saturated" water regime modifier is added to the bog classification example to indicate that the water table is at the surface of the substrate for much of the growing season.

3. WETLAND MAPPING

Due to the magnitude of this effort, wetland mapping by the NWI is primarily a remote sensing task. High altitude aerial photography is the basic information source. Since 1980, the NWI has regularly utilized 1:58,000 scale color infrared photography acquired for the U.S. Geological Survey's National High Altitude Photography Program. The use of satellite images is periodically investigated by the NWI and others and may eventually prove useful for monitoring wetland changes, updating NWI maps, and for producing maps in unmapped areas (Wilen and Pywell, 1992). At present, aerial photographs are the preferred tool for wetland mapping (Federal Geographic Data Committee, 1992).

The interpretation of the aerial photographs is performed by skilled photointerpreters following detailed guidance (conventions) developed by the NWI (U.S. Fish and Wildlife Service, 1990). The interpreters look at stereo-paired photographs through 4 to 6 power mirror stereoscopes. Viewing the images in stereo provides a 3 dimensional image that enables the interpreters to distinguish vegetation heights and to discern topographic relief to facilitate delineation of wetlands. Colors, textures, tones, and topographic position are among the characteristics of wetland signatures recognized by the interpreters. The delineations are made using 4X0 to 6X0 pen tips in waterproof black ink on clear stabalene overlays attached directly to the photographs. Field-checks and quality control reviews are conducted at specific intervals throughout the interpretation process. Tiner (1990) lists eleven steps performed in every NWI mapping project. Careful attention is paid to collateral information, especially county soil surveys and topographic maps.

When delineations are complete and have received a satisfactory review by NWI project personnel, the linework and classifications are transferred from the 1:58,000-scale aerial photographs to 1:24,000 scale base maps using zoom transfer scopes. Wetland delineations are superimposed over and composited with the corresponding topographic quadrangle (Figure 1.). The composited maps are distributed to a variety of Federal and State agencies for review and field checking. Editorial comments are compiled, maps are corrected, and final maps are prepared. The entire process takes 2 to 3 years from photo acquisition to final map production.

The process of preparing wetland maps through the interpretation of high altitude aerial photography has inherent limitations related to 1) the skill and experience of the photointerpreters, 2) the scale, quality, and acquisition date of the photography, and 3) the specific types of wetlands being classified and delineated. The NWI strives to train and employ individuals with a special aptitude for photointerpretation and an interest in wetland ecology. Remote sensing specialists and biologists from the Tennessee Valley Authority assisted the NWI with wetland photointerpretations for large areas in Tennessee, Kentucky, Alabama, and Georgia. However, most of the photointerpretations along with the cartographic tasks required in map preparation have been provided by a service support contractor associated with the NWI Central Control Group in St. Petersburg, Florida. The photointerpreters have degrees in the natural or biological sciences and are specifically trained in wetland classification and delineation. They perform wetland photointerpretation exclusively on a full time basis thereby continually maintaining and improving their skill. Service personnel from the Central Control Group and Regional Offices accompany the photointer- preters on ground-truthing field trips and provide feedback on the delineations, further fine-tuning the contractor's abilities.

Even though high altitude aerial photography has proven to be a useful and cost effective remote sensing tool for mapping wetlands, it is not without limitations. The scale of the photography determines the size of wetlands which can be delineated. Color infrared photography at a scale of 1:58,000 permits the NWI to delineate wetlands as small as 0.5 hectares and linear wetlands as narrow as 8 meters. Color infrared photography of optimum quality can record thousands of colors, shades, hues, and textures which can be interpreted as wetlands or other land cover. However, color infrared film is sensitive and requires careful handling, processing, and duplication. Photographs which are darker, lighter, bluer, or redder than normal can obscure standard wetland signatures. Poor quality imagery necessitates increased field checking and greater reliance on collateral information.

The aerial photography utilized by the NWI is usually taken during the period from late fall to early spring when deciduous trees are without leaves. This permits the interpreters to see beneath the forest canopy. Observations of moist to flooded substrates provide visual clues to the extent of wetlands. However, photographs inadvertently taken during periods of extreme drought or unusual flooding can be misleading. Compensation is again achieved by additional field checking and reliance on collateral information.

Some types of wetlands are inherently difficult to recognize and delineate regardless of the technology selected. Wetlands that are flooded or saturated for relatively short periods, wetlands that are vegetated by species common to the adjacent upland, and wetlands that have had alterations to the hydrology are especially difficult to photointerpret. For example, floodplain and riparian wetland habitats associated with small watersheds flood regularly but briefly during periods of heavy precipitation. Because flooding is of short duration, the probability is high that the photography will be obtained when the wetlands show no sign of flooding or saturation. Furthermore, floodplain and riparian wetlands are usually vegetated by species common to the surrounding uplands. For instance, the alluvial wetland forests of western North Carolina are vegetated by species with wide ecological tolerances, such as Tsuga canadensis, Liriodendron tulipifera, Platanus

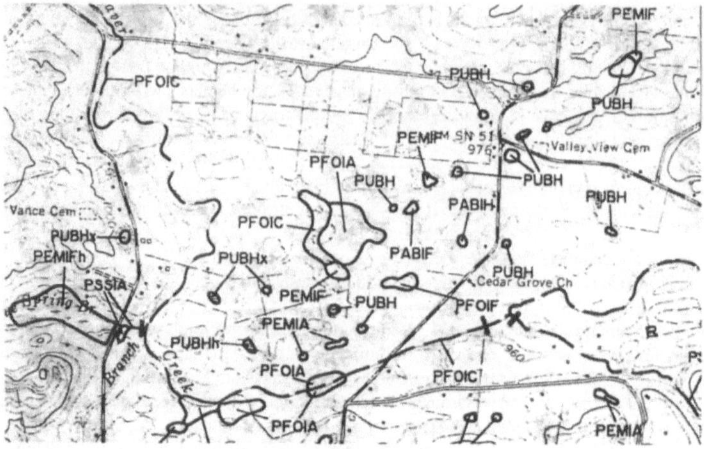

Fig. 1. West-central section of
the New Market, Tennessee, National
Wetlands Inventory Map (1:24,000 scale).

occidentalis, Betula alleghaniensis, Quercus alba, and Acer rubrum (Schafale and Weakley 1990). Even the edges of mountain bogs can be dominated by broadly tolerant tree species, such as Acer rubrum, T. canadensis, Pinus strobus, P. rigida and Picea rubens, which obscure the wetland perimeter. In cases when hydrologic indicators are not visible or when the wetlands grade nearly imperceptibly into upland habitats, the reliance on the aerial photography is augmented by the use of collateral information, intensified fieldwork, use of topographic maps, and whenever possible, the opinions of local experts.

Among the most difficult wetland delineation problems is the accurate determination of the extent of wetlands in areas where natural riparian overbank flooding has been controlled or altered as the result of channel modifications or run-of-the-river dams. These conditions are prevalent throughout the Appalachians and adjoining piedmont. Intensive site specific hydrologic studies beyond the scope of the NWI are required to accurately map wetlands in these situations. Therefore, the NWI should be used as only an approximation of wetland locations when extensive hydrologic modifications have occurred.

The employment of skilled full-time photointerpreters dedicated exclusively to wetland delineation assisted with collateral information and ground-truth data normally results in the preparation of good quality draft wetland maps. Constant oversight by Service biologists and reviewers from outside agencies, enable the NWI to produce high quality finished map products. Studies in New England by Swartout et al. (1981) and Crowley et al. (1988) determined the maps to be extremely accurate for their study areas. However, these results reveal little regarding the quality of the NWI elsewhere. A subjective gauge of NWI map reliability and utility is the willingness of outside agencies to sponsor NWI mapping activities. In the Southern Appalachian Region, nearly all work by the NWI has been accomplished cooperatively. Major contributions of State funds enabled Tennessee and Kentucky to be among the first States in the Nation to be mapped by the NWI. The Army Corps of Engineers provided cooperative funding for mapping major portions of Alabama, Georgia, South Carolina, and North Carolina. The State of South Carolina is currently providing funding to the NWI to prepare digital maps which include not only wetland classification and delineation but also upland land cover and land use information. West Virginia, Virginia, and North Carolina have sponsored the conversion of NWI maps from traditional cartographic products to digital data. Since 1983, North Carolina has had a formal agreement with the NWI to supply logistical support and local expertise for ground-truth acquisitions, as well as provide editorial review of draft maps.

4. MAPPING STATUS AND AVAILABILITY

The NWI had prepared maps, draft or final, for nearly 90% of the Southern Appalachian Region by early 1994. Final wetlands inventory maps are available for all of West Virginia, Virginia, Kentucky, and Tennessee. Nearly all of Georgia has been mapped, although most NWI maps are in the draft stage of preparation. Maps are available for most of North Carolina and Alabama. Complete coverage for these States will be available by the end of 1994. Wetland mapping has lagged in South Carolina, with only about half of the State having been completed. This is due in part to the State's desire to develop a database which includes both wetland and upland coverage, which is now underway.

An important strength of the NWI is the accessibility of its products. Maps are routinely distributed to the U.S. Army Corps of Engineers, the U.S. Environmental Protection Agency, and the U.S.D.A. Soil Conservation Service, as well as State agencies which have cooperated in their preparation. The Service alone distributes about 150,000 copies of NWI maps annually throughout the 50 States.

Maps are available for purchase from several sources. The Earth Science Information Centers (ESIC) of the U.S. Geological Survey (USGS) cooperate with the Service in the distribution of NWI maps. Mylar map copies can be purchased for $5.25 and paper copies for $3.50 from USGS. Details regarding the ordering process and information on product availability can be obtained by calling USGS/ESIC toll free at 1-800-USA-MAPS. The NWI has also established State-run distribution centers across the country. Locations and telephone numbers of distribution centers in the Southern Appalachian States are found on Table 1.

The NWI is mandated by the Emergency Wetlands Resources Act of 1986 (P.L. 99-645) as amended by the Wild Exotic Bird Conservation Act of 1992 (P.L. 102-440) to convert traditional cartographic products to digital data by September 30, 2004. Approximately 10,000 maps nationwide have been digitized and incorporated into the georeferenced NWI digital database. Complete Statewide digital wetlands databases will be available for Virginia and West Virginia in 1994. The digital database for North Carolina and South Carolina are also in progress but are several years away from completion. NWI digital data files can be purchased through USGS/ESIC. The data are stored on magnetic tape in MOSS export, DLG3 optional, or GRASS formats, written to 9 track tape, 1/4 inch or 8 mm cartridge in ASCII or UNIX-TAR. Tennessee and Kentucky are independently digitizing wetlands maps. The NWI is working with the States in order that the data might be shared and eventually made available for distribution.

A deliberate effort has been made by the NWI to make its products available to the greatest number of people possible. However, new NWI map users are sometimes deterred by the seemingly complex classifications displayed on the maps. To overcome this, NWI personnel located in each Regional Office of the Service are available to provide assistance in understanding the maps. In addition, formal training sessions in wetland classification and mapping procedures are regularly scheduled.

TABLE 1.

Sources of National Wetlands Inventory Products for the Southern Appalachian Region

Alabama	Alabama Geological Survey Post Office Box O Tuscaloosa, AL 35486	(205) 349-2852
Georgia	Georgia Geologic Survey Room 406A 19 Martin Luther King, Jr. Drive, S.W. Atlanta, GA 30334	(404) 656-3214
Kentucky	Natural Resources and Environmental Protection	(502) 564-5174

	Cabinet Division of Administrative Services Data Processing Branch 14th Floor, Capitol Plaza Tower Frankfort, KY 40601	
North Carolina	North Carolina Department of Environment, Health, and Natural Resources Division of Soil and Water Conservation Post Office Box 27687 512 North Salisbury Street Raleigh, NC 27611	(919) 733-2302
South Carolina	State of South Carolina Land Resources Conservation Commission Cartographic Information Center 2221 Devine Street Suite 222 Columbia, SC 29205	(803) 734-9100
Tennessee	ESIC/USGS* National Headquarters 507 National Center Reston, VA 22092	1-800-USA-MAPS (703) 648-6045
Virginia	ESIC/USGS* National Headquarters 507 National Center Reston, VA 22092	1-800-USA-MAPS (703) 648-6045
West Virginia	National Heritage Program of Wildlife Resources West Virginia Division of Natural Resources Post Office Box 67 Ward Road Elkins, W. VA 26241	(304) 637-0245

*Earth Science Information Center, United States Geological Survey

5. WETLANDS STATUS AND TRENDS REPORTS

Recognizing that maps are a static representation of wetland conditions, the Service conducts periodic studies to determine wetland gains and losses nationwide. The first wetland trends study was completed in the early 1980's and evaluated wetland changes from the mid-1950's to the mid-1970's (Frayer et al. 1983; Tiner, 1984). A second study (Dahl and Johnson, 1991) developed trend information for the mid-1970's to the mid-1980's. In accordance with the Emergency Wetlands Resources Act, the NWI will continue these studies at 10-year intervals.

A stratified random sampling design was used. Aerial photographs taken at the start and the end of each study period were interpreted and wetland acreages measured for 3,629 four-square mile sample plots nationwide. Estimates of wetland acreages were then generated through statistical analysis of the data obtained from the plots.

The study design was such that sample plots were concentrated in areas recognized as having high wetland densities such as the Atlantic and Gulf of Mexico Coastal Plains, and the Lower Mississippi Valley. Conversely, samples were sparsely distributed in areas of anticipated low wetland densities. For example, only 50 sample plots were assigned to the combined area of Tennessee and Kentucky while Florida and Louisiana each were allocated over 600 plots. In some areas such as the Southeastern United States (Hefner and Brown, 1984) and Florida (Hefner, 1986; Frayer and Hefner, 1991) where sample sizes were large, it has been possible to employ the statistical procedures of the national studies to develop localized wetland status and trends information. Conversely, the subset of samples in the southern Appalachians is so small that statistical analyses would result in extremely large standard deviations in the estimates and little useful information would be obtained.

6. SUMMARY

The NWI is nearly completed with wetland mapping in the Southern Appalachian States. The maps represent the only uniform, accessible, and extensive source of cartographic information related to wetlands in the region. The NWI not only locates wetlands but also describes them in terms useful for a variety of evaluation purposes. Wetland maps have been prepared through the interpretation of high altitude aerial photography augmented by field checking, use of collateral information, and quality control reviews by the Service and NWI cooperators.

Wetland delineation from high altitude aerial photography has inherent limitations related to scale, quality, date, and type of photography; the types of wetlands to be delineated; and the skill and experience of the photointerpreters. Tiner (1990) described the special considerations for using high altitude aerial photography for inventorying forested wetlands, the predominant wetlands of the Southern Appalachians and adjoining piedmont. The NWI relies heavily on field checking and the use of collateral information to overcome photointerpretive limitations.

The NWI periodically conducts studies to determine wetland gains and losses. However, trend information has not been developed for the region. Therefore, follow-up studies should be designed not only to monitor changes in the areal extent of the wetlands but also to evaluate the functional health of the resource.

7. REFERENCES

Cowardin, L.M. Carter, V. Golet, F.C., and LaRoe, E.T. 1979.: Classification of Wetlands and Deepwater Habitats of the United States, USDI Fish and Wildlife Service, Washington, DC. FWS/OBS-79/31, pp. 103.

Crowley, S. O'Brien, C., and Shea, S.: 1988. Results of the Wetland Study and the 1988 Draft Wetland Rules. Report by the Agency of Natural Resources Divisions of Water Quality, Waterbury, Vermont. pp. 33.

Dahl, T.E. and Johnson, C.E.: 1991. Wetlands Status and Trends in the Conterminous United States, Mid-1970's to Mid-1980's. U.S. Fish and Wildlife Service, National Wetlands Inventory Project, Washington, D.C. pp. 28.

Federal Geographic Data Committee: 1992. Application of Satellite Data for Mapping and Monitoring Wetlands - Fact Finding Report; Technical Report 1. Wetlands Subcommittee, FGDC. Washington, D.C. pp. 32 plus Appendices.

Frayer, W.E. and Hefner, J.M.: 1991. Florida Wetlands: Status and Trends, 1970's to 1980's. U.S. Fish and Wildlife Service, National Wetlands Inventory Project, Atlanta, Georgia. pp. 31.

Frayer, W.E. Monahan, T.J. Bowden, D.C. and Graybill, F.A.: 1983. Status and Trends of Wetlands and Deepwater Habitats in the Conterminous United States, 1950's to 1970's. Fort Collins: Colorado State University. pp. 32.

Hefner, J.M.: 1986. Wetlands of Florida, 1950's to 1970's. In: Estevez, E.D., Miller, J.; Morris, J.; Hamman, R. eds., Proceedings, Conference on Managing Cumulative Effects in Florida Wetlands; 1985 October 17-19; Sarasota, FL. New College Environmental Studies Prog. Publ. No. 37. Madison, WI: Omnipress. pp 23-31.

Hefner, J.M. and Brown, J.D.: 1984. Wetland trends in the Southeastern United States. Wetlands 4:1-11.

Reed, P.B. Jr.: 1988. National List of Plant Species that Occur in Wetlands: National Summary. U.S. Fish and Wildlife Service, National Wetlands Inventory Project, Washington, D.C., Biological Report 88 (24). pp. 244.

Schafale, M.P. and Weakley, A.S.: 1990. Classification of the Natural Communities of North Carolina. Third approximation. North Carolina Natural Heritage Program, Department of Environment, Health, and Natural Resources, Raleigh, North Carolina pp. 325.

Shaw, S.P. and Fredine, C.G.: 1956. Wetlands of the United States, USDI Fish and Wildlife Service, Circular 39. pp. 67.

Swartwout, D.J. MacConnell, W.P. and Finn, J.T.: 1981. An Evaluation of the National Wetlands Inventory in Massachusetts. In: Proc. In-Place Resource Inventories Workshop, 9-14 August 1981, University of Maine, Orono. pp. 685-691.

Tiner, R. W. Jr.: 1984. Wetlands of the United States: Current Status and Recent Trends. U.S. Fish and Wildlife Service, National Wetlands Inventory, Washington D.C. pp. 59.

Tiner, R.T. Jr.: 1990. Use of High-Altitude Aerial Photography for Inventorying Forested Wetlands in the United States. For. Ecol. and Management, Vol. 33/34, pp. 593-604.

U.S. Fish and Wildlife Service: 1990. Photointerpretation Conventions for the National Wetlands Inventory. National Wetlands Inventory Project, St. Petersburg, Florida. Unpublished. pp. 45.

U.S. Soil Conservation Service. 1991. Hydric Soils of the United States. U.S. Department of Agriculture Soil Conservation Service. Washington. D.C. (not paginated).

Wilen, B.O. and Pywell, H.R.: 1992. Remote Sensing the Nation's Wetlands: The National Wetlands Inventory. In: Greer, J.D. editor. Proceedings of the Fourth Forest Service Remote Sensing Applications Conference. 1992 April 6-11; Orlando, Florida. pp. 6-17.

IDENTIFICATION OF WETLANDS IN THE SOUTHERN APPALACHIAN REGION AND THE CERTIFICATION OF WETLAND DELINEATORS

JAMES S. WAKELEY
U.S. Army Engineer Waterways Experiment Station,
Vicksburg, Mississippi, USA

Abstract. According to the *Corps of Engineers Wetlands Delineation Manual*, wetlands are identified by the presence of field indicators of hydrophytic vegetation, hydric soils, and wetland hydrology. In the southern Appalachian region, situations that present problems for wetland delineators include (1) wetlands developed on recently deposited alluvial soils that may show little evidence of hydric conditions, (2) areas occupied by FAC-dominated plant communities, (3) wetlands affected by past or present drainage practices, (4) man-induced wetlands that may lack certain wetland field indicators, and (5) hydric soil units that are too small or narrow to be delineated separately on soil survey map sheets. In March 1993, under direction of Section 307(e) of the Water Resources Development Act of 1990, the Corps of Engineers initiated a Wetland Delineator Certification Program. A 1-year demonstration program has recently ended in Maryland, Florida, and Washington, with nationwide implementation scheduled for later in 1994. This voluntary program is designed to increase the quality of wetland delineations submitted with Section 404 permit applications, and reduce processing time by reducing the need for extensive field verification of wetland boundaries.

1. Introduction

Ecologists recognize that wetlands are transitional areas that occupy the gradient between permanently inundated aquatic habitats and well-drained uplands. The transition is sometimes very sharp, as in the case of an incised stream or a marsh that abuts steeply sloping hills. More often, however, wetlands occupy relatively flat terrain (e.g., broad floodplains) in which it is not readily apparent where the wetlands end and uplands begin.

Federal and state agencies have different responsibilities and purposes for wetland inventory, management, and regulation, and therefore have adopted different methods for wetland identification and delineation. The U.S. Fish and Wildlife Service National Wetlands Inventory (NWI), for example, is based on the wetland definition and classification system developed by Cowardin *et al.* (1979). Currently, the U.S. Army Corps of Engineers (USACE) and U.S. Environmental Protection Agency use the *Corps of Engineers Wetlands Delineation Manual* (hereafter, the Manual) (Environmental Laboratory, 1987, plus updates provided in memoranda and guidance letters to Corps districts by

Water, Air and Soil Pollution **77**: 217–226, 1994.

Headquarters, USACE) to identify and delineate wetlands for regulatory purposes under Section 404 of the Clean Water Act.

The purpose of this paper is to present an overview of wetland identification according to the Manual, to discuss selected problem situations that increase the difficulty of wetland determinations in the southern Appalachian region, and describe a certification program for wetland delineators that is scheduled for nationwide implementation in 1994.

2. Overview of Wetland Identification

According to the Manual, wetlands are identified and delineated in the field by the presence of indicators of each of three essential wetland parameters -- hydrophytic vegetation, hydric soils, and wetland hydrology. To the extent practicable, each parameter is evaluated separately at each sampling point, and the wetland boundary generally is defined as the highest point on the gradient where evidence of all three parameters is present.

Of course, the three parameters are not independent and are not equally amenable to investigation during a brief field visit. Hydrology is recognized as the driving force behind the establishment and maintenance of wetlands, whereas soil morphology and plant community composition generally reflect the long-term hydrologic regime. Because it is impossible to characterize the hydrology of a site during a single, brief visit that may occur at any time of the year, wetland determinations in practice are based mainly on vegetation and soil characteristics. Hydrologic field indicators (e.g., observed saturation, water marks, oxidized root channels) are used in a supporting role to provide evidence that vegetation and soil features are not relics of a previous hydrologic regime.

Problems arise when vegetation and soil indicators are weak or problematic, and direct hydrologic evidence is scanty or nonexistent. Under such conditions, wetland delineators must rely on their previous experience in the local area and best professional judgment, a situation that sometimes results in different wetland calls by different investigators.

2.1. Hydrophytic Vegetation

The Manual defines *hydrophyte* as a plant "that grows in water or on a substrate that is at least periodically deficient in oxygen as a result of excessive water content" [Appendix, p. A6]. Furthermore, for a plant community to be considered hydrophytic, it must be dominated by species that are "adapted for life in anaerobic soil conditions" [p. 17]. Tiner (1991) discussed the concept of a hydrophyte and its relevance to wetland delineation.

The first step in evaluating a plant community is to divide the vegetation into strata (i.e., trees, saplings/shrubs, herbs, and woody vines) and to select dominant species independently from each stratum present. The Manual recommends that the three most abundant species (by percent cover, basal area, density, or other recognized measure) be selected from each stratum. However, recent guidance from Headquarters USACE allows the use of the method for selecting dominant species (sometimes called the "50/20 rule") given in the 1989 *Federal Manual for Identifying and Delineating Jurisdictional Wetlands* (Federal Interagency Committee for Wetland Delineation, 1989):

> "For each stratum ... in the plant community, dominant species are the most abundant plant species (when ranked in descending order of abundance and cumulatively totaled) that immediately exceed 50 percent of the total dominance measure ... for the stratum, plus any additional species comprising 20 percent or more of the total dominance measure for the stratum" [p. 5-6].

After dominant species have been selected from each stratum, they are combined across strata into a single list, and the wetland indicator status for each species is determined by looking it up on the appropriate regional version of the *National List of Plant Species that Occur in Wetlands* (hereafter, the plant list) (Reed, 1988). The plant list categorizes each species as either obligate (OBL), facultative wetland (FACW), facultative (FAC), facultative upland (FACU), or obligate upland (UPL) based on its presumed frequency of occurrence in wetlands under natural conditions. Finally, for a plant community to be considered hydrophytic according to the Manual, more than 50% of the dominant species from all strata must be OBL, FACW, or FAC.

2.2. Hydric Soils

"A hydric soil is a soil that is saturated, flooded, or ponded long enough during the growing season to develop anaerobic conditions in the upper part" (U.S.D.A. Soil Conservation Service, 1991). Hydric soils are defined by a set of four criteria, three of which describe hydrologic regimes that are thought to be sufficient to bring about anaerobic conditions in the field. Current hydric soil criteria and the national hydric soil list (U.S.D.A. Soil Conservation Service, 1991) supersede those printed in the Manual.

The first criterion for hydric soils states that all organic soils (Histosols) are hydric, except for the Folists, which in the United States are widespread only in Alaska and Hawaii. Thus, virtually all organic soils (i.e., soils consisting of at least 16 inches of organic material in the top 32 inches of profile, or any thickness of organic material over bedrock) in the conterminous United States are considered to have developed under saturated conditions and, therefore, are hydric.

The second criterion for hydric soils describes soils with high water tables, and specifies that the water table must be within 6, 12, or 18 inches of the surface continuously for at least 2 weeks during the growing season in most years, the required depth varying with soil drainage class and permeability. The third and fourth criteria deal with ponded and flooded soils, respectively, and require that the soil be inundated for at least 1 week during the growing season in most years.

The criteria for hydric soils were developed originally by the Soil Conservation Service (SCS) to compile a list of potentially hydric soils in the United States by selecting appropriate series and phases from a nationwide database of soils interpretation records. In practice, an investigator rarely has enough information about the hydrologic regime of a site to evaluate hydric soils based on these criteria. Instead, hydric soil decisions are usually based on soil morphology. The most widely used morphological indicators of hydric soils are (1) color and (2) organic matter accumulation.

The typical brownish or reddish colors of most upland mineral soils are mainly due to ferric iron (Fe^{+3}) oxides and oxyhydroxides (e.g., FeOOH) that coat soil particles. During prolonged or repeated soil saturation, microbial metabolism under anaerobic conditions results in the chemical reduction of iron to the ferrous (Fe^{+2}) form. Ferrous iron itself imparts a grayish color to the soil. Furthermore, unlike ferric iron compounds, ferrous iron is soluble and can move in the groundwater. Areas in the soil from which iron has been removed (called redox depletions) become more gray, due to exposure of the natural color of the uncoated mineral grains. Ferrous iron can re-oxidize in areas of the soil where oxygen may be available, such as along ped faces or in macropores, forming bright mottles or ferromanganese concretions (collectively called redox accumulations). Therefore, hydric soils are predominantly gray, and may or may not be mottled.

Soil color is determined by matching samples to color chips on Munsell Soil Color Charts (Kollmorgen Corporation, 1992) and reported in the Munsell system of hue, value, and chroma. Chroma is a measure of the strength or purity of the color, and ranges in most soils from 0 (neutral colors of the photographic gray scale) to 8 (richly pigmented colors).

According to the Manual, mineral soils (except sands) are considered to be hydric if the predominant (matrix) color has a chroma of 1 or less in an unmottled soil, or a chroma of 2 or less in a mottled soil. In general, soil colors are measured in the horizon immediately below the A-horizon (i.e., below the zone in which accumulated organic matter is the primary coloring agent).

The second most widely used indicator of hydric soils is the accumulation of large amounts of organic matter on the soil surface. In general, organic matter accumulates abundantly only in wet areas where anaerobic conditions retard its decomposition and oxidation. The Manual considers soils with more than 8 inches of organic matter on the surface (i.e., histic epipedon) to be hydric. Furthermore, in sands, hydric soils are indicated by a surface mineral layer several inches thick in which the sand grains are coated and darkly stained by accumulated organic matter. Recently the SCS has proposed revised lists of hydric soil indicators that greatly expand on those given in the Manual.

2.3. Wetland Hydrology

Although hydrology creates and maintains all wetlands, it is the most difficult of the three parameters to investigate in the field. It is far easier to infer the influence of hydrology from the site's vegetation and soils. The Manual acknowledges that "[h]ydrology is often the least exact of the parameters, and indicators of wetland hydrology are sometimes difficult to find in the field. However, it is essential to establish that a wetland area is periodically inundated or has saturated soils during the growing season" [p. 34], in order to distinguish current wetlands from areas that may have been drained recently.

Perhaps reflecting the difficulties involved in dealing with the hydrology parameter, the Manual does not give an unequivocal definition of wetland hydrology. Based on work in bottomland hardwood wetlands in the Southeast, the Manual indicates that an area has wetland hydrology if it is inundated or saturated to or near the surface continuously for at least 5% of the growing season in most years. Because of the lack of direct long-term hydrologic information on most sites, decisions about the presence of wetland hydrology usually are made on the basis of field indicators. Acceptable hydrologic indicators include direct observation of inundation or soil saturation at the time of the site visit, water marks, debris lines, sediment deposits, water-stained leaves, and oxidized rhizospheres around living roots and rhizomes.

3. Wetland Delineation Problems in the Southern Appalachian Region

At times, wetland identification and delineation is made more difficult by human disturbances that impact wetland parameters, or by naturally occurring situations in which evidence of one or more parameters is absent or misleading. The Manual calls these Atypical Situations and Problem Areas, respectively. The following section discusses situations that may cause uncertainty in wetland delineations in the southern Appalachian region.

3.1. Misleading Color Indicators in Alluvial Soils

Many of the wetlands in the southern Appalachian region occur within the floodplains of rivers and streams. In general, only the wettest portions of floodplains satisfy wetland requirements. Floodplain wetlands are most likely to exist in (1) backwater areas that retain water for long periods following flood events, (2) areas lying below the elevation of the 2-year flood, or (3) areas whose hydrology is augmented by groundwater input from the surrounding uplands.

Widespread land clearing since the late 19th century has resulted in extensive loss of highly erodible soils from valley slopes and increased deposition of new alluvium along streams and in floodplains (W. Nutter, this symposium). In poorly drained areas, these recently deposited soils may support hydrophytic plant communities but have not had time to develop the gray colors usually associated with hydric mineral soils. However, the soils are hydric because they meet hydric soil criteria for frequency and duration of flooding or saturation. Thus the soils are wetter than they appear to be, and wetlands must be identified primarily on the basis of vegetation and field indicators of wetland hydrology.

3.2. FAC-Dominated Plant Communities

Floodplain plant communities in the southern Appalachian region are often dominated by FAC species (e.g., red maple [*Acer rubrum*], sweetgum [*Liquidambar styraciflua*], water oak [*Quercus nigra*]). The Manual states that more than 50% of dominant species must be FAC, FACW, or OBL for the community to be considered hydrophytic. Therefore, a community in which most or all the dominants are FAC easily satisfies hydrophytic vegetation criteria. However, because a FAC species is equally likely to occur in nonwetlands as in wetlands, a FAC-dominated community alone is not strong evidence of wetland conditions.

Usually the wetland determination is based on the combination of soil morphology, vegetation, and wetland hydrology; but when the hydric soil determination is problematic (e.g., recent alluvial soils), the vegetation parameter takes on increased importance. In that case, a FAC-dominated plant community probably should be investigated further. One option is to tally subdominant species, particularly perennial species in the understory, which may either support or contradict the conclusion based on dominants alone. In addition, one can increase the level of confidence in the determination of hydrophytic vegetation by use of the FAC-neutral test (in which the FAC category is ignored and the hydrophytic vegetation decision is based on the number of dominant species rated OBL and FACW versus FACU and UPL).

3.3. Past and Present Wetland Drainage

Throughout the southern Appalachian region, obvious hydrologic modifications (e.g., ditches, subsurface drains, levees, channelized streams) present problems for the wetland delineator because they cast doubt on the reliability of evidence derived from current soil and vegetation conditions. Hydrophytic vegetation may persist for years, and hydric soil colors indefinitely, after a wetland has been drained. In agricultural areas, the problem is often compounded by the absence of natural vegetation due to cultivation.

On the other hand, the presence of ditches and drains does not necessarily mean that an area no longer meets wetland criteria. Often only the immediate vicinity of a ditch is effectively drained, and a levee may prevent flooding but have no influence on groundwater. Agricultural improvements may permit earlier planting of wet areas but have little influence on wetness during the critical first few weeks of the growing season.

In areas with native vegetation, the plant community may offer the first clues that a wetland has been effectively drained. Invasion by upland-adapted perennial plants in the understory may occur long before any changes in overstory species composition. In addition, SCS drainage specialists can be consulted to estimate the zone of influence of any ditches or tile drains. There may be no alternative, however, to direct long-term hydrologic monitoring (e.g., with shallow groundwater wells) in controversial cases.

3.4. Man-Induced Wetlands

The opposite problem to wetland drainage is the deliberate or inadvertent creation of wetlands in areas that formerly were uplands. These areas also may exhibit soil colors and plant communities that are not indicative of current hydrology. In the southern Appalachian region, wetlands are often created during highway construction, due to unintentional blockage of surface flows. In addition, abandoned settling ponds and man-made depressions in abandoned mined lands soon exhibit wetland characteristics.

Man-induced wetlands often have obvious hydrology but may lack hydric soil indicators for a number of years. Although it may take time for plant community composition to change appreciably, the first indication of wetter conditions may be stress on or mortality of nonadapted plants. Most man-induced wetlands can be identified and delineated based primarily on vegetation and hydrology.

3.5. Scale of Soil Mapping

The Manual describes various levels of effort and reliability in wetland determinations. The lowest level involves the use of available office data without an on-site inspection. Generally, these off-site determinations are used for preliminary mapping of potential wetland areas for planning purposes; off-site determinations may not be sufficient for regulatory purposes.

Off-site wetland determinations are based on information in SCS soil survey reports, aerial photographs, NWI maps, and any other available maps or imagery. Soil survey map sheets are often the best source of information for mapping potential wetland areas. Soil map units have a high probability of being wetlands if they are listed on the national or local hydric soils list, and they correspond to areas mapped as wetlands by NWI or appear on aerial photos to support hydrophytic vegetation.

In the southern Appalachian region, however, most wetlands are either small and isolated (e.g., mountain bogs) or are narrow and linear (e.g., riparian wetlands). At the typical mapping scale used in SCS soil survey reports (i.e., second order surveys with a minimum delineation of approximately 3-5 acres), small or narrow hydric soil units often cannot be mapped separately and are either listed as dissimilar inclusions within larger, generally nonhydric mapping units, or appear as spot symbols on map sheets. Therefore, soil survey maps are not as useful for off-site wetland determinations in the southern Appalachian region as they are in other areas, and even preliminary wetland inventories may require time-consuming on-site inspections.

4. The Wetland Delineator Certification Program

In March of 1993, the US Army Corps of Engineers initiated the first phase of a Wetland Delineator Certification Program (WDCP) in response to Section 307(e) of the Water Resources Development Act of 1990. The purpose of the WDCP is to improve the accuracy of wetland delineations submitted by individuals as part of Section 404 permit applications, and facilitate permit processing by minimizing the time that Corps field personnel must spend verifying wetland boundaries. The WDCP is strictly a voluntary program. Non-certified individuals and consultants may continue to submit delineations to the Corps for verification and approval, and landowners may still choose to have Corps personnel do the wetland determination. However, the latter two options may result in longer processing times due to heavy workloads in most Corps regulatory offices.

The WDCP was initiated as a demonstration project in three states -- Florida, Maryland, and Washington -- administered by the Corps' Jacksonville, Baltimore, and Seattle Districts, respectively. For the demonstration phase, applicants for certification must pass a written examination and a separate field practicum, which test knowledge of basic wetland concepts, delineation criteria and procedures described in the Manual, and problem wetland situations encountered in that state or region. The WDCP is scheduled for nationwide implementation in 1994, following review of the demonstration projects. At that time, new applicants for certification will also be required to take a course in wetland delineation, taught by a certified delineator. The course must include both classroom and field training.

A package of training materials, including lesson plans and detailed lecture notes, is available to individuals or firms who wish to provide training under the WDCP. The training package is based on the 5-day course that the Corps of Engineers provides to its regulatory personnel, and is available from the Wetland Research and Technology Center of the US Army Engineer Waterways Experiment Station (Attn: CEWES-EP-W), 3909 Halls Ferry Road, Vicksburg, MS 39180-6199.

Acknowledgements

I thank Dave Baker, Steve Chapin, Cathy Elliott, Ray Hedrick, Susan Joy, Bob Lichvar, Tom Roberts, Ronnie Smith, Steve Sprecher, and Doug Winford for ideas concerning problem wetland situations in the southern Appalachian region. Mary Davis, Bob Lichvar, Steve Sprecher and two anonymous reviewers provided comments on the manuscript. The work was funded by the U.S. Army Corps of Engineers Wetland Research Program. Permission to publish this paper was granted by the Chief of Engineers.

References

Cowardin, L.M., Carter, V., Golet, F.C., and LaRoe, E.T.: 1979, *Classification of Wetlands and Deepwater Habitats of the United States.* U.S. Dept. of the Interior, Fish and Wildlife Service, FWS/OBS-79/31, Washington, D.C.

Environmental Laboratory.: 1987, *Corps of Engineers Wetlands Delineation Manual*. U.S. Army Engineer Waterways Experiment Station, Technical Report Y-87-1, Vicksburg, Miss.

Federal Interagency Committee for Wetland Delineation.: 1989, *Federal Manual for Identifying and Delineating Jurisdictional Wetlands*. U.S. Army Corps of Engineers, U.S. Environmental Protection Agency, U.S. Fish and Wildlife Service, and U.S.D.A. Soil Conservation Service, Cooperative Technical Publication, Washington, D.C.

Kollmorgen Corporation.: 1992, *Munsell Soil Color Charts*. Newburgh, N.Y.

Reed, P. B., Jr.: 1988, *National List of Plant Species that Occur in Wetlands: National Summary*. U.S. Dept. of the Interior, Fish and Wildlife Service, Biological Report 88(24), Washington, D.C.

Tiner, R. W.: 1991, The concept of a hydrophyte for wetland identification. *Bioscience* 41, 236-247.

U.S.D.A. Soil Conservation Service.: 1991, *Hydric Soils of the United States*. Miscellaneous Publication 1491, Washington, D.C.

PART III

BIOGEOCHEMICAL PROCESSES

HILLSLOPE NUTRIENT FLUX

DURING NEAR-STREAM VEGETATION REMOVAL

I. A MULTI-SCALED MODELING DESIGN.

J.A. YEAKLEY[1,3], J.L. MEYER[1] AND W.T. SWANK[2]

[1]*Institute of Ecology, University of Georgia, Athens, Georgia, 30602, USA.* [2]*Coweeta Hydrologic Laboratory, Southeastern Forest Experiment Station, USDA-Forest Service, Otto, North Carolina, 28763, USA.* [3]*Current address: Department of Environmental Sciences and Resources, Portland State University, Portland, Oregon, 97207, USA.*

Abstract. At the Coweeta Hydrologic Laboratory in the southern Appalachians of western North Carolina, a near-stream vegetation manipulation experiment is being conducted to determine the effect of removal of streamside *Rhododendron maximum* L. on the export of hillslope nutrients (K, Na, Ca, Mg, N, P, S) and organic matter. Experimental hillslope transects that span topographical flowpaths from a local highpoint to the stream have been instrumented with lysimeters and TDR rods at two depths, as well as with streambed and streambank piezometers. We present a review of studies of nutrient flux in the riparian zone of forested watersheds. In the southern Appalachians, we hypothesize that *R. maximum* is a keystone species at the interface between terrestrial and aquatic systems, with extensive near-stream thickets having a possible impact on carbon and nutrient transport into streams. We present the conceptual basis and initial implementation of a model-based experimental design to test the effect of *R. maximum* removal on hillslope nutrient and organic matter export in upland watersheds. The model is terrain-based and will be used to extrapolate elemental flux measurements both spatially from the hillslope to watershed scale and temporally for various climate regimes. The model consists of three modules: (1) objective terrain analysis (TAPES-C); (2) a dynamic interception canopy module; (3) a hillslope hydrology module (IHDM4) with a 2-D Richard's equation of subsurface moisture dynamics. Calibration and validation of the model will occur at two scales: at the hillslope scale, using well, lysimeter, and TDR data; at the watershed scale, using streamflow measurements across a variety of storm types. We show watershed terrain analysis for the experimental watershed (WS56) and discuss use of the model for understanding effects of watershed management of riparian zone processes.

1. Introduction

One of three major thrusts of the Coweeta Long Term Ecological Research (LTER) project is to investigate the role of riparian zone linkage between terrestrial and stream ecosystems in the southern Appalachian mountains (Franklin *et al.*, 1990; Van Cleve and Martin, 1991). Central to this investigation, a manipulation experiment is underway that seeks to determine the effect of removal of streamside *Rhododendron maximum* L. on the export of coarse particulate organic matter, of dissolved organic carbon (DOC) and of nutrients (Na, K, Ca, Mg, NO_3-N, NH_4-N, SO_4, PO_4-P). Information gained from the experiment will assist the U.S. Forest Service in understanding stream water quality effects of proposed management strategies to remove streamside *R. maximum* in the southern Appalachian mountains.

Water, Air and Soil Pollution 77: 229–246, 1994.

We review studies of nutrient flux in the riparian zone of forested watersheds. In the southern Appalachians, we hypothesize that *R. maximum* is a keystone species at the interface between terrestrial and aquatic systems, with extensive near-stream thickets having a possible impact on carbon and nutrient transport into streams. We present the conceptual basis and initial implementation of a model-based experimental design to test the effect of *R. maximum* removal on hillslope nutrient and organic matter export in upland watersheds. Our model has the capability to extrapolate elemental flux from ongoing plot-scale measurements to larger scales both spatially at the level of the watershed and temporally for various climate regimes, providing a tool for regional analysis and management. We show results from the first stage of model implementation, watershed terrain analysis, for an experimental watershed in the riparian component of the Coweeta LTER.

2. Literature Review

2.1. RIPARIAN ZONES AS NUTRIENT FILTERS

Riparian zones in forested watersheds have been defined as locations of direct interaction between aquatic and terrestrial ecosystems, with boundaries extending outward to the limits of flooding or near-surface saturation and upward into the canopy of streamside vegetation (Gregory *et al.*, 1991). In agricultural watersheds, forested riparian zones have been depicted as sinks for nutrients transported from upland sources that thereby buffer nutrient discharge from surrounding agro-ecosystems (Lowrance *et al.*, 1984a). For some time, it has been emphasized that riparian vegetation was important in improving water quality in agricultural watersheds (Schlosser and Karr, 1981). In a contrasting view, Omernik *et al.* (1981) hypothesized that mature riparian forests are not nutrient filters, because no net annual uptake would occur. They speculated that forest buffer strips can reach nutrient saturation and lose their nutrient filter capacity as they reach maturity, as has also been proposed in general for successional "climax" vegetation by Vitousek and Reiners (1975). It has been suggested, moreover, that ecosystem management be directed toward keeping riparian vegetation below mature stages by periodic selective harvesting (Lowrance *et al.*, 1983). Yet another view comes from research on the impact of afforestation in acid moorlands in the United Kingdom, where studies have shown streams draining forested areas are more acidic and contain larger aluminum concentrations compared with adjacent moorland catchments (Ormerod *et al.*, 1989; Waters and Jenkins, 1992; Reynolds *et al.*, 1992). A likely reason for these findings is increased scavenging of acid deposition by coniferous canopies (Anonymous, 1991); one modeling study showed afforestation in western Britain increasing total sulfur deposition by one third and total nitrogen deposition by a factor of two (Fowler *et al.*, 1989). Streams draining such clearfelled forests have shown proximal increases in nitrate and decreases in sulfate and chloride (Reynolds *et al.*, 1992).

Most empirical studies, however, have found that riparian forests generally act as nutrient sinks. For coastal agricultural watersheds, it has been shown that forested riparian zones removed total-N predominantly by subsurface uptake, whereas total-P removal in the riparian zone was equally divided between surface and subsurface losses

(Peterjohn and Correll, 1984). They later found that the riparian forest acted as an important sink for NO_3-N and significantly reduced the acidity of groundwater and precipitation that entered it (Peterjohn and Correll, 1986). In an agricultural watershed in the Georgia Coastal Plain, a riparian forest was found to act as a sink for NO_3-N, Ca, Mg, K and SO_4-S (Lowrance *et al.*, 1984b). It was projected that conversion of riparian forest to cropland would increase NO_3-N and NH_4-N loads by as much as 800% (Lowrance *et al.*, 1983). A study of riparian peatlands of a forested watershed in Minnesota found that 36-60% of all annual nutrient inputs were retained in the streamside zone (Verry and Timmons, 1982).

In coastal plain riparian forests with shallow underlying aquicludes, areas of near-surface saturation may remove much of the nitrate load through denitrification rather than uptake by vegetation (Jacobs and Gilliam, 1985; Jordan *et al.* 1993). Studies of riparian nutrient dynamics in other watersheds, however, have indicated either that significant nitrate removal occurs both through vegetation uptake and denitrification (Correll and Weller, 1989) or that vegetation processes were the primary form of removal (Lowrance, 1992). Additionally, nitrate retention by riparian zones has been shown to occur during winter months under either grass or poplar cover (Haycock and Pinay, 1993). They found that nitrate retention was greater under poplar and suggested that, although aboveground vegetation has no active role in retaining nitrate in the winter, poplar root systems contributed more carbon to denitrifying soil bacteria. In a northern hardwood forest in Turkey Lakes Watershed in Ontario, nitrate concentrations were highest in soil and stream water during the dormant season, peaking at the start of snowmelt. During the growing season, NO3-N and NH4-N increases were greatest in the Oe horizon, decreasing with depth (Foster *et al.*, 1989). It was recently shown at the Coweeta Hydrologic Laboratory in western North Carolina that biological and geochemical processes in the upper soil horizons in oak-hickory forests at Coweeta retained N and P, thus reducing inputs to streams (Qualls *et al.*, 1991). In the Walker Branch Watershed in eastern Tennessee, partitioning between biological uptake and leaching during a 12 year period was site specific and depended significantly on slope position in the watershed (Johnson and Todd, 1990).

Dynamics of nutrient export from forested watersheds are dependent on several factors. The geological setting largely determines both the supply of inorganic nutrients available to forest biota and the general character of streamwater chemistry (Velbel, 1988; Gilvear *et al.*, 1993; Mulholland, 1993; O'Brien *et al.*, 1993). Biological controls on nutrient export include type, maturity and extent of vegetation cover (Swank, 1988), as well as biological processes in soils (Swank, 1986; Correll and Weller, 1989). Geomorphic and climatic controls include channel morphology (Pinay *et al.*, 1992; Harvey and Bencala, 1993), distribution and depth of near-stream soil saturation (Gaskin *et al.*, 1989; Geyer *et al.*, 1992) as well as frequency and duration of precipitation events (Mulholland *et al.*, 1990; Schnabel *et al.*, 1993). The subsurface riparian zone has been depicted as an ecotone between two interfaces. Inland is a terrestrial boundary where transport of water and dissolved solutes is toward the channel and controlled by watershed hydrology. Streamside is an aquatic boundary, where exchange of surface water and dissolved solutes is bi-directional and flux is strongly influenced by channel hydraulics (Triska *et al.*, 1993a). Several recent studies have explored solute dynamics across this ecotone (Triska *et al.*, 1989, 1993a,b; Mulholland, 1992; Harvey and Bencala, 1993). Terrestrial water is often low in dissolved oxygen (DO) and therefore chemically reducing. In contrast, hyporheic zone, or stream subsurface, water may entrain DO laterally into near-stream hillslope areas causing a chemically oxidizing environment (Triska *et al.*, 1993b). Dynamics of hyporheic exchange have been shown to be a function of streambed topography (Harvey and Bencala, 1993) as well as channel roughness, gradient, sediment size and permeability, and pool-riffle sequence (Triska *et al.*, 1993b). For example, a study at Walker Branch indicated that the riparian zone was a potential source of NH_4^+ and P to the stream when dissolved oxygen concentrations in riparian groundwater were low, but a sink for P when dissolved oxygen concentrations were high (Mulholland, 1992). Such a fluctuating and reversible nutrient pathway in the riparian zone, where the direction is related to the redox state of the soil, indicates the need to understand the timing, duration and extent of saturated and unsaturated states of riparian zone soils.

2.2. NUTRIENT UPTAKE BY FOREST BIOMASS

Forest biomass accumulation of nutrients has been verified in a variety of forests, with studies indicating that understory species play a significant role in aboveground nutrient accumulation (Day and Monk, 1977b; Grove and Malajczuk, 1985; Kellman *et al.*, 1987; Turner *et al.*, 1992). On a whole watershed basis in second growth forests, extensive study in the Coweeta Basin (e.g., Day and Monk, 1974, 1977a,b; Yount, 1975; McGinty, 1976) has yielded nutrient cycling budgets for Ca, K, Mg, N, P for aggrading mixed oak-hickory forests (Monk and Day, 1988). They showed that elemental content of net annual biomass accretion, in kg ha^{-1} yr^{-1}, was Ca: 4.5, K: 4.9, Mg: 0.9, P: 3.0, and N: 6.7. As a percentage of standing soil-litter pools, to a depth of 60 cm, annual net nutrient assimilation by forest biomass was Ca: 0.7%, K: 1.1%, Mg: 0.2%, P: 8.0%, N: 0.1%. The effect of nutrient accumulation over an extended time period in an aggrading forest at Coweeta was shown recently by comparing soil-litter pools (0-20cm depth) in 1970 vs. 1990 (Knoepp and Swank, 1994). In all nutrients studied (Ca, K, Mg), there were large decreases in soil-litter concentration over the 20 year period due both to losses to the stream and to accumulation in forest biomass.

In a lowland riparian forest adjacent to an agricultural field, approximately 25% of nitrogen retained annually in the riparian zone was assimilated by trees (predominantly *Liquidambar styraciflua* and *Acer rubrum*; Correll and Weller, 1989). Detailed analysis of biomass uptake of nutrient within riparian zones in forested southern Appalachian uplands, however, is yet lacking.

2.3. NUTRIENT RESPONSES TO VEGETATION REMOVAL

Removal of forest vegetation generally causes transient increases in nutrient exports to streams (Johnson *et al.*, 1982; Hopmans *et al.*, 1987; Swank, 1988; Blackburn and Wood, 1990; Hornbeck *et al.*, 1990). Elevated nutrient export resulting from vegetation removal has been found to depend greatly on the method and extent of removal (Pye and Vitousek, 1985; Swank, 1988; Blackburn and Wood, 1990). In three northeastern forests and an eastern Tennessee forest, clearcutting and stem removal caused total ecosystem Ca depletion to a much greater extent than other elements (Johnson *et al.*, 1982; Hornbeck *et al.*, 1990), although Ca loss was primarily via removal of woody tissues rather than to leaching through the soil to the stream. A study in the root zone of four northern hardwood species at Hubbard Brook in New Hampshire following clearcutting showed that K and Mg were released while Ca was retained by decaying roots (Fahey *et al.*, 1988). They also found that N and P release from decaying roots was an important nutrient flux pathway to both stream outflow and vegetation regrowth for two years following harvest. A variety of vegetation removal experiments conducted at the Coweeta Hydrologic Laboratory showed only marginal increases in nutrient concentrations in streamwater after vegetation removal. The constituent most sensitive to vegetation disturbance was found to be NO_3-N, with elevated levels in streams draining clearcuts persisting as long as 20 years after cutting (Swank, 1988).

2.4. *RHODODENDRON MAXIMUM* AS A KEYSTONE RIPARIAN SPECIES

In southern Appalachian upland watersheds, near-stream understory vegetation is frequently dominated by the evergreen sclerophyllous ericaceous shrub *Rhododendron maximum* L. In many cases, this mesic shrub acts to completely close the understory canopy in pure stands over upland streams. We hypothesize that *R. maximum* is a keystone species on this landscape at the interface between terrestrial and aquatic systems, with near-stream thickets acting as hillslope debris dams and having an impact on organic matter processing in the riparian zone, on element transport into streams, and on stream ecosystem structure and function.

It was reported that *R. maximum* establishment at Bent Creek Experimental Forest in western North Carolina primarily occurred in the early 1900's, during a period that coincided with the cessation of fire and grazing disturbance in the region (McGee and Smith, 1967). It has been suggested that the long history of burning of forests in the southern Appalachians (Sharitz *et al.*, 1992) until the turn of this century prevented substantial establishment of *R. maximum* thickets in the area (Phillips and Murdy, 1985). Logging operations in the early twentieth century dramatically opened the canopy and stimulated the establishment and growth of understory species. Subsequent opening of the forest canopy in the 1930's by blight-induced decline of the American chestnut (*Castanea dentata*) has also been suggested as a means of *R. maximum* establishment at the Coweeta Hydrologic Laboratory (McGinty, 1972).

Vegetation analyses at Coweeta, both over long periods as well as following severe drought, indicate that *R. maximum* canopies can have significant impact on hardwood regeneration (Clinton, 1989). Significant hardwood suppression may have been occurring since the 1930's. A long-term study found that regeneration of *Quercus prinus* and *Q. alba* was significantly reduced in plots with high *R. maximum* densities, although other hardwood species (*Q. coccinea*, *Q. velutina*, *Acer rubrum*) were not affected (Phillips and Murdy, 1985).

In comparison with three common species (*Q. prinus*, *Tsuga canadensis*, *Cornus florida*), Day and McGinty (1975) found that *R. maximum* had the largest leaf biomass for WS 18 at Coweeta. *R. maximum* leaf turnover time ranges from 4 to 7 years (Nilsen, 1986). Although nutrient concentrations in *R. maximum* leaves are generally lower than that in deciduous leaves, such long-lived and abundant leaf mass creates a significant nutrient storage reservoir in the riparian zone (Monk *et al.*, 1985).

2.5. RIPARIAN ZONE HYDROLOGY

In the deep-soiled forested watersheds of the southern Appalachian mountains, the variable extent of the saturated source areas, or near-stream areas, primarily determine the timing and volume of streamflow. This variable source area mechanism of streamflow generation, described by hydrologists at the Coweeta Hydrologic Laboratory (Hewlett and Hibbert, 1963; Hewlett and Nutter, 1970), begins as precipitation infiltrates undisturbed forest soils and migrates downslope, accumulating at lower slope positions. These saturated or near-saturated areas maintain baseflow and readily contribute subsurface flow to storm flow as the zone of saturated soil surface expands during a storm event. The degree to which saturation and subsequent expansion would occur for a given hillslope varies as a function of antecedent soil moisture conditions, precipitation volume and duration of input (Hibbert and Troendle, 1988). The hydrological extent of riparian zones in the southern Appalachians vary temporally, then, in response to the frequency and amount of precipitation.

Spatially, the hydrological extent of riparian zones are constrained by watershed topography as well as distribution and depths of soils (Hewlett and Hibbert, 1966; Dunne *et al.*, 1975; Anderson and Burt, 1978; Beven *et al.*, 1988). Upland watersheds in the southern Appalachians generally have steep hillslopes that constrain riparian zones to relatively small near-stream areas. Relatively deep soils in these watersheds, however, provide drainage to near-stream areas that allow riparian zones to persist even during short-term drought periods. Within a watershed, extent of near-stream saturated zones varies with hillslope type (Anderson and Burt, 1978). Persistence of saturated areas varies with hillslope planform, with convergent hillslope planform resulting in riparian zones of greater width (Crabtree and Burt, 1983) and of greater variation in response to climate variation (Yeakley, 1993). Geomorphic differences in the structure of stream channels and floodplains can cause differences in whether riparian zones are effective sources or sinks of carbon, nitrogen and phosphorus (Pinay *et al.*, 1992). They suggest explicit analysis of geomorphic characteristics to determine the retention ability of a given reach.

Important processes of nutrient retention and cycling, however, occur at several hierarchical levels within valley floor landforms (Gregory *et al.*,1991), from stream unit scales on the order of a meter to reaches or sections on the order of 10^2-10^5 meters. Mechanistic investigation should begin at lower space and time scales compatible with generative processes (Allen *et al.*, 1984; Luxmoore *et al.*, 1991; Yeakley and Cale, 1991), to understand patterns produced at higher levels. To characterize the extent and dynamics of variable riparian zones in mountain watersheds over climatic time periods, we present a multi-scaled modeling approach that incorporates both topographic spatial variation at the plot scale and temporal variation at the scale of a storm event.

3. A Multi-scaled Modeling Approach

The remainder of this paper describes our experimental design featuring a terrain-based watershed hydrology model. We will use this model to extend elemental flux estimates from ongoing plot-scale measurements spatially to hillslope and watershed scales and temporally both for short-term storm responses and for long-term climate regimes. We present results from watershed terrain analysis, the first stage of model implementation, for the experimental watershed of the riparian component of the Coweeta LTER and describe potential use of the model for watershed management of riparian zone processes.

A terrain-based hillslope hydrology model has been developed from several existing models to characterize climate scale distributions of hillslope soil moisture in the southern Appalachians (Yeakley, 1993). The model consists of a dynamic canopy interception module (Rutter *et al.*, 1971, 1975) and a two-dimensional hillslope hydrology module (IHDM4: Institute of Hydrology Distributed Model, version 4; Beven *et al.*, 1987) having hillslope planes objectively delineated using contour-based terrain analysis (TAPES-C: Topographic Analysis Programs for the Environmental Sciences-Contour; Moore and Grayson, 1991). Calibration is performed at two scales: watershed and hillslope. Results are validated at both scales, using information from plot measurement studies of soil moisture and downstream measurement of watershed stream discharge over storm and baseflow periods (Yeakley, 1993).

3.1. CANOPY MODULE

The subsurface model (IHDM4) receives inputs from an aboveground module that accounts for canopy and litter fluxes. The aboveground model was specified by Rutter *et al.* (1971, 1975). The Rutter model follows a dynamic canopy storage (C) with input of a constant fraction of rainfall determined by leaf area index and vegetation type, and output as evaporation and drainage. The equations of the model are:

$$\frac{dC}{dt} = Q - K_C[\exp(b_C C)-1] \quad (1a)$$

$$Q = (1-p)R - E_P \cdot f(C) \quad (1b)$$

where K_C and b_C are drainage parameters, R is the total rainfall, E_P is potential evaporation (determined by a Penman-Monteith equation with stomatal resistance set to zero), p is the canopy throughfall fraction. If $C > S$, f(C) equals 1. Otherwise, if $C < S$, then f(C) equals C/S, where S is interception storage and corresponds to a completely wet canopy. The model allows for simultaneous evaporation and transpiration from a partially wet canopy ($C < S$).

The model is regulated by a water balance given as:

$$R = T + E + \Delta C \quad (2)$$

where T is throughfall and E is evaporative loss. Transpiration demand is calculated as E_p for that fraction of the canopy that is dry. An effective precipitation is then calculated that is the throughfall amount (which includes direct throughfall as well as drainage) minus the transpiration demand. In the absence of throughfall, effective precipitation at the soil surface is negative, which is input to the hillslope hydrology model (IHDM4) as a sink term at the surface. The sink is regulated by soil moisture availability times the fractional root distribution in a given layer in the hydrology model as given by Feddes *et al.* (1976). If positive, i.e. if rainfall is occurring, then input to the surface becomes a source term.

3.2. HILLSLOPE HYDROLOGY MODULE

For a given hillslope plane in IHDM4, cells are bounded by two vertically-layered sets of finite element nodes. The top surface of the soil (i.e., highest set of nodes) is treated as a flux boundary with fluxes controlled by applied input rates of effective precipitation unless the surface becomes saturated and overland flow develops. The surface boundary then changes to a fixed head boundary while saturation persists, with potentials fixed at atmospheric pressure. Change of boundary conditions at the soil surface can occur locally on the slope to enable simulation of a time-varying near-stream saturated area. Subsurface elements may be extended to beneath the mid-point of the channel, which is assumed to be a no-flow boundary. Other no-flow boundaries in the model are the upslope divide, the base of the soil/rock profile, hillslope plane sides (i.e. "streamtube" boundaries, see discussion below) and the unsaturated part of the seepage face (Beven *et al.*, 1987).

For a given hillslope plane, subsurface flow is given by a Richards equation expressed as

$$BC(\psi)\frac{\partial\phi}{\partial t} - \frac{\partial}{\partial x}[BK_x(\psi)\frac{\partial\phi}{\partial x}] - \frac{\partial\phi}{\partial z}[BK_z(\psi)\frac{\partial\phi}{\partial z}] = Q_s \quad (3)$$

where B is a streamtube width; ψ is capillary potential; x is horizontal distance downslope; z is gravity potential (measured vertically from some arbitrary datum); ϕ (= ψ + z) is total hydraulic potential; $C(\psi)$ is the specific moisture capacity of the soil (slope of relation between θ and ψ); θ is soil moisture content by volume; K_x, K_z are saturated hydraulic conductivities in the x, z directions; Q_s is a source/sink term (Q from equation (1b)); and t is time. Implementation of (3) requires several assumptions, including: (a) water is of constant viscosity and unit density; (b) flow occurs in an isothermal medium; (c) Darcy's law applies with time-invariant parameters; (d) only single phase water flow in response to hydraulic gradients is considered; (e) the relationship between θ and ψ is locally differentiable (Beven *et al.*, 1987; Calver, 1988).

If either the infiltration capacity of the soil surface is exceeded by input rates or the soil becomes fully saturated resulting in return flow, then overland flow occurs and is given by

$$B\frac{\partial Q}{\partial t} - c\frac{\partial [BQ]}{\partial y} - Bic = 0 \quad (4)$$

where Q is discharge, i is net lateral inflow rate per unit downslope length, y is distance downslope, c is kinematic wave velocity defined by dQ/dA where A is the cross-sectional area of flow (Calver, 1988). Solution of (4) requires a specification between discharge and cross-sectional area, which in IHDM4 is given as a power law function of the form

$$Q = f\, s^{0.5} A^{b} \quad (5)$$

where s is local slope angle, f is an effective roughness parameter (Chezy), such that the kinematic wave velocity is given by $c = Qb/(Q/fs^{0.5})^{1/b}$ (Beven *et al.*, 1987).

Soil moisture characteristics are determined using modified Campbell (1974) relationships with parameters based on soil textural differences (Clapp and Hornberger, 1978). Actual evapotranspiration (E_A) is given as a function of E_P and soil moisture based on Feddes *et al.* (1976):

$$E_A = W_R \cdot \alpha(\psi) \cdot E_P \quad (6)$$

where W_R is a weighting of proportion of root mass for a depth and $\alpha(\psi)$ is a linear scaling function: when $\psi_i < \psi < \psi_a$, $\alpha(\psi)$ equals 1; otherwise when $\psi_w < \psi < \psi_i$, then $\alpha(\psi)$ equals $(\psi-\psi_w)/(\psi_i-\psi_w)$; otherwise when $\psi < \psi_w$ or when $\psi_a < \psi$, then $\alpha(\psi)$ equals 0. Note here that ψ_a is anaerobiosis point (-0.05 bars), ψ_i is vegetation stress initiation point (-0.3 bars, Hewlett, 1962), and ψ_w is wilting point (-15.0 bars).

At the end of each subsurface timestep, inputs from each hillslope section to both overland flow and the channel are calculated. To compute channel flow, IHDM4 uses the same kinematic wave equation and power law flow relationship (4-5) as the overland flow solution on hillslope planes, except that each channel is assumed to be of uniform width. Four levels of timestep occur in IHDM4. The highest level is the input climate data timestep, which here is at one-hour intervals. The next level involves flux exchange between hillslope and channel at a timestep equal to or smaller than the climate step. Subsurface and channel flow is calculated at a finer time resolution; following (Calver and Wood, 1989), a one-half hour step is used. Finally, overland flow if it occurs is calculated a fixed number times in each subsurface flow timestep (Beven *et al.*, 1987).

3.3. TERRAIN ANALYSIS

There are three primary ways of structuring a network of topographic data: (1) triangulated irregular networks (TINs); (2) raster or grid networks; (3) vector or contour line based networks (Moore and Grayson, 1991). Of the three, contour-based networks provide more physical realism than grid based networks that restrict water flow from a given node to only one of eight possible directions. TINs provide physical realism, but require interpretive alignment of the elements, many times based on vector digital elevation maps (DEMs). Moore and Grayson (1991) provide an automated contour-based method (TAPES-C) for partitioning watersheds into natural units bounded by irregularly shaped polygons. These polygons are bounded by equipotential (or contour) lines on two sides and by streamlines, orthogonal to the contours, on the other two sides. Streamlines are assumed to be no-flow boundaries; thus groundwater flow is constrained to flow through a series of elements positioned along a natural gradient. Such a series of cells is termed a "streamtube." By orienting the flow equations of a distributed parameter model along streamtubes, spatial complexity in the equations may be reduced from three dimensions to two, while accomplishing a terrain-based model structure.

Contour-based terrain analysis as developed by Moore *et al.* (1988) and Moore and Grayson (1991) required three general steps. First, a contour map of the watershed was digitized, creating a vector DEM. Here the Arc/INFO geographic information system (GIS) was used to accomplish this task for watershed 56 (WS 56) at Coweeta, which is the primary experimental watershed for the riparian component of the Coweeta LTER (Figure 1, left). A preprocessing program (PREPROC) was used to transform the vector DEM for input to the program TAPES-C, which partitioned the watershed into streamtube, or hillslope, units using a constant offset between trajectory (i.e., stream tube boundary) starting points. Figure 1, center, shows results from a TAPES-C computation for WS 56 using a 30 meter offset, selected to correspond to the width of the experimental cut.

Further processing is then required to transform the streamtube output of TAPES-C into a structure suitable for IHDM4. Each hillslope plane in IHDM4 is represented by a two-dimensional vertical cross-section of finite-element nodes running longitudinally from watershed divide (or interior high point) to stream. At each vertical set of nodes in the cross-section, a constant width is assumed. From map view, a hillslope plane in IHDM4 is constrained to a series of adjacent trapezoids beginning at the stream and continuing to the divide (Yeakley, 1993). To fit TAPES-C output to IHDM4, the no-flow boundaries shown in Figure 1 (center) were extended to permanent stream locations using Arc/INFO to derive 21 hillslope planes (Figure 1, right).

3.4. INSTRUMENTATION AND DATA COLLECTION

The experimental transects for the planned *R. maximum* removal are shown in Figure 1 (right, planes #18 and #19). The lower 15 meters of each transect have been instrumented with tension lysimeters for the BA and B soil horizons for solute measurement (3 locations x 2 depths x 4 replicates per transect). The entire span of each transect, as well as upper slope positions continuing to the highpoint (HP1), has been instrumented with TDR measurement points for soil moisture measurement for the BA and B soil horizons (18 locations x 2 depths x 3 replicates). Installation of streambed and near-stream piezometers is in process to determine depth and slope of phreatic surfaces in the hillslopes (2 locations x 3 replicates per transect). These experimental transects correspond to streamtubes determined from the terrain analysis (Figure 1, right).

Sample collection frequency is weekly (composited to monthly) for lysimeters and bi-weekly for TDR and piezometer measurements. Storm samples for solute

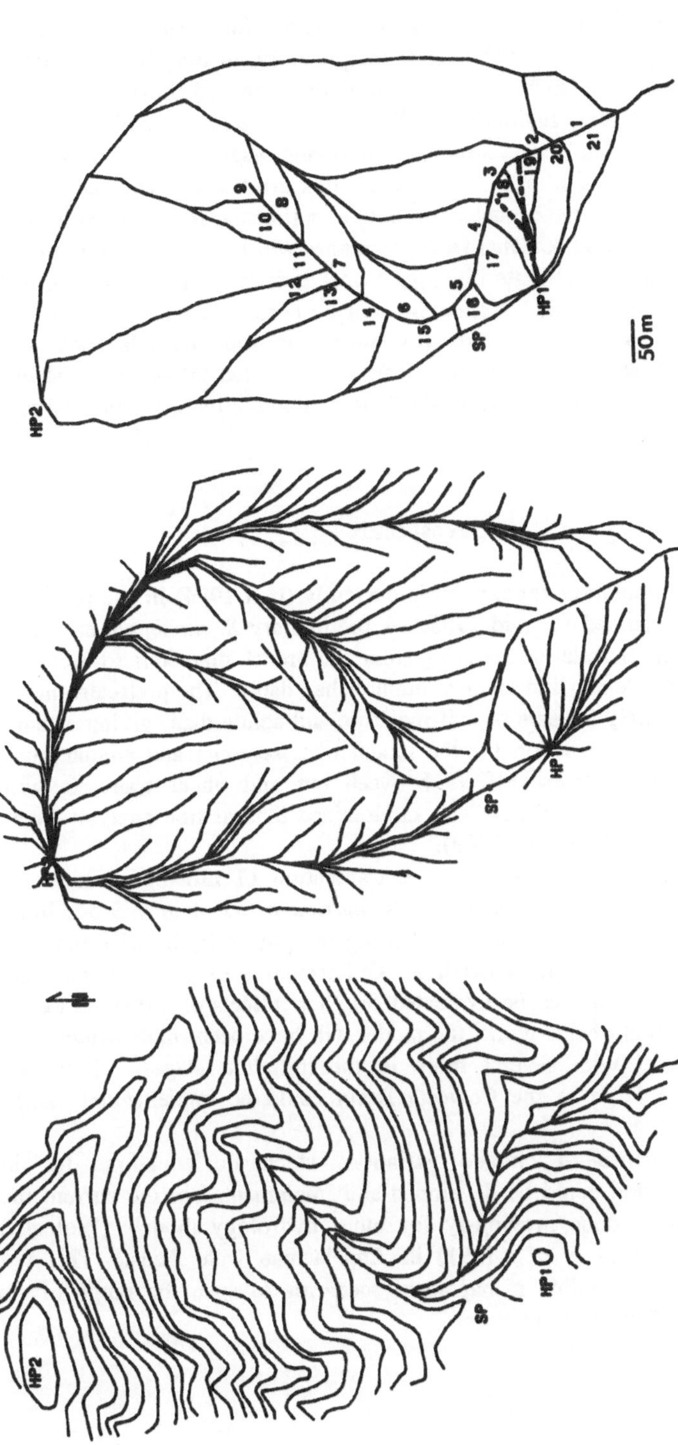

Figure 1. Terrain analysis of Watershed 56 at the Coweeta Hydrologic Laboratory. Shown at left is the original contour map with stream, two high points (HP1, HP2) and a saddle point (SP). Contours are from 725.4 m to 981.5 m M.S.L. (interval alternates between 6.10 m and 9.14 m). Shown center is "streamtube" delineation using TAPES-C (Moore and Grayson, 1991). Stream tube width interval was set at 30 meters. Shown at right are IHDM4 (Beven *et al.*, 1987) hillslope planes selected from TAPES-C analysis. Streamtube boundaries are specified as no flow boundaries. Shown are one headwater plane (#9) and twenty side planes. Also shown are control (in plane #18) and treatment (in plane #19) transects for the near-stream (< 15 m) vegetation removal experiment. Each transect (dotted lines from HP1 to stream) is the centerline of TDR, lysimeter and piezometer instrumentation.

concentrations, water content and flow are being taken for various storm types in throughfall collectors, flow-weighted stream samplers, lysimeters, TDR stations and piezometers throughout WS 56, with one spring storm collected thusfar (Webster and Yeakley, unpublished data). Estimates of depth to saturation from piezometers will provide further information for calibration, following suggestion from Wood and Calver (1992). Estimates of litterfall and detrital flux are being conducted with traps every 5 meters through the first 20 meters on both experimental transects. Soil moisture measurements over storm and seasonal response ranges will be used, along with WS 56 discharge measurements, to calibrate hillslope model parameters as was performed for nearby WS 2 in Yeakley (1993). Validation will be performed using both measured streamflow and hillslope soil moisture distribution on time periods successive to calibration periods. Water fluxes resulting from the validated hillslope model will be coupled with solute concentration measurements to compute hillslope elemental exports for the experimental transects.

4. Prospectus

The *R. maximum* manipulation experiment is currently (mid 1994) in the pretreatment phase, with removal planned for mid 1995. A preliminary *R. maximum* cut has been conducted, with allometric relations relating stem and branch diameters to biomass and leaf area in progress (Coleman and Haines, unpublished data). The pretreatment period will provide approximately 2 years of soil moisture and solute data, giving a temporal control and calibration information for the model over seasonal and annual response ranges. Analysis for treatment effects between cut and uncut transects will be conducted using a BACI approach, such as randomized intervention analysis (Stewart-Oaten *et al.*, 1986; Carpenter *et al.*, 1989).

The model presented here allows for extrapolation of hillslope results to the watershed scale. Coupled with estimates of *R. maximum* extent in WS 56, hillslope results from this experiment can be extended using the physically-based framework of the model to the remainder of the watershed. Cut scenarios for selected hillslopes or for the entire watershed can then be evaluated for their effects on hillslope export of solutes. With completion of this work, the model will have been implemented for two watersheds at Coweeta, with several more in the initiation stages. Management strategies for areas as large as the Coweeta Basin could be implemented using the model.

In terms of response to climate variation, stochastic equations for input meteorological information could be developed and implemented in the current model framework, as has been done previously for other hydrology models (Wolock and Hornberger, 1991; Yeakley *et al.*, 1991) in Appalachian watersheds. This would extend understanding of possible management scenarios to span the range of storm types (March *et al.*, 1979) and sequences.

Acknowledgements

This work is a part of the riparian component of the Coweeta LTER (National Science Foundation grant #BSR 90 11661) and has greatly benefited from ongoing discussion with B. Argo, D. Coleman and B. Haines. Helpful comments on this manuscript were received from C. Trettin and two anonymous reviewers. B. Argo and K. Saari conducted the survey of the experimental transect areas on WS 56. K. Saari provided raw DEM data for WS 56 analysis. The late I. Moore of Australian National University provided TAPES-C software. K. Beven of Lancaster University, U.K., provided IHDM4 software.

References

Allen, T.H.F., O'Neill, R.V. and Hoekstra, T.W.: 1984, *Interlevel Relations in Ecological Research and Management: Some Working Principles from Hierarchy Theory.* Gen. Tech. Rep. RM-110. USDA Rocky Mountain Forest and Range Experimental Station, Fort Collins, Co.

Anderson, M.G. and Burt, T.P.: 1978, The role of topography in controlling throughflow generation. *Earth Surface Proc.* 3, 331-344.

Anonymous: 1991, *Forests & Water Guidelines*. U.K. Forestry Commission, Forest Research Station, Wrecclesham, Farnham, Surrey, U.K.

Beven, K.J., Calver, A., and Morris, E.M.: 1987, *The Institute of Hydrology Distributed Model.* Institute of Hydrology Report No. 98, Wallingford, U.K.

Beven, K J., Wood, E.F. and Sivapalan, M.: 1988, On hydrological heterogeneity - Catchment morphology and catchment response. *J. Hydrol.* 100, 353-375.

Blackburn, W.H. and Wood, J.C.: 1990, Nutrient export in stormflow following forest harvesting and site-preparation in east Texas. *J. Environ. Qual.* 19, 402-408.

Calver, A.: 1988, Calibration, sensitivity and validation of a physically-based rainfall-runoff model. *J. Hydrol.* 103, 103-115.

Calver, A. and W.L. Wood.: 1989, On the discretization and cost-effectiveness of a finite element solution for hillslope subsurface flow. *J. Hydrol.* 110, 165-179.

Campbell, G.S.: 1974, A simple method for determining unsaturated conductivity from moisture retention data. *Soil Sci.* 117, 311-314.

Carpenter, S.R., Frost, T.M., Heisey, D. and Kratz, T.K.: 1989, Randomized intervention analysis and the interpretation of whole-ecosystem experiments. *Ecol.* 70, 1142-1152.

Clapp, R.B. and Hornberger, G.M.: 1978, Empirical equations for some soil hydraulic properties. *Water Resources Res.* 14, 601-604.

Clinton, B.D.: 1989, *Regeneration Patterns and Characteristics of Drought-Induced Canopy Gaps in Oak Forests of the Coweeta Basin.* M.S. Thesis, University of Georgia, Athens.

Correll, D.L. and Weller, D.E. 1989, Factors limiting processes in freshwater wetlands: an agricultural primary stream riparian forest, in R.R. Sharitz and J.W. Gibbons (eds.): *Freshwater Wetlands and Wildlife*. USDOE Office of Scientific and Technical Information, Oak Ridge, Tennessee, 9-23.

Crabtree, R.W. and Burt, T.P.: 1983, Spatial variation in solutional denudation and soil moisture over a hillslope hollow. *Earth Surface Proc. Landforms* 8, 151-160.
Day, F.P. and Monk, C.D.: 1974, Vegetation patterns on a southern Appalachian watershed. *Ecol.* 55, 1064-1074.
Day, F.P. and McGinty, D.T.: 1975, Mineral Cycling Strategies of Two Deciduous and Two Evergreen Tree Species on a Southern Appalachian Watershed, in F. G. Howell, J. B. Gentry and M. H. Smith (eds.): *Mineral Cycling in Southeastern Ecosystems*. NTIS, Springfield, Virginia, 736-743.
Day, F.P. and Monk, C.D.: 1977a. Net primary production and phenology on a southern Appalachian watershed. *Am. J. Bot.* 64, 1117-1125.
Day, F.P. and Monk, C.D.: 1977b. Seasonal nutrient dynamics in the vegetation on a southern Appalachian watershed. *Am. J. Bot.* 64, 1126-1139.
Dunne, T., Moore, T.R. and Taylor, C.H.: 1975, Recognition and prediction of runoff-producing zones in humid regions. *Hydrol. Sci. Bull.* 20, 305-327.
Fahey, T.J., Hughes, J.W., Pu, M., Arthur, M.A.: 1988, Root decomposition and nutrient flux following whole-tree harvest of northern hardwood forest. *Forest Sci.* 34, 744-768.
Feddes, R., Kowalik, P., Kolinska-Malinka, K. and Zaradny, H.:1976, Simulation of field water uptake by plants using a soil water dependent root extraction function. *J. Hydrol.* 31, 13-26.
Foster, N.W., Nicolson, J.A., and Hazlett, P.W.: 1989, Temporal variation in nitrate and nutrient cations in drainage waters from a deciduous forest. *J. Environ. Qual.* 18, 238-244.
Fowler, D., Cape, J.N. and Unsworth, M.H.: 1989, Deposition of atmospheric pollutants on forests. *Phil. Trans. Royal Soc. London* B324, 247-265.
Franklin, J.F., Bledsoe, C.S. and Callahan, J.T.: 1990, Contributions of the Long-Term Ecological Research Program. *BioSci.* 40:509-523.
Gaskin, J.W., Dowd, J.F., Nutter, W.L. and Swank, W.T.: 1989, Vertical and lateral components of soil nutrient flux in a hillslope. *J. Environ. Qual.* 18, 403-410.
Geyer, D.J., Keller, C.K., Smith, J.L. and Johnstone, D.L.: 1992, Subsurface fate of nitrate as a function of depth and landscape position in Missouri Flat Creek watershed, U.S.A. *J. Contam. Hydrol.* 11, 127-147.
Gilvear, D.J., Andrews, R., Tellam, J.H., Lloyd, J.W. and Lerner, D.N.: 1993, Quantitification of the water balance and hydrogeological processes in the vicinity of a small groundwater-fed wetland, East Anglia, UK. *J. Hydrol.* 144, 311-344.
Gregory, S.V., Swanson, F.J., McKee, W.A., and Cummins, K.W.: 1991, An ecosystem perspective of riparian zones. *BioSci.* 41, 540-551.
Grove, T.S. and Malajczuk, N.: 1985, Nutrient accumulation by trees and understorey shrubs in an age-series of *Eucalyptus diversicolor* F. Muell. stands. *For. Ecol. Manage.* 11, 75-95.
Harvey, J.W. and Bencala, K.E.: 1993, The effect of streambed topography on surface-subsurface water exchange in mountain catchments. *Water Resources Res.* 29, 89-98.
Haycock, N.E. and Pinay, G.: 1993, Groundwater nitrate dynamics in grass and poplar vegetated riparian buffer strips during the winter. *J. Environ. Qual.* 22, 273-278.
Hewlett, J.D.: 1962, *Internal Water Balance of Forest Trees on the Coweeta Watershed*. Ph.D. Dissertation, Duke University, Durham, North Carolina.

Hewlett, J.D. and Hibbert, A.R.: 1963, Moisture and energy conditions within a sloping soil mass during drainage. *J. Geophys. Res.* 68, 1080-1087.

Hewlett, J.D. and Hibbert, A.R.: 1966. Factors Affecting the Response of Small Watersheds to Precipitation in Humid Areas, in *International Symposium on Forest Hydrology*. Pergamon Press, New York, 275-290.

Hewlett, J.D. and Nutter, W.L.: 1970. "The Varying Source Area of Streamflow from Upland Basins" in *Interdisciplinary Aspects of Watershed Management*, American Society of Civil Engineers, Bozeman, Montana, 65-83.

Hibbert, A.R. and Troendle, C.A.: 1988, Streamflow Generation by Variable Source Area, in W.T. Swank and D.A. Crossley (eds.): *Forest Hydrology and Ecology at Coweeta*. Springer-Verlag, New York, 111-127.

Hopmans, P., Flinn, D.W. and Farrell, P.W.: 1987, Nutrient dynamics of forested catchments in southeastern Australia and changes in water quality and nutrient exports following clearing. *For. Ecol. Manage.* 20, 209-231.

Hornbeck, J.W., Smith, C.T., Martin, Q.W., Tritton, L.M. and Pierce, R.S.: Effects of intensive harvesting on nutrient capitals of three forest types in New England. 1990, *For. Ecol. Manage.* 30, 55-64.

Jacobs, T.C. and Gilliam, J.W.: 1985, Riparian losses of nitrate from agricultural drainage waters. *J. Environ. Qual.* 14:472-478.

Johnson, D.W. and Todd, D.E.: 1990, Nutrient cycling in forests of Walker Branch Watershed, Tennessee: roles of uptake and leaching in causing soil changes. *J. Environ. Qual.* 19, 97-104.

Johnson, D.W., West, D.C., Todd, D.E. and Mann, L.K.: 1982, Effects of sawlog vs. whole-tree harvesting on the nitrogen, phosphorus, potassium, and calcium budgets of an upland mixed oak forest. *Soil Soc. Sci. Am. J.* 46, 1304-1309.

Jordan, T.E., Correll, D.L. and Weller, D.E.: 1993, Nutrient interception by a riparian forest receiving inputs from adjacent cropland. *J. Environ. Qual.* 22:467-473.

Kellman, M., Miyanishi, K. and Hiebert, P.: 1987, Nutrient sequestering by the understorey strata of natural *Pinus caribaea* stands subject to prescription burning. *For. Ecol. Manage.* 21, 57-73.

Knoepp, J.D. and Swank, W.T.: 1994: Long-term chemistry changes in aggrading forest ecosystems. *Soil. Sci. Soc. Am. J.* 58, 325-332.

Lowrance, R.: 1992, Groundwater nitrate and denitrification in a coastal plain riparian forest. *J. Environ. Qual.* 21:401-405.

Lowrance, R.R., Todd, R.L. and Asmussen, L.E.: 1983, Waterborne nutrient budgets for the riparian zone of an agricultural watershed. *Agric., Ecosyst. Environ.* 10, 371-384.

Lowrance, R., Todd, R., Fail, J., Hendrickson, O., Leonard, R. and Asmussen, L.: 1984a, Riparian forests as nutrient filters in agricultural watersheds. *BioSci.* 34, 274-277.

Lowrance, R.R., Todd, R.L. and Asmussen, L.E.: 1984b, Nutrient cycling in an agricultural watershed: I. Phreatic movement. *J. Environ. Qual.* 13, 22-27.

Luxmoore, R.J., King, A.W., and Tharp, M.L.: 1991, Approaches to scaling up physiologically based soil-plant models in space and time. *Tree Physiol.* 9, 281-292.

March, W.J., Wallace, J.R. and Swift, L.W.: 1979, An investigation into the effect of storm type on precipitation in a small mountain watershed. *Water Resources Res.*, 15, 298-304.

McGee, C.E. and Smith, R.C.: 1967, Undisturbed *R. maximum* thickets are not spreading. *J. Forestry* 65, 334-336.

McGinty, D.T.: 1972, *The Ecological Roles of* Kalmia latifolia *and* Rhododendron maximum *in the Hardwood Forest at Coweeta.* M.S. Thesis, University of Georgia, Athens.

McGinty, D.T.: 1976, *Comparative Root and Soil Dynamics on a White Pine Watershed in the Hardwood Forest in the Coweeta Basin.* Ph.D. Dissertation, University of Georgia, Athens.

Monk, C.D., McGinty, D.T. and Day, F.P.: 1985, The ecological importance of *Kalmia latifolia* and *Rhododendron maximum* in the deciduous forest of the southern Appalachians. *Bull. Torrey Bot. Club* 112, 187-193.

Monk, C.D. and Day, F.P.: 1988, Biomass, primary production, and selected nutrient budgets for an undisturbed watershed, in W.T. Swank and D.A. Crossley (eds.): *Forest Hydrology and Ecology at Coweeta*, Springer-Verlag, New York, 151-159.

Moore, I.D. and Grayson, R.B.: 1991, Terrain-based catchment partitioning and runoff prediction using vector elevation data. *Water Resources Res.* 27, 1177-1191.

Moore, I.D., O'Loughlin, E.M. and Burch, G.J.: 1988, A contour-based topographic model for hydrological and ecological applications. *Earth Surface Proc. Landforms* 13, 305-320.

Mulholland, P.J.: 1992, Regulation of nutrient concentrations in a temperate forest stream: roles of upland, riparian, and instream processes. *Limnol. Oceanogr.* 37, 1512-1526.

Mulholland, P.J.: 1993, Hydrometric and stream chemistry evidence of three storm flowpaths in Walker Branch Watershed. *J. Hydrol.* 151, 291-316.

Mulholland, P.J., Wilson, G.V. and Jardine, P.M.: 1990, Hydrogeochemical response of a forested watershed to storms: effects of preferential flow along shallow and deep pathways. *Water Resources Res.* 26, 3021-3036.

Nilsen, E.T.: 1986, Quantitative phenology and leaf survivorship of Rhododendron maximum in contrasting irradiance environments of the southern Applachian mountains. *Am. J. Bot.* 73, 822-831.

O'Brien, A.K., Rice, K.C., Kennedy, M.M. and Bricker, O.P.: 1993, Comparison of episodic acidification of mid-Atlantic upland and coastal plain streams. *Water Resources Res.* 29, 3029-3040.

Omnerik, J.M., Abernathy, A.R. and Male, L.M.: 1981, Stream nutrient levels and proximity of agricultural and forest land to streams: some relationships. *J. Soil Water Conserv.* 36, 227-231.

Ormerod, S.J., Donald, A.P. and Brown, S.J.: 1989, The influence of plantation forestry on the pH and aluminium concentration of upland Welsh streams: A re-examination. *Environ. Pollut.* 62, 47-62.

Peterjohn, W.T. and Correll, D.L.: 1984, Nutrient dynamics in an agricultural watershed: observations on the role of a riparian forest. *Ecol.* 65, 1466-1475.

Peterjohn, W. T. and Correll, D. L.: 1986, The Effect of Riparian Forest on the Volume and Chemical Composition of Baseflow in an Agricultural Watershed, in

D.L. Correll (ed.): *Watershed Research Perspectives*. Smithsonian Institution Press, Washington, D.C., 244-262.

Phillips, D.L. and Murdy, W.H.: 1985, Effects of Rhododendron (*R. maximum maximum* L.) on regeneration of southern Appalachian hardwoods. *Forest Sci.* 31, 226-233..

Pinay, G., Fabre, A., Vervier, Ph. and Gazelle, F.: 1992, Control of C,N,P distribution in soils of riparian forests. *Landscape Ecol.* 6, 121-132.

Pye, J.M. and Vitousek, P.M.: 1985, Soil and nutrient removals by erosion and windrowing at a southeastern U.S. Piedmont site. *For. Ecol. Manag.* 11, 145-155.

Qualls, R.G., Haines, B.L. and Swank, W.T.: 1991, Fluxes of dissolved organic nutrients in a deciduous forest. *Ecol.* 72, 254-266.

Reynolds, B., Stevens, P.A., Adamson, J.K., Hughes, S. and Roberts, J.D.: 1992, Effects of clearfelling on stream and soil water aluminium chemistry in three UK forests. *Environ. Pollut.* 77, 157-165.

Rutter, A.J., Kershaw, K.A., Robbins, P.C. and Morton, A.J.: 1971, A predictive model of rainfall interception in forests. I. Derivation of the model from observations in a plantation of Corsican pine. *Agric. Meteorol.* 9, 367-384.

Rutter, A.J., Morton, A.J. and Robbins, P.C.: 1975, A predictive model of rainfall interception in forests. II. Generalization of the model and comparison with observations in some coniferous and hardwood stands. *J. Appl. Ecol.* 12, 367-380.

Schlosser, I.J. and Karr, J.R.: 1981, Water quality in agricultural watersheds: impact of riparian vegetation during base flow. *Water Resources Bull.* 17, 233-240.

Schnabel, R.R., Urban, J.B. and Gburek, W.J.: 1993, Hydrologic controls in nitrate, sulfate, and chloride concentrations. *J. Environ. Qual.* 22:589-596.

Sharitz, R.R., Boring, L.R., Van Lear, D.H., and Pinder, J.E.: 1992, Integrating ecological concepts with natural resource management of southern forests. *Ecol. Applic.* 2:226-237.

Stewart-Oaten, A, Murdoch, W.W. and Parker, K.R.: 1986, Environmental impact assessment: "Pseudoreplication" in time? *Ecol.* 67,929-940.

Swank, W.T.: 1986, Biological control of solute losses from forest ecosystems, in S.T. Trudgill (ed.): *Solute Processes*. John Wiley & Sons, New York,

Swank, W.T.: 1988, Stream chemistry responses to disturbance, in W.T. Swank and D.A. Crossley (eds.): *Forest Hydrology and Ecology at Coweeta*, Springer-Verlag, New York, 339-357.

Triska, F.J., Kennedy, V.C., Avanzino, R.J., Zellweger, G.W. and Bencala, K.E.: 1989, Retention and transport of nutrients in a third order stream: hyporheic processes. *Ecol.* 70, 1877-1892.

Triska, F.J., Duff, J.H., and Avanzino, R.J.: 1993a, The role of water exchange between a stream channel and its hyporheic zone in nitrogen cycling at the terrestrial-aquatic interface. *Hydrobiol.* 251, 167-184.

Triska, F.J., Duff, J.H., and Avanzino, R.J.: 1993b, Patterns of hydrological exchange and nutrient transformation in the hyporheic zone of a gravel-bottom stream: examining terrestrial-aquatic linkages. *Freshwater Biol.* 29, 259-274.

Turner, J., Lambert, M.J. and Holmes, G.: 1992, Nutrient cycling in forested catchments in southeastern New South Wales. 1. Biomass accumulation. *For. Ecol. Manage.* 55, 135-148.

Van Cleve, K. and Martin, S. (eds.): 1991, *Long-term Ecological Research in the United States*, LTER Publication No. 11, Long-Term Ecological Research Network Office, Seattle, Washington.

Velbel, M.A.: 1988, Weathering and soil-forming processes, in W.T. Swank and D.A. Crossley (eds.): *Forest Hydrology and Ecology at Coweeta*, Springer-Verlag, New York, 93-102.

Verry, E.S. and Timmons, D.R.: 1982, Waterborne nutrient flow through an upland-peatland watershed in Minnesota. *Ecol.* 63, 1456-1467.

Vitousek, P.M. and Reiners, W.A.: 1975, Ecosystem succession and nutrient retention: a hypothesis. *BioSci.* 25, 376-381.

Waters, D. and Jenkins, A.: 1992, Impacts of afforestation on water quality trends in two catchments in mid-Wales. *Environ. Pollut.* 77, 167-172.

Wolock, D.M. and Hornberger, G.M.: 1991, Hydrological effects of changes in levels of atmospheric carbon dioxide. *J. Forecast.* 10, 105-116.

Wood, W.L. and Calver, A.: 1992, Initial conditions for hillslope hydrology modelling. *J. Hydrol.* 130, 379-397.

Yeakley, J.A.: 1993, *Hillslope Soil Moisture Gradients in an Upland Forested Watershed.* Ph.D. Dissertation, University of Virginia, Charlottesville.

Yeakley, J.A. and Cale, W.G.: 1991, Organizational levels analysis: a key to understanding processes in natural systems. *J. Theor. Biol.* 149:203-216.

Yeakley, J.A., Swank, W.T., Hayden, B.P., Hornberger, G.M., Vose, J.M. and Shugart, H.H.: 1991, Variability of hydrologic components in a forested watershed during temperature change, in Preprint Vol. of *Amer. Met. Soc. Special Session on Hydrometeorol.* Boston, 195.

Yount, J.D. 1975, Forest-floor nutrient dynamics in southern Appalachian hardwood and white pine plantation ecosystems. in F.G. Howell and M.H. Smith (eds.): *Mineral Cycling in Southeastern Ecosystems.* ERDA Symposium Series, 598-608.

PLANT COMMUNITY COMPOSITION AND SURFACE WATER CHEMISTRY OF FEN PEATLANDS IN WEST VIRGINIA'S APPALACHIAN PLATEAU

MARK R. WALBRIDGE
Department of Biology, George Mason University, Fairfax, VA 22030-4444, U.S.A.

Abstract. I analyzed plant community composition, surface water chemistry, soil saturation, landscape position, and disturbance history in 4 small peatlands in WV's Allegheny Plateau, to determine vegetational differences among communities and identify environmental variables associated with community patterning. Thirty-four plant communites were identified, representing 5 physiognomic types: forest, tall and low shrub, herbaceous, and bryophyte. Of 138 species, only 34 were common to all sites; 56 were unique to single sites. Principal components analysis identified a major physiognomic separation between forest and tall shrub communities with less acid surface waters (pH 4.6 - 5.0) dominated by base cations (Ca^{++}, Mg^{++}, Na^{+}, K^{+}), vs. low shrub and bryophyte communities with more acidic surface waters (pH 4.0 - 4.4). Much of the variation in community composition resulted from changes in the distributions of *Hypericum densiflorum*, *Rubus hispidus*, *Polytrichum commune*, and *Sphagnum fallax*, with changes in soil saturation. Community distribution reflected an underlying pattern of basin geomorphology modified by beaver disturbance.

1. Introduction

Approximately 36,000 ha of forested, shrub, and emergent wetlands occur in WV's Appalachian Highlands (Tiner 1987). Many are fen peatlands that receive nutrients from both precipitation and groundwater flow, including several large complexes--Canaan Valley (Fortney 1975), Cranberry Glades (Darlington 1943, Edens 1973), and Cranesville Swamp (Robinette 1964)--and numerous smaller wetlands (e.g., Gibson 1970, Wieder et al. 1981). Similar systems occur throughout the southern Appalachians, though in decreasing number and size outside WV (Weakley and Schafle 1994).

WV fen peatlands share a common geomorphology, occurring in small watersheds at the headwaters of mountain streams, exhibit a characteristic pattern of plant community distribution, and have disturbance histories that include clearcutting (often followed by fire) associated with late 19th/early 20th century timber harvests, and periodic beaver colonization (Darlington 1943, Wieder et al. 1981, Fortney 1975).

Both North American and European peatlands are frequently classified along a gradient from bog to poor fen to rich fen (Sjors 1950, Gorham 1967, Heinselman 1970, Glaser et al. 1981). Ombrotrophic bogs exist on topographic highs, with water tables elevated above the regional groundwater table, and receive water and nutrients solely by precipitation. Minerotrophic fens receive additional nutrient inputs in groundwater, and are characterized as poor vs. rich depending on whether groundwater inputs arise within the immediate watershed, or are more extensive. Differences in the degree of mineral soil influence are reflected in both vegetation and surface water chemistry ("rich" and "poor"

Water, Air and Soil Pollution 77: 247–269, 1994.

TABLE I
Study site locations and baseline physical and chemical characteristics

	site			
	Big Run	Tub Run	Laurel Run	Cupp Run
elevation (m)	981	950	1000	823
latitude	39° 7' 0" N	39° 7' 15" N	39° 13' 25" N	39° 30' 32" N
longitude	79° 34' 55" W	79° 33' 5" W	79° 19' 27" W	79° 31' 10" W
WV county	Tucker	Tucker	Grant	Preston
stream order	1st	2nd	1st	1st
outflow stream pH	4.5	4.5	4.6	5.6
wetland area (ha)	15	23	6	21
watershed area (ha)	280	277	103	365
wetland:watershed ratio (%)	5.2	8.2	5.8	5.8
underlying geologic formation	Pottsville	Upper Freeport	Upper Mahoning	Catskill
underlying rock type	sandstone	sandstone/ coal	sandstone	shale/ sandstone

originally referred to fen flora (Sjors 1950)); the bog--poor fen--rich fen gradient represents a gradient of increasing pH and base cation concentrations (calcium (Ca^{2+}), magnesium (Mg^{2+}), sodium (Na^{+}), and potassium (K^{+})) in surface waters (Gorham 1967). Various Ca^{2+} concentrations have been suggested as a boundary between bogs and fens, but a more important characteristic signalling this change may be the shift in dominant cation from hydrogen (H^{+}) in bogs, to Ca^{2+} in fens (Moore and Bellamy 1974).

North American and European peatlands often exhibit characteristic patterns of plant community distribution (e.g., Vitt and Slack 1975, Glaser et al. 1981, Moore and Bellamy 1974). In contrast, plant community distribution in WV peatlands has been described as an 'irregular mosaic' (i.e., having no readily apparent pattern) (Wieder et al. 1981). My objectives were to analyze plant community composition in 4 WV peatlands, and identify underlying environmental factors associated with community patterning.

2. Study Sites

Study sites (Table I) were selected from 57 wetlands (2.5 to 225 ha in area) included in an air photo survey of WV's Allegheny Plateau. The survey was conducted October 15 - 22, 1979, at peak fall coloration, and covered portions of 9 WV counties and a small section of Garrett Co., MD. Complete coverage of each site was provided at an approximate scale of 1:6000 (Walbridge 1982).

WV fen peatlands occur in regions of high mean annual precipitation (139 - 154 mm) and low mean annual temperatures (8 - 9 °C) (U.S. Dept. of Commerce 1980, Walbridge and Lang 1982). Regional climatic conditions are intensified by local topography, creating 'frost pockets' that receive cold air drainage from surrounding uplands and are frequently impacted by fog (Wieder et al. 1981, Walbridge 1982, Walbridge and Lang 1982).

Logging and fire histories for Big Run (BR), Tub Run (TR), and Laurel Run (LR) are typical for Tucker Co., WV, during the turn-of-the-century logging boom (Thompson 1974, Allard and Leonard 1952) (Table II). Logging at Cupp Run (CR) occurred about 40 years earlier, and was not accompanied by post-logging wildfire (Brooks 1910).

All sites showed evidence of past beaver activity, but no active colonies were observed during this study (Table II). WV's beaver population was extirpated by 1825. Restocking began in 1933 (Bailey 1954). BR, LR, and TR lie in the range for beaver estimated by Swank (1949); CR lies in the 1952 range (Bailey 1954). Agricultural Stabilization and Conservation Service air photos show that beaver colonized CR between 1953 and 1967, and were established at TR by 1958. Beaver activity ceased at BR after 1974 (Wieder et al. 1981), but was renewed in 1981 and has continued to the present (Table II).

3. Methods

3.1. VEGETATIONAL ANALYSIS

Air photos were used to delineate polygons of similar vegetation in each wetland based on infrared color, texture, and vegetation height (Table III). Working vegetation maps were prepared by outlining polygons on mylar overlays while viewing photo pairs under a stereoscope. Maps were then used to locate polygons in the field. During summer 1980, the vegetational composition of each polygon was estimated using a series ($n \geq 5$ per polygon) of randomly placed nested quadrats (2 x 10 m for trees; 2 x 5 m for shrubs; 1 x 2 m for herbs; 0.5 x 1 m for bryophytes). Only polygons large enough to accomodate at least 5 quadrats were sampled (Walbridge 1982). Most samples were collected over a 1-2 week period (LR - 68.8 %, 6/22 to 7/3; CR - 88.1 %, 7/9 to 7/17; TR - 88.2%, 7/23 to 7/29; BR - 79.0%, 9/10 to 9/16), with remaining samples collected in late summer/early fall (CR - 8/19, 8/22; TR - 8/20; LR - 8/21, 8/27; BR - 10/8). A total of 466 quadrats were sampled (BR = 95; TR = 133; LR = 70; CR = 163).

Percent cover was estimated visually for each species in each quadrat. Species not readily identifiable in the field were collected for later lab identification. Vascular plant nomenclature follows Strasbaugh and Core (1977). Bryophyte nomenclature follows Welch (1957), with specimens previously identified as *Sphagnum recurvum* revised to *Sphagnum fallax* (Andrus 1980).

Following field checking, working maps were revised to produce vegetation maps for each site to an exact scale of 1:3000. Wetland and polygon areas were estimated from these maps using a Kent compensating polar planimeter (Walbridge 1982).

3.2. LANDSCAPE POSITION, SOIL SATURATION, AND WATER CHEMISTRY

Each polygon was assigned to one of 6 landscape positions (stream margin, wetland margin, glade, old beaver pond, old beaver dam, entire wetland). Soil saturation was estimated qualitatively in each 1 x 2 m quadrat using a 5 point scale (1 = extremely dry; 2 = moderately dry; 3 = damp; 4 = saturated (water table at/near the surface); 5 = flooded). Plots with intermediate conditions were given half values (Walbridge and Lang 1982).

On June 8, 1981, one surface water sample was collected from each polygon. Samples were collected in acid-rinsed Nalgene bottles from standing water when present, or from 50

TABLE II

Known disturbance histories of study sites

	site			
	Big Run	Tub Run	Laurel Run	Cupp Run
present disturbance in watershed	none	inactive coal surface mine	none	agriculture
logging	clearcut @ 1900	clearcut @ 1900	clearcut @ 1900	clearcut @ 1860
fire	@1900 (post-logging)	@1900 (post-logging)	@1900 (post-logging)	periodically burned (lower portion only)
beaver				
last colonized*	between 1950 and 1974	by 1958	unknown	between 1953 and 1967
abandoned*	after 1974**	between 1958 and 1979	unknown	between 1967 and 1979

*Determined from available ASCS aerial photography

**Evidence of renewed activity during summer 1981 and continued occupation to the present

TABLE III
Identification codes describing infrared color, texture, and vegetation height of vegetation polygons

identification code	description
DGS	dark green shrub
DK	dark
DKR	dark red
DIST	disturbed
DRB	dark red-brown
GCL	green coarse low
GCLRC (composite)	green coarse low/red coniferous
GG	gray-green
GN	green
LB	light brown
LG	light green
LRB	light red-brown
M	mixed
MAG	magenta
O	orange
PK	pink
R	red
RB	red-brown
RC	red coniferous
RS	red shrub
RSF	red swamp forest
RT	red trees
RW	red and white
RY	Red-Yellow
TAN	tan
W	white
YG	yellow-green

cm diameter wells allowed to equilbrate for ≥ 30 min before collection. Samples were kept on ice until returning to the lab, analyzed for pH and conductivity, and then frozen. Samples were analyzed for Ca^{2+}, Mg^{2+}, and dissolved Fe using a Varian AA6 atomic absorption spectrophotometer, with excess lanthanum added prior to Ca^{2+} and Mg^{2+} analysis. Sodium and K^+ were analyzed by flame photometry; ammonium (NH_4^+) and nitrate (NO_3^-) were determined on an autoanalyzer (Technicon Industrial Systems 1977).

3.4. STATISTICAL ANALYSIS

For each site, mean percent cover values were determined for each species in each polygon, and polygons were grouped by agglomerative cluster analysis (unweighted arithmetic mean method) (Sokal and Sneath 1963). Clusters (communities) were defined as polygons separated by Euclidean distances ≤ 50 or 55, depending on the site. Major lines of variation in species composition were determined by principal components analysis (PCA) of mean percent cover values (covariance matrix), and correlations between original and component variables were determined (SAS 1982). Polygons were plotted in component space (ordination) and grouped by communities identified by cluster analysis.

Major lines of variation in species composition across sites were also analyzed by cluster analysis and PCA as described above, but using only species (n=78) with either a frequency $\geq$ 60%, or a mean percent cover $\geq$ 10%, in at least one polygon.

For communities represented by multiple polygons, significant differences in surface water chemistry were analyzed by ANOVA and Duncan's NMR test (SAS 1982). Major lines of variation in surface water chemistry among plant communities were analyzed by PCA (correlation matrix), and relationships based on surface water chemistry were compared with relationships based on vegetational composition.

4. Results

4.1. VEGETATIONAL ANALYSIS AND CLASSIFICATION

Vegetational analysis identified 138 species--13 trees, 25 shrubs, 85 herbs, and 15 bryophytes (mosses) (Table IV). Only 34 species were common to all sites; 56 were unique to single sites. The total number of species per site ranged from 63 to 99.

Nearly all species in Table IV have ranges that extend northward beyond WV (Fernald 1970). At least 48 species in Table IV (37.2%) also occur in peatlands in the northern U.S. and Canada (Heinselmann 1970, Vitt et al. 1975, Vitt and Slack 1975, Richardson et al. 1976, Schwintzer 1978a, Schwintzer 1978b, Horton et al. 1979, Slack et al. 1980, Glaser et al. 1981, Schwintzer and Tomberlin 1982, Chee and Vitt 1989). Only 3 vascular species have limited northern ranges (*Hypericum densiflorum, Ilex montana,* and *Smilax glauca*) (Fernald 1970).

Cluster analysis (Figure 1) identified 34 plant communities (7 to 12 per site) representing 5 physiognomic types: forest, tall shrub, low shrub, herbaceous, and bryophyte (Table V).

Forest communities were dominated by *Tsuga canadensis* or *Picea rubens*, with an understory of *Rhododendron maximum*, and *Acer rubrum* and *Betula alleghaniensis* as common canopy associates. Forest communities were found along wetland margins or over the entire wetland, and had average soil moistures of 3.2 to 4.2 (Table Va, c-d).

Tall shrub communities were dominated by *Alnus rugosa, Salix sericea*, or *Viburnum recognitum.* Both *A. rugosa* and *S. sericea* communities were found along wetland margins with 'wet' and 'dry' variants that differed in understory vegetation. *A. rugosa* communities were also found along stream margins. *V. recognitum* communities were found lining old beaver dams, with average soil moistures of 3.3 - 3.6 (Table Vb-d).

Low shrub communities dominated by *Hypericum densiflorum* occurred along both stream and wetland margins, within a narrow range of average soil moisture (3.0 - 3.2). Low shrub communities dominated by *Rubus hispidus* occupied stream margins, wetland margins, and old beaver ponds, with average soil moistures of 3.1 to 3.8 (Table V).

Herbaceous communities were found at all sites, but no single herbaceous community occurred at all sites. Dominants included *Leersia oryzoides, Typha latifolia, Carex canescens, Carex stricta, Juncus subcaudatus, Glyceria canadensis, Calamagrostis canadensis*, and *Dulichium arundinaceum.* Herbaceous communities had average soil moistures of 3.1 - 5.0, and were found primarily in senescent beaver ponds (Table V).

Bryophyte communities were dominated by *Polytrichum commune, Sphagnum fallax*, or a mixture of these two species. Although landscape position varied, these communities

TABLE IV

Vascular and bryophyte (moss) species found at Big Run, Tub Run, Laurel Run, and Cupp Run. Lines separate species unique to a single site or common to all sites.

		site			
species	type	Big	Tub	Laurel	Cupp
Hamamelis virginiana	shrub	+			
Ilex montana	shrub	+			
Nemopanthus mucronata*	shrub	+			
Vaccinium microcarpon	shrub	+			
Aster acuminata	herb	+			
**Carex baileyi*	herb	+			
Carex canescens*	herb	+			
Graminaceae (unid.)	herb	+			
Oxalis sp.	herb	+			
Rhyncospora alba*	herb	+			
*Drosera rotundifolia***	herb	+		+	
Sphagnum magellanicum*	moss	+		+	
Betula alleghaniensis*	tree	+		+	+
*Vaccinium angustifolium***	shrub	+		+	+
**Carex gynandra*	herb	+		+	+
**Carex incomperta*	herb	+		+	+
Chelone glabra	herb	+		+	+
Sphagnum papillosum*	moss	+		+	+
Fraxinus americana	tree	+			+
**Pyrus americana*	shrub	+			+
*Polytrichum juniperinum***	moss	+			+
Polytrichum ohioense	moss	+			+
*Sphagnum subsecundum***	moss	+			+
**Picea rubens*	tree	+	+		
**Danthonia compressa*	herb	+	+	+	
Pteridium aquilinum	herb	+	+	+	
**Carex folliculata*	herb	+	+		+
Panicum clandestinum	herb	+	+		+
**Dennstaedtia punctilobula*	herb	+	+		+
Acer rubrum*	tree	+	+	+	+
Tsuga canadensis*	tree	+	+	+	+
**Amelanchier laevis*	shrub	+	+	+	+
**Hypericum densiflorum*	shrub	+	+	+	+
Ilex verticillata*	shrub	+	+	+	+
**Kalmia latifolia*	shrub	+	+	+	+
**Pyrus arbutifolia*	shrub	+	+	+	+
**Rhododendron maximum*	shrub	+	+	+	+
**Salix sericea*	shrub	+	+	+	+
Vaccinium myrtilloides*	shrub	+	+	+	+
Viburnum cassinoides*	shrub	+	+	+	+
**Rubus hispidus*	shrub	+	+	+	+
**Agrostis hyemalis*	herb	+	+	+	+
**Agrostis perennans*	herb	+	+	+	+
*Carex leptalea***	herb	+	+	+	+
**Carex scoparia*	herb	+	+	+	+
Carex trisperma*	herb	+	+	+	+
Eriophorum virginicum*	herb	+	+	+	+
**Gentiana linearis*	herb	+	+	+	+
**Glyceria melicaria*	herb	+	+	+	+
**Juncus brevicaudatus*	herb	+	+	+	+
**Juncus effusus*	herb	+	+	+	+
**Leersia oryzoides*	herb	+	+	+	+
Polygonum saggitatum*	herb	+	+	+	+
Scirpus cyperinus*	herb	+	+	+	+
**Solidago rugosa*	herb	+	+	+	+
Solidago uliginosa*	herb	+	+	+	+
**Sparganium chlorocarpum*	herb	+	+	+	+
**Viola sp.*	herb	+	+	+	+
Osmunda cinnamomea*	herb	+	+	+	+
Polytrichum commune*	moss	+	+	+	+
Sphagnum fallax*	moss	+	+	+	+
Sphagnum girgensohnii*	moss	+	+	+	+
**Sphagnum imbricatum*	moss	+	+	+	+
Prunus serotina	tree		+	+	+
**Sambucus canadensis*	shrub		+	+	+
**Viburnum recognitum*	shrub		+	+	+
**Aster puniceus*	herb		+	+	+
**Carex lurida*	herb		+	+	+
**Carex stipata*	herb		+	+	+
**Eleocharis obtusa*	herb		+	+	+
**Eupatorium perfoliatum*	herb		+	+	+
Galium tinctorum*	herb		+	+	+
Glyceria striata*	herb		+	+	+
**Hypericum mutilum*	herb		+	+	+
Lycopus uniflorus*	herb		+	+	+
**Sphagnum fimbriatum*	moss		+	+	+
Ostrya virginiana	tree		+		
*Pinus strobus***	tree		+		
**Spiraea tomentosa*	shrub		+		
Callitriche heterophylla	herb		+		
Dulichium arundinaceum*	herb		+		
Mimulus sp.	herb		+		
*Epilobium leptophyllum***	herb		+	+	
Viburnum lentago	shrub		+		+
Calamagrostis canadensis*	herb		+		+
Carex flexuosa	herb		+		+
**Eleocharis acicularis*	herb		+		+
**Impatiens capensis*	herb		+		+
Panicum microcarpon	herb		+		+
Crataegus sp.	tree			+	
Gaultheria hispidula*	shrub			+	
**Aster dumosus*	herb			+	
*Aster umbellatus***	herb			+	
*Carex brunnescens***	herb			+	
Clematis virginiana	herb			+	
*Epilobium glandulosum***	herb			+	
Epilobium sp.	herb			+	
**Glyceria canadensis*	herb			+	
Habernia clavellata	herb			+	
**Juncus subcaudatus*	herb			+	
Poa palustris	herb			+	
Potentilla simplex	herb			+	
Rosa sp.	herb			+	
Solidago sp.	herb			+	
Calliergon schreberi	moss			+	
Sphagnum capillaceum*	moss			+	
Acer saccharum	tree			+	+
*Arisaema triphyllum***	herb			+	+
Hypericum virginicum*	herb			+	+
*Scirpus validus***	herb			+	+
**Solidago graminifolia*	herb			+	+
Typha latifolia*	herb			+	+
*Drypoteris cristata***	herb			+	+
**Onoclea sensibilis*	herb			+	+
*Aulacomnium palustre***	moss			+	+
Magnolia acuminata	tree				+
Nyssa sylvatica	tree				+
*Quercus rubra***	tree				+
Alnus rugosa*	shrub				+
Pyrus melanocarpa	shrub				+
Sassafras albidum	shrub				+
Smilax glauca	shrub				+
Caltha palustris	herb				+
*Carex rostrata***	herb				+
**Carex stricta*	herb				+
Cinna arundinacea	herb				+
Danthonia spicata	herb				+
**Eleocharis smallii*	herb				+
Juncus canadensis	herb				+
Mitchella repens	herb				+
Potamogeton epihydrus	herb				+
Potentilla norvegica	herb				+
**Symplocarpus foetidus*	herb				+
Thalictrum polygamum	herb				+
**Vernonia novaboracensis*	herb				+
*Dryopteris spinulosa***	herb				+
Atrichum crispum	moss				+
Atrichum undulatum	moss				+

*species chosen for across-site analysis

**species identified in literature review of 12 northern U.S. and Canadian wetlands

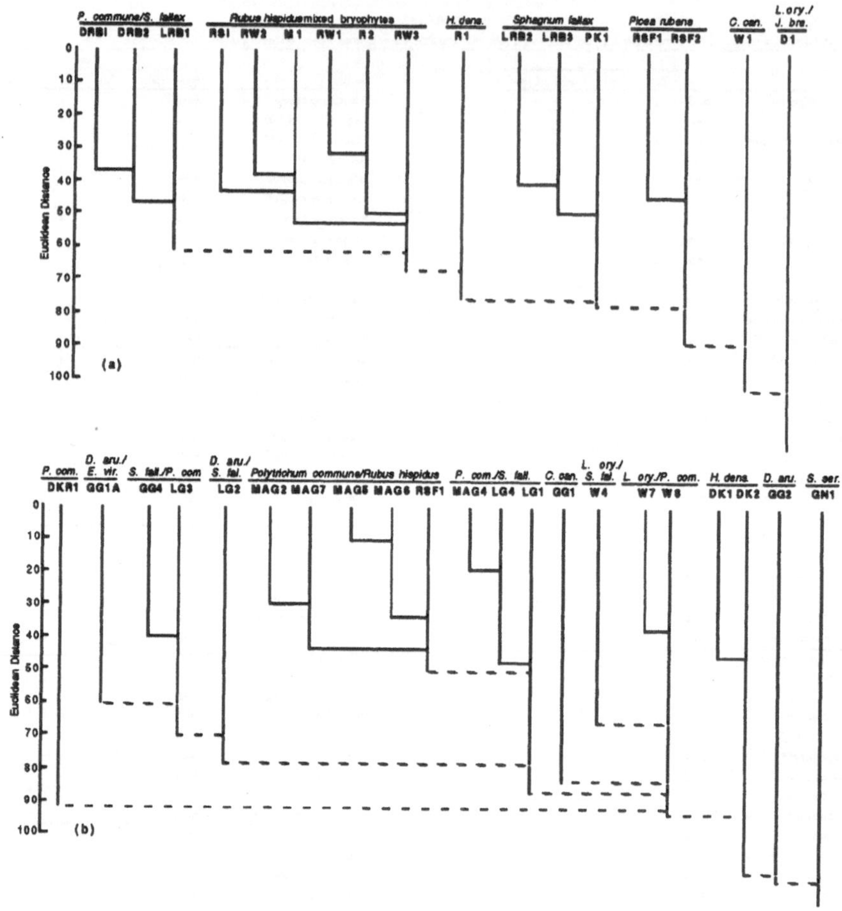

Fig. 1. Individual site cluster diagrams for Big Run (a) and Tub Run (b). Polygons are represented by vertical lines. Solid horizontal lines link community clusters; dashed lines show linkages at Euclidean distances > 50 or 55 (see Table V for community composition).

were primarily centered in 'glades', areas of deeper peat lying between stream and wetland margins. Average soil moistures ranged from 2.7 to 4.5 (Table Va-c).

Across-site cluster analysis identified polygon groupings similar to individual site analyses with two important differences (Figure 2). One was a group of 12 polygons, 11 of which represented herbaceous senescent pond communities at CR. These polygons fell out as 2 closely related groups in the across-site analysis, possibly representing 2 successive stages of senescent pond revegetation. A second was a group of 28 polygons from all 4 sites that included bryophyte, low shrub, and herbaceous communities with varying relative proportions of *H. densiflorum, R. hispidus, P. commune,* and *S. fallax.* Covering the full range of landscape positions and exhibiting varying soil moistures (2.7 - 4.4), these polygons were classified into 3 groups: 1) *P. commune/S. fallax* (mean soil moisture = 3.8 ± 0.2); 2) *P. commune/R. hispidus*(mean soil moisture = 3.4 ± 0.1), and 3) *R. hispidus* (mean soil moisture = 3.3 ± 0.1), which in combination with the *H.*

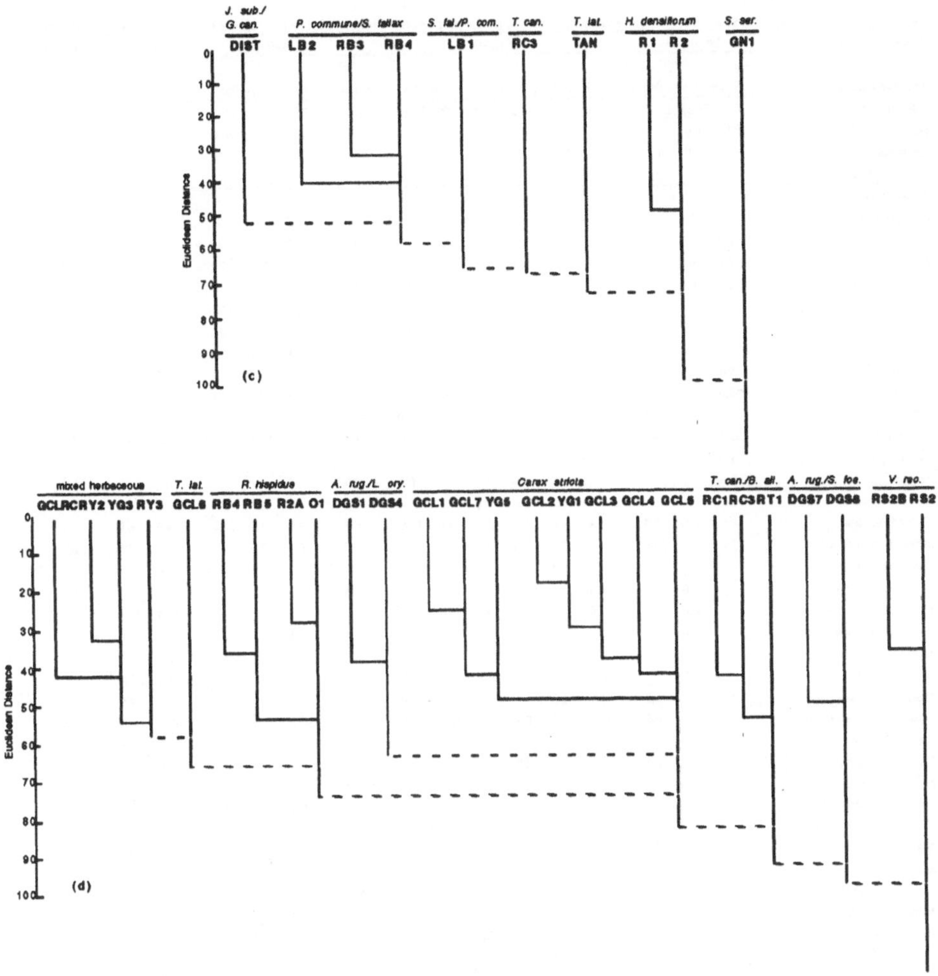

Fig. 1 (cont.). Individual site cluster diagrams for Laurel Run (c) and Cupp Run (d).

densiflorum and *S. fallax* communities, comprises a gradient, from *H. densiflorum* (mean soil moisture = 3.1 ± 0.1) to *S. fallax* (mean soil moisture = 4.2 ± 0.2).

4.2. WATER CHEMISTRY

Within sites, significant differences in surface water chemistry among plant communities were found only for Mg^{2+} and NO_3^-, and only at BR, LR, and CR (Walbridge 1982). Across sites, significant differences were observed in Ca^{2+}, Mg^{2+}, Na^+, NO_3^-, and pH (Table VI). Surface waters at CR and LR were generally higher in Ca^{2+}, Mg^{2+}, Na^+, and pH than at BR and TR. Highest Ca^{2+} concentrations occurred in *T. canadensis* forests at CR and LR. Base cation concentrations were generally higher in forest and tall shrub vs.

TABLE Va

Big Run vegetation data (means of % cover for 78 'important' species, n = 5 or greater per polygon, '+' indicates % cover < 0.1); see Walbridge (1982) for full data set

physiognomic type	bryophyte			low shrub							bryophyte			forest		herbaceous	
plant community	*P. commune/S.fallax*			*Rubus hispidus* /mixed bryophytes						*H.den.*	*Sphagnum fallax*			*Picea rubens*		*C.ca.*	*Lor./J.br.*
polygon code	DRB1	DRB2	LRB1	RS1	RW2	M1	RW1	R2	RW3	R1	LRB2	LRB3	PK1	RSF1	RSF2	W1	D1
polygon area (ha)	0.9	0.9	0.7	1.6	1.3	1.1	0.7	0.3	0.4	0.3	1.3	0.4	0.3	3.2	0.9	1.0	0.5
soil moisture class	4.0	3.9	4.4	3.4	3.5	3.8	3.4	3.1	3.2	3.2	4.3	4.5	3.9	3.7	4.2	3.4	3.8
landscape position*	E**	E	E	S,W	S	S,W	S	S	S,W	S	G	G	G	W	W	P	P
plant species																	
Carex trisperma	+	2.0	1.4								4.0			5.6			
Viburnum cassinoides	0.2	1.4	+	7.0		0.4	4.4	14.0		0.2	1.0		0.4		+		
Eriophorum virginicum	14.0	10.0	22.0	2.0	0.8	0.6	0.8				16.6	16.3	2.6			0.6	
Carex canescens	1.0	0.4		4.0	1.0	3.0		11.6	28.0	5.2		1.0				66.0	
Rhyncospora alba	3.6	6.0				0.6					14.2	24.6	10.0				
Pyrus arbutifolia	+	26.0	3.2	1.0	2.6	5.2		0.2	0.2	2.0	4.7	2.3	0.2	+			
Solidago uliginosa	12.2	5.4	10.2		11.0	4.0					4.8	9.6	7.4		0.2		
Ilex verticillata	5.0	3.2	1.8	10.2	15.2	22.0	+	0.2	2.6	1.2	5.2	1.4		5.7	9.0		
Polytrichum commune	65.0	49.2	34.0	4.2	22.2	9.0	46.0	43.0	25.0	16.0	0.1	4.6	1.8	0.4	2.0	48.0	
Rubus hispidus	37.0	44.0	38.0	55.0	52.0	50.0	71.0	66.0	53.0	46.0	31.2	25.0	54.0	26.5	23.0	20.0	
Sphagnum fallax	36.8	25.2	60.0	13.0	28.0	16.0	10.0	4.0	27.0	6.4	54.4	83.9	67.6	26.3	7.4	44.0	
Agrostis perennans	1.0						2.2						0.2	0.6			
Juncus effusus	3.0				0.8	1.0	5.2		0.4	1.2	1.0			0.3		1.0	
Agrostis hyemalis	2.0				0.2	5.6	9.0	+		6.6		0.4	0.2				2.0
Carex folliculata	4.0	6.0	0.6	1.4	8.0	3.4	12.0	5.0	17.0	4.6	1.0	1.0	30.0	3.4	4.0	0.2	0.2
Juncus brevicaudatus	2.0	0.8	13.4		5.4	7.4	1.2		+	2.0	0.6	7.7	7.0		1.0	0.2	21.6
Amelanchier laevis		+			0.4				+					0.6			
Hypericum densiflorum		1.0		0.4				17.0	5.2	41.4						1.4	
Nemopanthus mucronata		0.6		12.0			+				1.3			10.4	12.4		
Sphagnum magellanicum		5.0	5.0			7.0					19.3	5.0	1.0		30.8		
Scirpus cyperinus		0.4			2.0		6.4			0.4				0.3			0.2
Sphagnum imbricatum		10.0		0.4	3.4	4.0	6.0		10.0		2.0	0.3	6.0	27.6	20.8	1.0	
Osmunda cinnamomea			1.6	3.4	1.2				1.0						4.0		
Rhododendron maximum			0.6	18.4	5.0	0.8	1.2	10.0	0.2		3.8	0.3	1.0	39.5	34.0		
Betula alleghaniensis			0.6		1.0	0.8						+		0.7	1.6		
Tsuga canadensis			1.0			0.8								6.4	+		
Gentiana linearis				0.4			+	0.8	0.4	1.0							
Vaccinium myrtilloides				1.6	0.2						0.3		1.0				
Dennstaedtia punctilobula				11.0				+						1.5			
Sphagnum girgensohnii				2.0	6.0		1.0		20.0					13.1	10.0		
Acer rubrum				0.2			+						+	+	3.6		
Carex incomperta					2.4	3.4	0.6			9.0							
Sphagnum papillosum					4.0	29.0					15.0						
Sparganium chlorocarpum					1.0	1.6	3.0					3.3					4.0
Glyceria melicaria					0.2									10.9			
Kalmia latifolia					2.0	1.4			+					6.9	17.2		
Polygonum saggitatum					7.0	3.4	1.0			16.0				0.6			3.0
Leersia oryzoides					16.2	3.0	4.6			27.0		0.3		0.6			66.4
Salix sericea						+											
Picea rubens						1.4					0.2			3.5	5.4		
Solidago rugosa							1.4	+									
Danthonia compressa									0.2								
Pyrus americana									0.6					0.5	+		
Viola sp.										0.2				1.3	0.4		0.2
Carex baileyi														12.5			
Carex gynandra															4.0		
Carex scoparia																+	

*P=senescent beaver pond, E=entire wetland, G=glade, S=stream margin, W=wetland margin

**primarily centered in glades

low shrub and bryophyte communities. Lowest pH values occurred in bryophyte-dominated communities in general, and in all BR communities.

Characteristic of fens, Ca^{2+} was generally the dominant cation in surface waters. However, Ca^{2+}/H^{+} ratios < 1.0 were occasionally observed (Table VI).

TABLE Vb

Tub Run vegetation data (means of % cover for 78 'important' species, n = 5 or greater per polygon, '+' indicates % cover < 0.1); see Walbridge (1982) for full data set

physiognomic type	bryophyte	herbaceous D.er./	bryophyte		herbaceous D.er./	bryophyte								herbaceous L.or./				low shrub		herbaceous	tall shrub
plant community	P.ca	E.vir.	S.fal./P.com.		S.fal	P. commune/R. hispidus					P. comm./S. fall			C.ca.	S.fal	L.or./P.ca		H. densifl.		D.aru.	S.ser.
polygon code	DKR1	GG1A	GG4	LG3	LG2	MAG2	MAG7	MAG5	MAG6	RSF1	MAG4	LG4	LG1	GG1	W4	W7	W8	DK1	DK2	GG2	GN1
polygon area (ha)	0.2	0.7	0.7	0.5	0.1	0.2	0.9	1.6	0.5	3.2	0.4	0.1	0.3	0.8	0.2	0.2	0.2	0.2	0.4	0.1	0.3
soil moisture class	4.4	3.9	4.1	3.9	4.2	3.4	3.0	3.2	3.2	3.6	3.3	4.1	3.9	4.2	4.9	4.0	3.8	3.0	3.0	5.0	4.2
landscape position*	G	P	G	G	P	W	S,W	S	S,W	S,W	G	G	G	P	P	P	P	W	W	P	W
plant species																					
Eriophorum virginicum	14.0	25.0	20.6	12.6		10.2		0.7		0.3	1.0	7.8	3.2	0.6	0.3						
Viburnum cassinoides	0.4	2.1	1.4	3.3		1.0	+	1.4		10.7	3.0	5.2	0.6								
Gentiana linearis	0.4	0.6	0.4	0.3					0.2	1.2						1.0					1.2
Solidago uliginosa	39.0	5.6	20.7	22.1			1.0	0.8	1.6	7.9	4.0	4.4	4.0					5.0			7.4
Carex scoparia	2.0		0.1	0.3	0.2	1.0	0.6	0.2	0.8	1.3						12.2	16.0	3.2			17.4
Agrostis hyemalis	0.6	0.3	0.1	0.9	5.6	0.8	0.6	3.4	0.6	1.6		0.6	0.6		0.6	10.0	5.0	1.0	1.2		
Hypericum densiflorum	2.4	0.1	1.4	6.4	2.4	2.0	17.4	18.3	14.8	12.5	14.0	2.4	2.4			13.0	9.8	75.0	79.0		2.0
Scirpus cyperinus	3.0		0.7	0.1	20.0	2.4		3.5	9.8	1.4	0.4		4.2		2.3	16.6	6.6	3.2	3.2		
Polytrichum commune	91.0	53.8	22.9	47.5	9.0	44.0	34.0	63.8	64.4	65.0	34.6	38.0	56.6	13.0	2.1	46.0	39.0	29.0	69.0		2.0
Glyceria melicaria	3.4																				1.0
Solidago rugosa	0.4						1.2			2.7								2.4	0.4		
Carex folliculata		1.9	2.1							0.4											
Amelanchier laevis		0.5	0.1	0.1			+			2.2	1.0										
Carex trisperma		2.5	2.9				4.0			0.4		2.0									
Pyrus arbutifolia		10.9	1.3	2.5		0.6		1.8		0.5	0.8	+									
Sphagnum fallax		9.4	62.9	46.3	64.0	3.0	4.0	5.4	4.0	5.9	37.0	35.4	20.0	12.0	27.9	2.0		4.0	1.0		
Dulichium arundinaceum		26.9		15.6	45.0	3.4	0.2	2.6		0.6	10.2	9.6	39.0	16.0	6.9		0.6			85.0	
Juncus brevicaudatus		10.9	4.7	5.4	13.0	2.0				0.6	4.0	12.4	0.6	4.0	0.4	4.4	4.0			1.3	
Rubus hispidus		15.0	14.3	18.8	5.4	83.6	78.0	66.5	63.0	36.3	45.0	52.0	65.0			15.6	2.0	19.0	36.0		32.0
Leersia oryzoides		5.0		0.3	10.8	0.2		1.0		0.6		0.4	8.0	0.4	66.4	24.0	48.0	0.6		7.0	
Carex lurida		0.3					1.0									3.0					2.0
Acer rubrum			+																		
Danthonia compressa			0.1							6.3		0.2									
Vaccinium myrtilloides			1.4				1.1		7.0	0.1	4.2	+						0.6			
Juncus effusus			11.3	2.1	0.2		3.6	1.5	1.6	2.3	3.2	2.6				6.6	5.6	8.2	0.4		1.0
Salix sericea			13.6																		80.0
Viola sp.			0.4																		4.0
Kalmia latifolia				+																	
Picea rubens				0.1			0.6			0.7	0.8										
Agrostis perennans				+	0.2		1.0	0.1			0.4					0.6	1.0				
Viburnum recognitum						+			0.2	0.6											
Calamagrostis canadensis						8.2								55.0							
Sphagnum fimbriatum							9.0														
Tsuga canadensis							6.0			2.2											
Dennstaedtia punctilobula							3.0			4.8		0.2									
Spiraea tomentosa							8.4					0.8							1.0		
Osmunda cinnamomea							2.0														0.4
Lycopus uniflorus										0.2											
Ilex verticillata										0.1											
Rhododendron maximum												0.2									
Eleocharis acicularis													10.0								
Sphagnum imbricatum													2.0								
Polygonum saggitatum													3.6			2.0					11.0
Sparganium chlorocarpum														1.0		17.0	3.0				
Galium tinctorum																11.0					
Eleocharis obtusa																3.0					
Carex stipata																0.2					
Hypericum mutilum																8.0					6.0
Glyceria striata																		0.6			8.4
Sphagnum girgensohnii																					19.0
Impatiens capensis																					7.6
Aster panicaus																					3.0
Eupatorium perfoliatum																					2.4
Sambucus canadensis																					0.4

*P=senescent beaver pond, G=glade, S=stream margin, W=wetland margin

4.3. GRADIENT ANALYSIS (ORDINATION)

PCA revealed similar major lines of variation in species composition at each site. About half (46.8 - 60.1 %) of the variation was explained by the first two dimensions (Walbridge 1982). When forest communities were present, a bog-vs.-forest gradient was represented by either axis 1 or 2 (Figure 3a, c-d). Soil saturation gradients were represented by axis 1 at LR and TR (Figure 3b, c), and as a complex gradient in two-dimensional space at BR (Figure 3a). Stand-age gradients were concomitant with the axis 1 hydrologic gradient at TR (Figure 3b), and were represented by axis 2 at BR and CR (Figure 3a, d). Stand age gradients separated senescent beaver ponds from other vegetation (TR), or separated younger ponds from both older ponds (deduced from aerial photography) and mature forest and tall shrub communities (CR). The stand age gradient at BR distinguished early-successional mixed *P. commune/S. fallax* (Fortney 1975) and aggrading *R. hispidus* shrub communities from more mature *S. fallax* and *P. rubens* forest communities. Differences in landscape position occurred concomitantly with the separation of polygons based on physiognomy,

TABLE Vc

Laurel Run vegetation data (means of % cover for 78 'important' species, n = 5 or greater per polygon, '+' indicates % cover < 0.1); see Walbridge (1982) for full data set

physiognomic type	herb-aceous	bryophtye				forest	herb-aceous	low shrub		tall shrub
plant community	*J.su./G.can.*	*P. comm./S. fall.*			*S.fal./P.com.*	*T.can.*	*T.lat.*	*H. densifl.*		*S.ser.*
polygon code	DIST	LB2	RB3	RB4	LB1	RC3	TAN1	R1	R2	GN1
polygon area (ha)	0.3	0.3	1.4	0.7	0.6	0.3	0.6	0.1	0.1	0.4
soil moisture class	4.4	4.2	2.7	3.6	3.9	3.7	4.0	3.0	3.0	3.0
landscape position*	P	E**	E**	E**	E**	W	P	S	S,W	W
plant species										
Sparganium chlorocarpum	1.7									
Carex lurida	1.5									
Sphagnum papillosum	0.8									
Leersia oryzoides	4.2			0.2						
Glyceria canadensis	16.7				0.3					
Eleocharis obtusa	7.2				0.5					
Scirpus cyperinus	0.2			0.1	1.0	0.2				
Juncus subcaudatus	18.3	6.4	0.1	5.2	5.0	1.4		0.2		
Hypericum mutilum	7.5			0.0				8.0		
Pyrus arbutifolia	0.2	7.4	6.8	6.8		0.2	2.0	0.2		
Sphagnum fallax	25.9	46.3	10.7	21.9	64.6	19.4	48.8	16.4		
Sphagnum imbricatum	0.2	0.1		1.7		11.4	37.0	5.0		
Polytrichum commune	6.0	18.6	21.7	33.6	23.3	7.2		13.0	0.2	
Hypericum virginicum	9.3				1.4				3.0	
Juncus effusus	1.0	1.6	0.8	2.1	14.9	0.4		3.0	0.8	5.0
Solidago rugosa	3.0		0.8	0.2	0.1				13.4	8.0
Solidago uliginosa	1.0	17.7	19.7	27.8	21.5	22.0	16.0	36.0	15.4	35.0
Rubus hispidus	0.8	14.3	37.8	15.6	10.1	10.0	14.0	56.0	38.0	4.4
Salix sericea	0.2		1.7	0.3		6.0	11.2		16.4	76.0
Glyceria striata	2.5					0.4				1.2
Viola sp.	3.2	0.1		0.2	1.5	4.4	+	1.0	0.6	3.5
Carex scoparia	0.7			0.4	0.7	1.0			7.8	4.0
Agrostis perennans	6.0	1.9			0.3	0.6		0.2	1.2	4.0
Carex gynandra	1.0		1.1	3.6	1.2	1.0	0.6	1.0	1.0	0.4
Galium tinctorum	2.3				+	0.4			7.0	3.2
Polygonum saggitatum	5.9								3.6	10.0
Eupatorium perfoliatum	1.3								1.0	2.4
Carex trisperma		5.1								
Vaccinium myrtilloides		1.7	2.3	0.1						
Eriophorum virginicum		5.3		0.2	0.1					
Sphagnum capillaceum		0.4		1.9		11.0				
Drosera rotundifolia		2.1		1.9	0.2	5.0				
Acer rubrum		0.1	+	+		0.4	+			
Sphagnum magellanicum		10.0					4.0			
Amelanchier laevis		0.6	1.7	0.2		1.8			0.2	
Sambucus canadensis		+							0.2	
Hypericum densiflorum		1.3	11.2	3.1	32.2	0.8	0.2	42.0	37.0	8.4
Danthonia compressa			3.1	0.3						
Viburnum recognitum			0.1	0.1			3.0			
Betula alleghaniensis			0.8			4.4	0.6			
Rhododendron maximum			0.3			29.6	10.0			
Osmunda cinnamomea			0.2			18.0	6.4			
Typha latifolia			0.1			2.0	30.0			
Gentiana linearis			1.8	0.3	0.7	2.0	1.2	5.2		
Sphagnum girgensohnii			0.3						1.0	
Glyceria melicaria				0.3		4.0				
Gaultheria hispidula				+			8.0			
Viburnum cassinoides				0.3			1.2		1.0	
Ilex verticillata				0.1		6.4		0.4	0.4	
Juncus brevicaudatus					0.5					
Agrostis hyemalis					0.2					
Tsuga canadensis						30.8				
Sphagnum fimbriatum						4.0				
Kalmia latifolia						3.4	2.4			
Onoclea sensibilis						7.4				18.0
Carex incomperta								0.6		
Aster dumosus									14.2	
Solidago graminifolia									15.2	24.0
Aster pumicans									2.4	0.4
Carex stipata										6.0
Lycopus uniflorus										0.4
Epilobium glandulosum										0.2

*P=senescent beaver pond, E=entire wetland, S=stream margin, W=wetland margin

**primarily centered in glades

TABLE VI

Cupp Run vegetation data (means of % cover for 78 'important' species, n = 5 or greater per polygon, '+' indicates % cover < 0.1); see Walbridge (1982) for full data set

physiognomic type	herbaceous					low shrub				tall shrub		herbaceous								forest			tall shrub			
plant community	mixed herbaceous				*T. lat.*	*R. hispidus* meadow				*A.rug./L.ory.*		*Carex stricta*								*T. can./B. all.*			*A.rug./S.fos.*		*V. recog.*	
polygon code	GCLRC	RY2	YG3	RY3	GCL6	RB4	RB5	R2A	O1	DGS1	DGS4	GCL1	GCL7	YG5	GCL2	YG1	GCL3	GCL4	GCL5	RC1	RC3	RT1	DGS7	DGS8	RB2B	RB2
polygon area (ha)	0.5	0.3	0.5	0.2	0.1	0.6	0.4	0.1	0.1	0.3	2.0	0.1	4.4	1.0	0.2	0.4	0.3	0.1	0.7	2.1	1.0	1.0	0.2	0.5	0.1	0.1
soil moisture class	4.2	3.7	4.0	3.2	4.0	3.2	3.2	3.1	3.4	4.0	4.0	3.2	3.1	3.5	3.6	3.1	3.8	4.2	5.0	3.8	3.3	3.2	3.3	3.1	3.3	3.6
landscape position	P	P	P	P	P	W	W	P	P	S,W	S,W	P	W	E	P	P	P	P	P	E	E	E	W	W	D	D
plant species																										
Carex trisperma	6.2		0.4	3.0																						
Eleocharis acicularis	5.0		0.9		2.0																					
Solidago uliginosa	0.6					16.6	13.0				8.0															
Amelanchier laevis	1.0	0.2					0.2												0.2							
Hypericum densiflorum	0.4		0.1	0.8	9.2	7.0	19.0	1.4				2.0	3.2	6.7	0.8											
Juncus brevicaudatus	6.4	0.2	1.6		4.2	1.0		1.0	0.4				0.3	6.0	+	0.6	1.3		12.8							
Carex stipata	0.4	0.2	0.1			0.2	1.0			1.4	0.2		0.2	0.8					2.8				10.0			
Carex gynandra	9.2	0.8	0.7	3.0	5.0						3.0			3.0					0.8	0.4		1.7				
Carex incomperta	5.2	0.2	1.1	14.0	5.4	4.0	4.0			1.0	1.6			3.2		+			0.2				0.4	4.0		
Osmunda cinnamomea	2.6			0.6		0.2	4.2		2.0			0.4								+				6.0		
Sphagnum fallax	29.0	27.8	23.2	31.0	33.0	3.0	1.0			3.4				6.5					23.8	9.0		0.4	4.4			
Juncus effusus	3.2		3.1	1.2		8.6	13.6		7.0	8.6	1.0		2.8	5.3	4.2	4.9	6.6					0.1				
Rhododendron maximum	12.8	10.2	0.9	14.0	2.4					0.8									0.9	62.0	55.0	24.7				
Tsuga canadensis	8.0	0.2	0.9	3.4	3.4					0.4									1.4	15.0	7.2	9.4				
Viburnum cassinoides	0.2				0.6						0.2															5.0
Salix sericea	7.0				7.0								12.2											2.0		
Betula alleghaniensis	3.8	6.2	1.2	1.2	0.6					1.2		0.2							0.9	7.2	5.8	7.5	2.4			
Acer rubrum	0.6	0.2	0.1	0.8	0.2			1.6	0.6	+		0.2							0.4		5.0	1.1			1.2	1.8
Sphagnum fimbriatum	6.0	5.0	8.0	2.0				9.0		2.4	1.2			19.8			1.3			8.0	4.0	6.7				2.0
Solidago rugosa	0.6	2.5	0.5	2.0	1.4	3.4	1.6			1.4	2.0	0.4	0.3	0.3		0.9	0.1		1.3			2.5	2.0	5.8	0.2	
Solidago graminifolia	2.4		1.3			2.0	0.2	0.4		9.2		6.0		2.8	10.0	0.6	0.5		4.8			1.3		0.6		
Agrostis perennans	1.4	0.3	1.5			1.0	+	0.4		3.4	6.0			1.3		0.1		3.6		2.4		1.4		0.2	1.4	
Carex folliculata	0.6	0.2	1.8	13.0		8.6	13.0	1.0		1.4	1.0					0.4			0.7	0.4			0.4	3.2	0.8	3.0
Rubus hispidus	28.0	42.0	33.2	72.0	31.0	54.0	65.0	63.0	43.0	0.2	3.0	20.0	15.6			3.8	0.6		3.5	0.6	1.0	15.0	6.0	37.0	13.6	21.0
Sphagnum imbricatum	5.0		4.2	4.0	3.0	12.0	26.0		0.4	2.8	9.0		0.1		0.8	3.1	0.6	0.8	5.0	3.0	10.0	16.4		2.2	13.0	
Lycopus uniflorus	10.0	1.7	2.4		2.0		0.2		0.2	4.2	2.8	2.6	14.3	11.5	7.7	1.4	8.8	3.2	7.2	6.2	1.2	3.7		4.6	1.4	
Polygonum sagittatum	6.0	9.2	16.0		3.4		0.2	1.8	11.0	8.4	6.6	8.0	6.3	12.5	10.8	10.4	12.4	6.8	11.2	5.0	5.0	2.2	3.0	0.4	+	
Symplocarpus foetidus	23.4	0.8	0.4	5.6			2.0		2.4	14.0	14.6		2.8					0.6		8.4	7.0	31.7	33.0	34.0	30.0	15.0
Leersia oryzoides	8.0	10.2	21.4		1.4			3.0	1.2	51.0	28.0		5.8	2.7	27.5	28.5	25.0	20.0	10.8	7.2	2.0	0.8	5.0	4.0	3.0	
Galium tinctorum	2.0	2.2	1.2						6.2	17.2	7.0	1.4	6.5	4.5	11.7	3.6	15.6	4.0	6.3	4.0	1.2	2.8		1.0	0.4	
Alnus rugosa	1.0	2.5	0.9				4.4	0.2		29.0	39.0	0.6	2.3						0.7	6.2	30.0	26.8	82.0	61.0	6.0	13.0
Viola sp.	11.6	1.3	3.6		12.6		7.0		2.0		7.0	0.4	1.3	11.7		1.8	0.6	2.0	3.2	9.0	7.0	6.7	11.2	9.0	+	0.4
Glyceria striata	10.6		5.1					0.2			0.6			5.3		0.3		0.2	0.8	9.0		1.7	8.0	7.4		
Eupatorium perfoliatum	0.2									0.6			0.1			0.3	0.4		1.7		0.2	0.5				
Glyceria melicaria	2.4									1.6	2.0								0.3		16.0	9.0	19.0			0.6
Eleocharis smallii		11.7												0.2			6.5		5.8							
Hypericum mutilum		11.7	1.1							9.2		0.4	2.8	3.3	3.0	3.1	6.5	3.0		0.4		1.8				
Calamagrostis canadensis		5.8	2.1									9.0	18.6	15.0		1.0								0.4		
Polytrichum commune		4.0	5.4	1.0		10.0	0.4			0.4					1.8	4.4	1.3	1.0				5.4				
Sparganium chlorocarpum		1.7	1.7						+		3.0			1.3	0.5	0.4	15.6	2.0	6.3	0.6						
Agrostis hyemalis		1.0	2.9							0.4			0.8	0.5					3.5							
Pyrus americana		0.3	0.1	4.4																1.2						
Viburnum recognitum		0.2	3.8	5.2				0.4	+	0.6	1.4	0.2		2.0			+	16.0	0.3	1.2	3.6	13.3	+	4.0	90.0	67.0
Carex stricta			11.8			23.0	2.0	49.0	43.6		10.2	50.0	45.3	30.2	30.7	25.0	11.3	31.0	10.0					15.0	3.0	3.4
Ilex verticillata			2.9	4.0		0.6	1.2	0.2	0.6	2.6	1.0							23.0	0.5			1.7		+	4.4	3.0
Carex scoparia			1.9	1.0			1.0			0.8		1.0	1.6	1.7		2.3	1.6		0.8							
Eleocharis obtusa			1.1									0.2	0.7		3.7	0.6		0.2								
Sphagnum papillosum			1.1																	14.0						5.8
Sambucus canadensis			0.4	+				0.4	1.0	0.8	6.2		1.8		0.3	0.1			0.7	1.0	3.0	1.0	2.2	2.4		0.6
Scirpus cyperinus			0.1			1.0	3.4							0.5												
Vaccinium myrtilloides				0.2			1.6	0.4																		
Typha latifolia					36.6														3.3							
Vernonia noveboracensis						11.4	10.6						0.6													
Eriophorum virginicum						2.0	6.4																			
Pyrus arbutifolia							2.0																	1.6		
Gentiana linearis							1.0						0.4													
Impatiens capensis									6.0		7.6	+	3.3	1.2	3.7	3.3	12.4	8.0	4.2	21.0	6.0	10.9	15.0	3.0	5.0	
Hypericum virginicum													0.8													
Carex lurida													0.2													
Aster puniceus														12.5									0.4			
Sphagnum girgensohnii																						2.9				
Dennstaedtia punctilobula																						0.1				
Kalmia latifolia																						+				
Onoclea sensibilis																							0.6			

*P=senescent beaver pond, D=senescent beaver dam, E=active wetland, G=glade, S=stream margin, W=wetland margin

hydrology, and stand age (cf., Table V, Figure 3).

A PCA of the vegetational relationships across sites identified the physiognomic separation of low shrub and bryophyte vs. forest and tall shrub communities as the most important line of variation among the 74 polygons, with herbaceous communities scattered throughout the space determined by the first two axes (Figure 4). Low shrub and bryophyte vegetation were separated along axis 2, along a gradient of decreasing soil moisture (Figure 4), similar to the hydrologic gradients identified by across-site cluster analysis (Figure 2) and within-site PCA (Figure 3a-c).

The relationship between the relative predominance of *H. densiflorum*, *R. hispidus*, *P. commune*, and *S. fallax* and soil moisture status was further explored by determining average percent cover values for each species as a function of soil moisture class across the full data set (n=466). *H. densiflorum* reached maximum cover at soil moistures of 2.5 to 3.0. *R. hispidus* reached maximum cover at soil moistures of 3.0 to 3.5. *P. commune* exhibited a bimodal distribution with maxima at soil moistures of 2.5 to 3.0 and 4.0 to

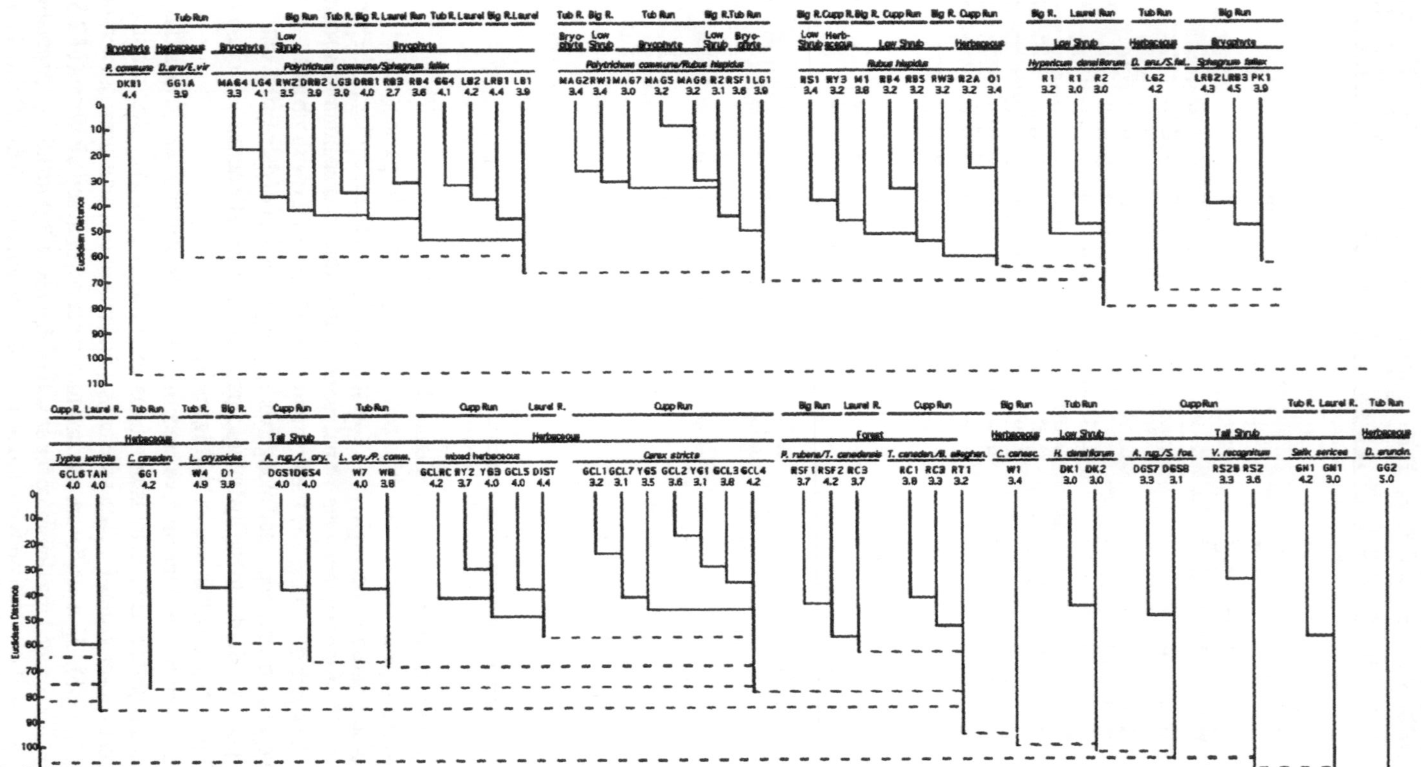

Fig. 2. Across-site cluster analysis of individual polygons (n=74) based on mean % cover values of 78 species that had either a frequency ≥ 60%, or a mean percent cover ≥ 10%, in at least 1 polygon (see Table IV for species identification). Each polygon is represented by a vertical line. Horizontal lines indicate linkages between polygons, or groups of polygons. Clusters are linked by solid lines; dashed lines show polygon/cluster linkages at Euclidean distances > 55. Polygons are identified by site, physiognomic type, and average soil moisture (see text for further explanation). Community names are based on vegetational composition (Table V).

TABLE VI

Comparison of surface water chemistries of plant communities across the four study sites (lower case letters indicate statistically similar means based on ANOVA and Duncan's NMRT (P=.95))

			(mg/L)								(uS/cm2)	(eq/L)
site	community	n	Ca	Mg	Fe	Na	K	NH4	NO3	pH	Cond.	Ca:H
Laurel Run	*Tsuga canadensis*	2	6.8a	1.9a	0.8	1.2a	1.0	0.2	0.0c	5.5a	20.2	107.3
Cupp Run	*T. can./B. all.*	2	6.6a	1.4b	0.5	1.0ab	0.8	0.2	0.0c	5.0ab	49.6	32.9
Laurel Run	*Typha latifolia*	1	6.1	1.4	0.3	0.9	1.5	0.1	0.0	5.5	28.4	96.3
Laurel Run	*J. sub./G. can.*	1	6.0	1.4	0.7	0.8	0.6	0.1	0.0	5.2	23.3	47.5
Cupp Run	*A. rug./S. foe.*	2	4.6a	1.1bc	1.1	0.7bcd	0.7	0.5	0.0c	4.9ab	35.0	18.2
Laurel Run	*Salix sericea*	1	4.3	0.8	0.4	0.4	0.8	0.1	0.0	4.9	24.1	17.0
Cupp Run	*V. recognitum*	2	4.2b	1.1bc	1.0	1.0ab	0.7	0.4	0.8bc	5.0ab	32.4	21.0
Laurel Run	*H. densiflorum*	2	4.1bc	0.9bcde	0.9	0.5cd	0.4	0.1	0.0c	4.5abc	46.1	6.5
Cupp Run	*A. rug./L. ory.*	2	3.9bc	1.4b	0.5	1.0ab	0.7	0.2	0.0c	4.9ab	30.7	15.5
Laurel Run	*P. com./S. fal.*	3	3.7bc	0.9bcd	0.6	0.7bcd	0.7	0.1	0.0c	4.3abcd	29.3	3.7
Cupp Run	*Carex stricta*	8	3.2bcd	0.8cde	0.9	0.8bc	0.6	0.2	0.1c	4.8ab	23.2	10.1
Tub Run	*Salix sericea*	1	3.1	0.7	0.3	0.6	0.5	0.2	0.0	5.9	16.4	122.9
Laurel Run	*S. fal./P. com*	1	3.0	0.7	0.4	0.5	0.5	0.1	0.0	4.1	34.1	1.9
Cupp Run	*Rubus hispidus*	4	2.8bcd	0.6cde	1.0	0.7bcd	1.0	0.3	0.0c	4.4abc	30.3	3.5
Tub Run	*P. com./S. fal.*	3	2.7bcd	0.7cde	0.4	0.7bcd	0.6	0.2	0.0c	4.1cd	29.6	1.7
Cupp Run	*Typha latifolia*	1	2.6	0.8	0.4	0.9	0.6	0.0	0.0	4.8	28.2	8.2
Cupp Run	mixed herbaceous	4	2.6bcd	0.7cde	0.7	0.8bc	0.8	0.2	0.0c	4.6ab	26.1	5.2
Tub Run	*D. aru./E. vir.*	1	2.5	0.7	0.2	0.4	0.4	0.3	0.0	4.4	25.1	3.1
Tub Run	*C. canadensis*	1	2.4	0.6	0.3	0.3	0.6	0.3	0.0	3.7	80.1	0.6
Tub Run	*L. ory./P. com.*	2	2.4	0.6	0.4	0.9	0.9	0.2	0.3	4.5	29.2	3.8
Tub Run	*H. densiflorum*	2	2.3bcd	0.5de	0.5	0.4d	0.5	0.2	0.3bc	4.4abc	24.2	2.9
Tub Run	*P. com./R. his.*	5	2.0cd	0.5e	0.7	0.4d	0.5	0.2	0.2c	4.2bcd	25.7	1.6
Tub Run	*S. fal./P. com.*	2	1.9cd	0.6cde	0.9	0.4d	0.5	0.2	0.0c	4.3abcd	21.7	1.9
Big Run	*L. ory./J. bre.*	1	1.8	0.6	1.6	0.4	0.6	0.1	2.3	4.4	23.5	2.3
Big Run	*Picea rubens*	2	1.7d	0.7cde	0.3	0.4d	0.6	0.3	3.1a	4.1cd	30.1	1.1
Big Run	*Rubus hispidus*	6	1.7d	0.5e	0.7	0.4d	0.5	0.2	1.3b	4.1bcd	33.4	1.1
Tub Run	*L. ory./S. fal.*	1	1.7	0.5	0.4	0.5	0.5	0.0	0.0	4.2	25.2	1.3
Tub Run	*D. arundinaceum*	1	1.7	0.5	1.1	0.6	0.5	0.2	0.0	4.0	37.0	0.8
Tub Run	*D. aru./S. fal.*	1	1.7	0.4	0.3	n.d.*	n.d.*	0.1	0.0	4.1	29.4	1.1
Tub Run	*P. commune*	1	1.7	0.4	0.9	0.4	0.4	0.3	0.5	4.2	23.6	1.3
Big Run	*Sphagnum fallax*	3	1.6d	0.4e	0.4	0.4d	0.3	0.2	0.0c	4.1cd	31.2	1.0
Big Run	*P. com./S. fal.*	3	1.6d	0.4e	0.2	0.4d	0.5	0.2	0.0c	4.0d	37.1	0.8
Big Run	*H. densiflorum*	1	1.5	0.5	0.7	0.4	0.4	0.2	4.2	4.2	26.2	1.2
Big Run	*Carex canescens*	1	1.4	0.4	0.4	0.4	0.3	0.3	0.0	3.8	46.2	0.4

*no data

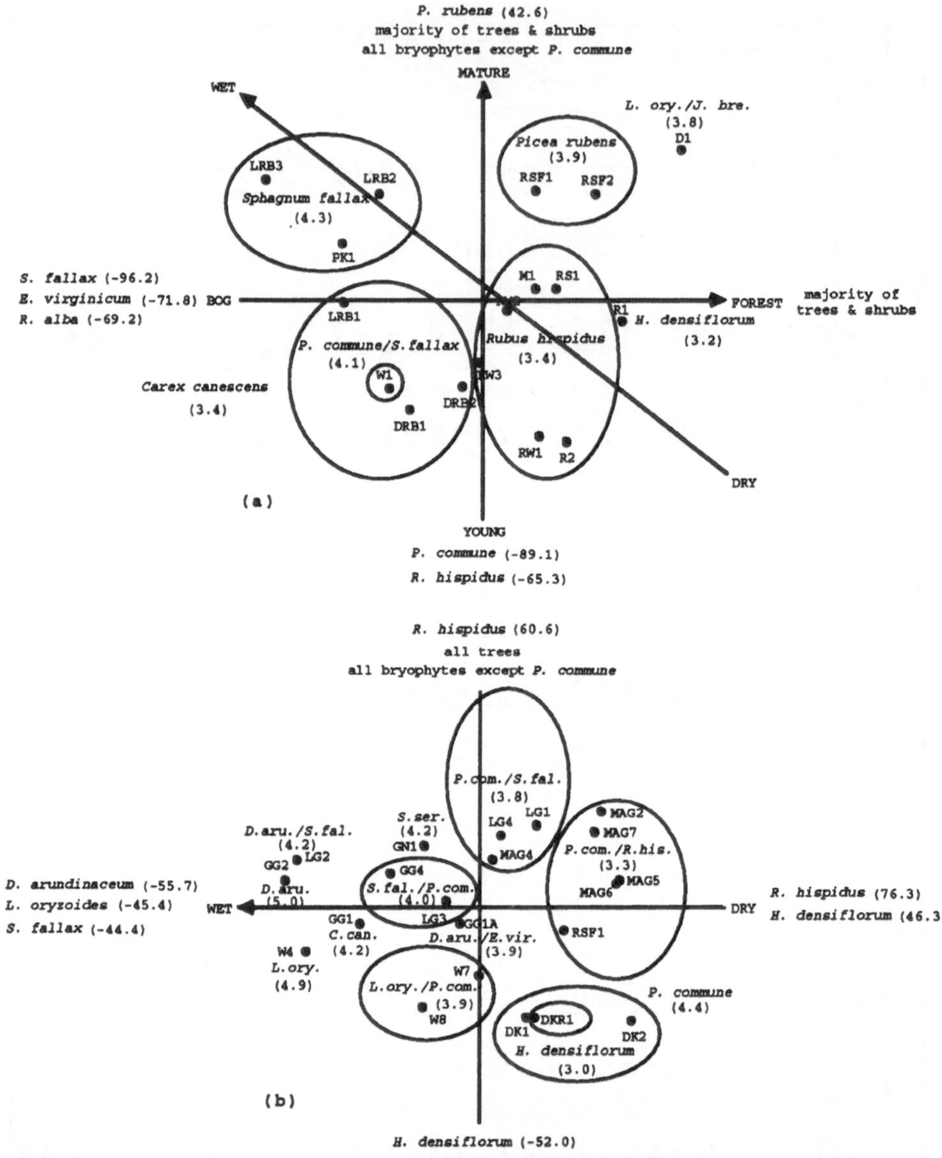

Fig. 3. Principal components analysis of the vegetational relationships among sampled polygons at Big Run (a) and Tub Run (b), based on mean percent cover values of all species present. Polygons are plotted in the first two component dimensions (Axis 1 = horizontal dimension; Axis 2 = vertical dimension), and identified using symbols given in Table V. Community clusters (see Figure 1) are enclosed by ellipses, all communities are labeled using community names given in Table V, and the weighted average soil moisture index value for each community is given (Walbridge 1982). Major gradients are indicated on each figure, as are correlation coefficients (expressed as %) for species highly correlated with each major axis.

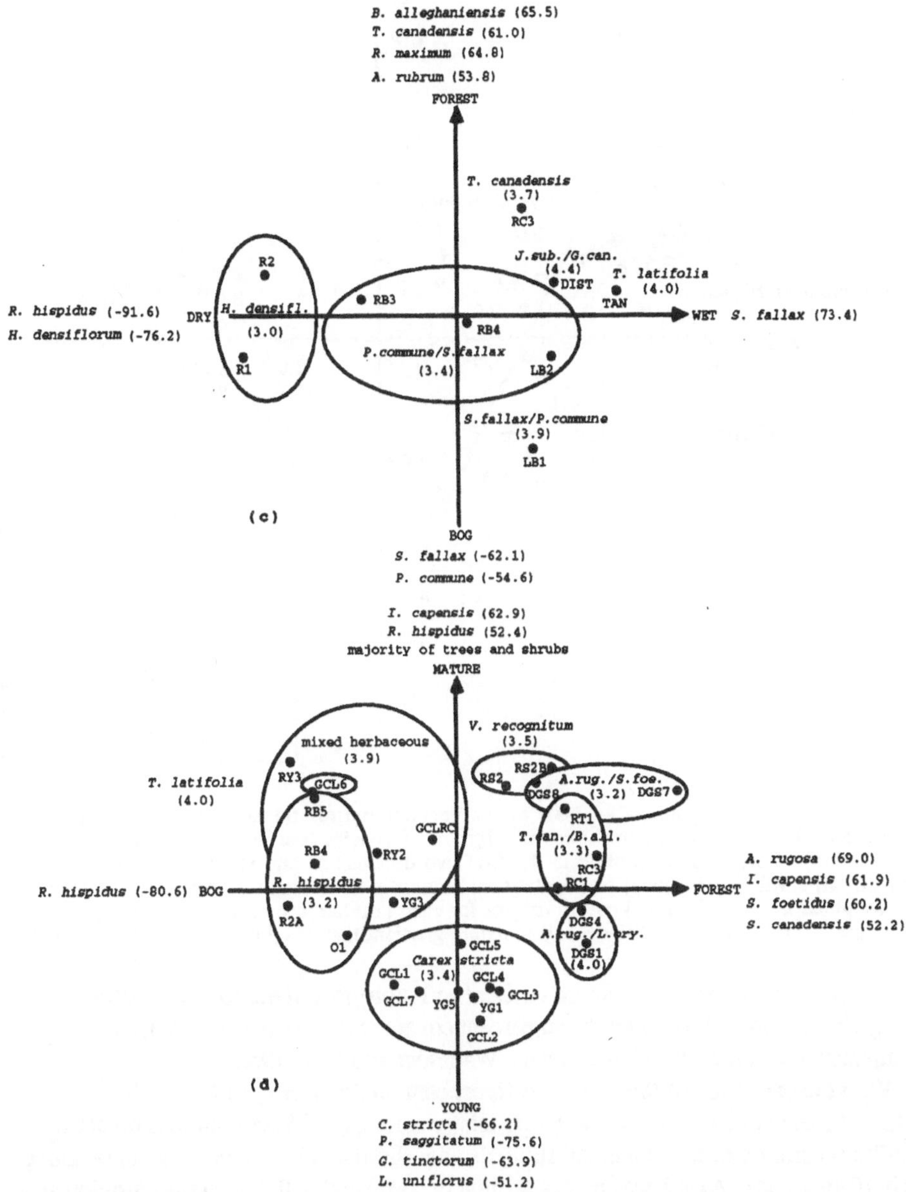

Fig. 3 (cont.). Principal components analysis of the vegetational relationships among sampled polygons at Laurel Run (c) and Cupp Run (d), based on mean percent cover values of all species present. Polygons are plotted in the first two component dimensions (Axis 1 = horizontal dimension; Axis 2 = vertical dimension), and identified using symbols given in Table V. Community clusters (see Figure 1) are enclosed by ellipses, all communities are labeled using community names given in Table V, and the weighted average soil moisture index value for each community is given (Walbridge 1982). Major gradients are indicated on each figure, as are correlation coefficients (expressed as %) for species highly correlated with each major axis.

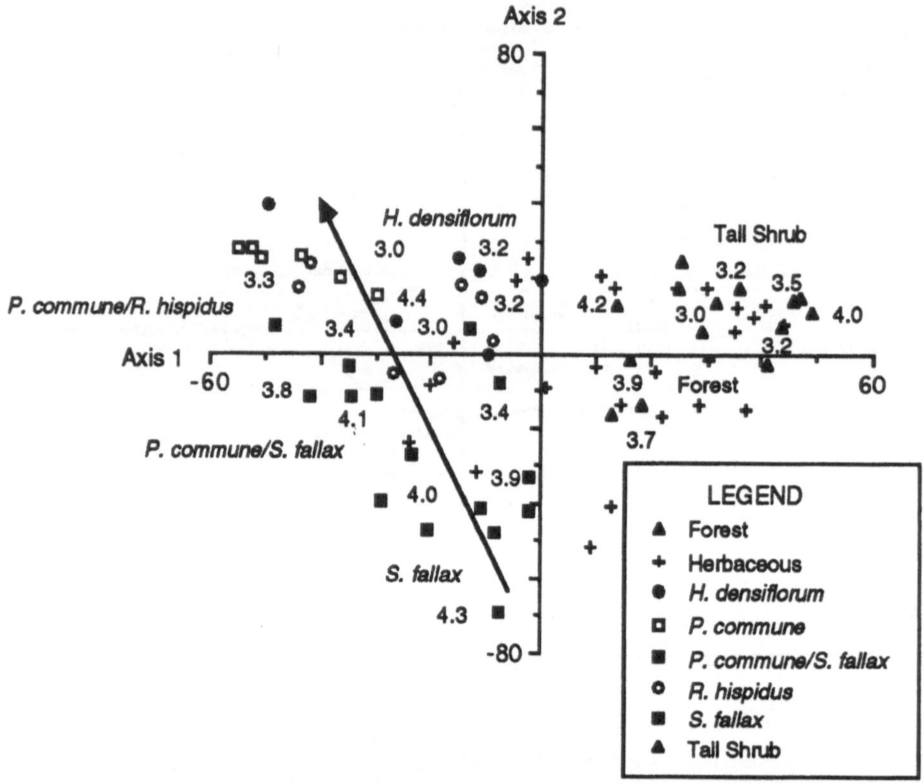

Fig. 4. Principal components analysis of the vegetational relationships among polygons across sites, based on mean percent cover values of 78 species (see Figure 2 for further explanation). Polygons are plotted in the first two dimensions and identified by physiognomy (forest, tall shrub, herbaceous) or vegetational composition (low shrub, bryophyte). Numbers are weighted average (by area) soil moistures for each non-herbaceous community type identified in Figure 1. The arrow identifies a soil moisture gradient that roughly coincides with axis 2.

4.5. *S. fallax* reached its maximum cover at soil moistures of 4.0 to 4.5 (Walbridge and Lang 1982). Plots of weighted mean percent cover values for these 4 species in component space revealed similar trends (Walbridge and Lang 1982).

With the exception of the *P. rubens* forest community at BR, a PCA of the relationships in average surface water chemistry among the 34 communities revealed a similar separation of forest and tall shrub vs. low shrub and bryophyte vegetation along axis 1(Figure 5a). Axis 1 was highly positively correlated with Ca^{2+} concentration and highly negatively correlated with pH (Figure 5b). Each site appeared to have a somewhat unique surface water chemistry (Figure 5c).

5. Discussion

Only 34 species (24.8%) were common to all sites, while 55 (40.2%) were unique to individual sites. To the extent that vegetational analysis produced a complete species list for each site, each wetland has a somewhat unique flora, a trait apparently shared by North and South Carolina Appalachian fens (Weakley and Schafle 1994). If each site is truly floristically unique, loss of any site would compromise regional biodiversity.

Despite this floristic uniqueness, there were two common vegetational trends among the 4 wetlands--a similarity in forest composition and the ubiquity of the *H. densiflorum/R. hispidus/P. commune/S. fallax* association. Both vegetation types are similar to communities described for other WV peatlands (Darlington 1943, Robinette 1964, Gibson 1970, Fortney 1975). Forest composition agrees with early WV settlers' descriptions of peatland vegetation (Allard and Leonard 1952, Fortney 1975); wetland forests probably covered a larger portion of the pre-settlement landscape than they do today.

The vegetational similarity of WV peatlands to their more northern counterparts has frequently been observed (Gibson 1970, Wieder et al. 1981). There also appears to be a general geochemical similarity between these two groups of wetlands (Walbridge 1982). Surface water chemistries of WV peatlands are comparable with other North American bogs and poor fens, but are lower in Ca^{2+} and pH than richer fens and swamps. These WV peatlands are correctly classified as fens--basin wetlands that receive nutrients in groundwaters draining surrounding uplands. They are extremely poor fens however, that occasionally exhibit chemical similarities to ombrotrophic bogs (Table VI). Dominance of H^+ in the surface waters of these poor fens could be due to: 1) patterns of surface and groundwater flow that isolate some communities from mineral soil influence (Wieder 1985); 2) low ionic contributions from the weathering of sandstone substrates (Table I);

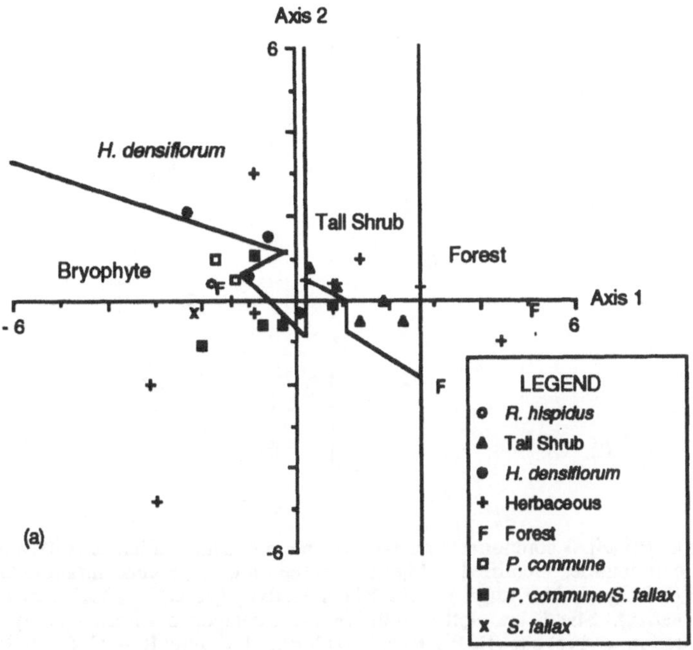

Fig. 5. PCA of chemical similarities among the 34 communities identified in Figure 1, based on average concentrations of Ca^{2+}, Mg^{2+}, Na^+, K^+, NH_4^+, NO_3^-, and H^+ (mg/L), and total conductivity ($\mu S/cm^2$) in surface waters (correlation matrix). Stands are plotted in the first two component dimensions by : (a) vegetation-type. Solid lines separate zones dominated by forest, tall shrub, low shrub (*H. densiflorum*), and bryophyte communities (legend identifies physiognomic community types after Figure 4).

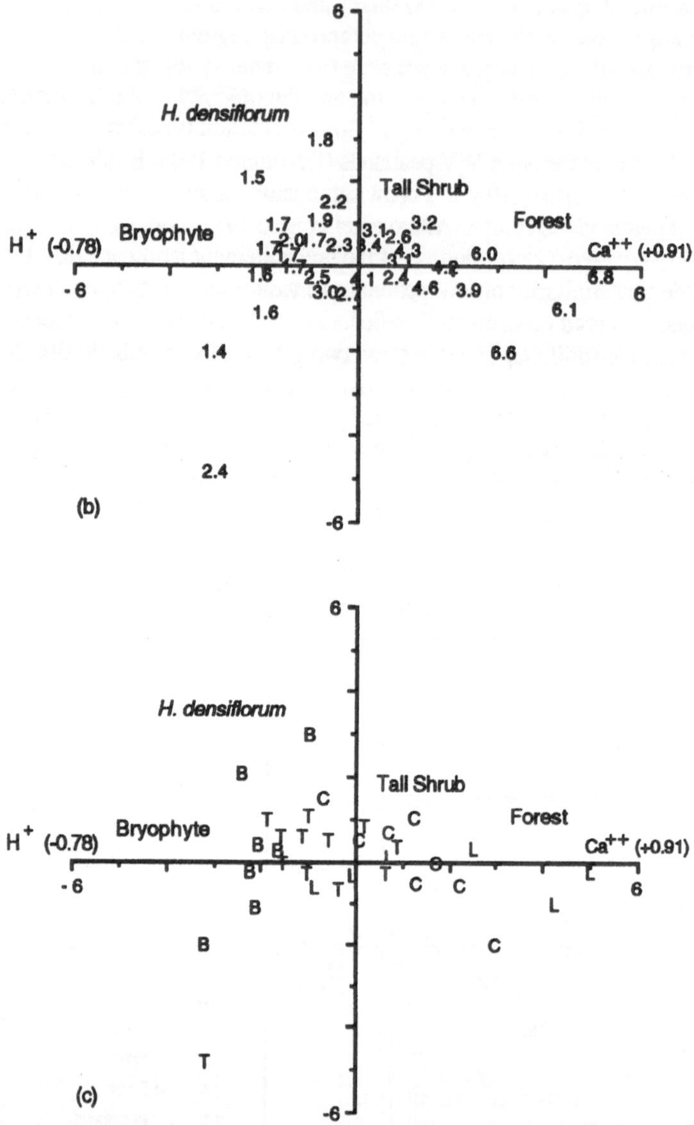

Fig. 5 (cont.). Principal components analysis of the chemical similarities/differences among the 34 plant communities identified in Figure 1, based on average concentrations of Ca^{2+}, Mg^{2+}, Na^{+}, K^{+}, NH_4^{+}, NO_3^{-}, and H^{+} (mg/L), and total conductivity (μS/cm²) in their surface waters (correlation matrix). Stands are plotted in the first two component dimensions by: (b) average Ca^{2+} concentration; and (c) site (B=Big Run; T=Tub Run; L=Laurel Run; C=Cupp Run).

and 3) inputs of sulfuric and nitric acids in acidic deposition (Gorham 1967). At TR, which has an abandoned surface mine in its watershed (Walbridge 1982), H^+ dominance could also be due to acid mine drainage.

Gradient analyses indicated a major physiognomic separation between forest and tall shrub vs. low shrub and bryophyte vegetation (Figures 3,4), consistent with a gradient of increasing pH and base cation concentrations in surface waters (Table VI, Figure 5b). This suggests a classic bog--poor fen--rich fen gradient, frequently described as a major line of vegetational variation in both European and North American peatlands (Gorham 1967, Heinselman 1970, Moore and Bellamy 1974, Vitt and Slack 1975, Glaser et al. 1981).

A second major line of vegetational variation separated plant communities dominated by *H. densiflorum*, *R. hispidus*, *P. commune*, and *S. fallax* (Figure 3a-c, 4). Both Daniels (1978) and Slack et al. (1980) found soil saturation to be an important line of vegetational variation in British mires and western Alberta fens, respectively. The *H. densiflorum/R. hispidus/P. commune/S. fallax* association also appears to be distributed along a soil moisture gradient. Variations in soil saturation were consistent with changes in landscape position, from drier stream and wetland margins to wetter glade communities (Table Va-c, Figure 3a-c). This gradient appears to be driven by differences in the hydrologic regimes of different landscape positions, and the hydrologic tolerances of component species.

Darlington (1943) described an underlying pattern of basin geomorphology at Cranberry Glades, identifying two landscape positions--stream margins, characterized by shallow peats and the formation of small levees, and glades--peat-filled basins lying between stream margins and the wetland/upland boundary. Wieder (1985) observed a similar geomorphology at BR. The spatial distributions of plant communities observed in this study suggest that this geomorphology is a recurring characteristic of this wetland type.

All sites showed evidence of past beaver colonization, which played an important role in determining community patterning. Beaver disturbances: 1) initiated successional gradients at TR and CR identified as major lines of vegetational variation (Figure 3b, d); 2) increased community diversity within wetlands, adding 2 to 6 communities per site (Table V); and 3) may have increased species diversity. Between 2 and 17 species per site were found only in senescent beaver ponds (Table V).

Raised water tables behind beaver dams kill trees, creating gaps in previously forested areas. Beaver effects on pre-historical wetland community dynamics are speculative, but historical air photos allow examination of their effects on the modern landscape.

The upper end of CR was entirely forested prior to initial beaver disturbance between 1953 and 1967. The upper portion of present-day CR consists of bands of wetland forest alternating with open areas of senescent beaver ponds (Walbridge 1982). Polygons behind old beaver dams are wider than forest polygons (Walbridge 1982), while senescent pond vegetation suggests the post-disturbance development of the *H. densiflorum/R. hispidus/P. commune/S. fallax* association (Figure 2).

Conclusions

The irregular mosaic of plant community patterning characteristic of WV fen peatlands can be explained by 3 factors: 1) basin geomorphology, 2) responses of key species to soil moisture gradients, and 3) beaver disturbance. Basin morphology produces lateral zonation, as component species respond to variations in hydrologic regime. Plant communities that develop following beaver disturbance follow longitudinal axes, because beaver dams alter water tables in a downstream-to-upstream direction. As breaches in dams occur following abandonment, water tables theoretically move towards pre-disturbance levels. Beaver pond senescence initiates a secondary successional sequence that follows the longitudinal gradients dictated by underlying basin geomorphology, but adds a temporal component to comunity patterning. The characteristic irregular mosaic results from the superimposition of community variation along lateral, longitudinal, and temporal axes. Observations at CR suggest that beaver disturbance may result in the replacement of forest vegetation by the *H. densiflorum/R. hispidus/P. commune/S. fallax* association.

Acknowledgments

Research was supported by the U.S. Army Corps of Engineers and the WV Dept. of Natural Resources, via a grant to G. E. Lang at West Virginia University (WVU). Research was done while the author was a graduate student at WVU, under the direction of G.E. Lang, without whose support this work would not have been possible. C. Bennett provided field and laboratory assistance, including vascular plant identification. A.M. Bartuska, E. Clark, A. McCormick, B.R. McDonald, M. Nebiolo, L. Pittman, and G. Shellito assisted with field and/or laboratory portions of this research. Grateful acknowledgment is also given to R. K. Wieder, C.H. Baer, A. Whitehouse, and R. Backhaus. G.E. Lang, K.L. Carvell, R.B. Clarkson, and two anonymous reviewers provided helpful comments on an earlier draft of this manuscript.

References

Allard, H. A. and Leonard, E. C.: 1952, *Castanea* **17,** 1.
Andrus, R. E.: 1980, *Sphagnaceae (Peat Moss Family) of New York State,* New York State Museum, Albany, NY.
Bailey, R. W.: 1954, *J. Wildl. Mgmt.* **18,** 184.
Brooks, A. B.: 1910, *West Virginia Geological Survey Volume 5 - Forestry and Wood Industries,* Acme Publishing Company, Morgantown, WV.
Chee, W. L. and Vitt, D. H.: 1989, *Wetlands* **9,** 227.
Daniels, R. E.: 1978, *J. Ecol.* **66,** 773.
Darlington, H. C.: 1943, *Bot. Gaz.* **104,** 371.
Edens, D. L.: 1973, 'The Ecology and Succession of Cranberry Glades, West Virginia', Ph.D. Dissertation, North Carolina State University, Raleigh.
Fernald, M. L.: 1970, *Gray's Manual of Botany,* 8th edition, D. Van Nostrand, New York.
Fortney, R. H.: 1975, 'The Vegetation of Canaan Valley, West Virginia, a Taxonomic and Ecological Study', Ph.D. Dissertation, West Virginia University, Morgantown.
Gibson, J. R.: 1970, *Castanea* **35,** 81.
Glaser, P. H., Wheeler, G. A., Gorham, E., and Wright, H. E., Jr.: 1981, *J. Ecol.* **69,** 575.
Gorham, E.: 1967, *Some Chemical Aspects of Wetland Ecology,* National Research Councial of Canada, p. 20.
Heinselmann, H. L.: 1970, *Ecol. Monogr.* **40,** 235.

Horton, D. G., Vitt, D. H., and Slack, N. G.: 1979, *Can. J. Bot.* **57**, 2283.
Moore, P. D. and Bellamy, D. J.: 1974, *Peatlands*, Springer-Verlag, New York.
Richardson, C. J., Kadlec, J. A., Wentz, A. W., Chamie, J. M., and Kadlec, R. H.: 1976, 'Background Ecology and the Effects of Nutrient Additions on a Central Michigan Wetland', in M. W. Lefor, W. C. Kennard, and T. B. Helfgott (eds.), *Proceedings Third Wetland Conference*, Institute of Water Resources, The University of Connecticut, Storrs.
Robinette, S. L.: 1964, 'Plant Ecology of an Allegheny Mountain Swamp', M.S. Thesis, West Virginia University, Morgantown.
Statistical Analysis System: 1982, *SAS User's Guide*, SAS Institute, Cary, NC.
Schwintzer, C. R.: 1978a, *Am. Mid. Nat.* **100**, 441.
Schwintzer, C. R.: 1978b, *Can. J. Bot.* **56**, 3044.
Schwintzer, C. R. and Tomberlin, T. J.: 1982, *Amer. J. Bot.* **69**, 1231.
Sjors, H.: 1950, *Oikos* **2**, 241.
Slack, N. G., Vitt, D. H., and Horton, D. G.: 1980, *Can. J. Bot.* **58**, 330.
Sokal, R. R. and Sneath, P. H. A.: 1963, *Principles of Numerical Taxonomy*, W. H. Freeman, San Francisco.
Strasbaugh, P. D. and Core, E. L.: 1977, *Flora of West Virginia*, 2nd edition, West Virginia University Bulletin, Morgantown.
Swank, W. G.: 1949, *Beaver Ecology and Management in West Virginia*, Consumer Commission of West Virginia Division of Game Management Bulletin No. 1.
Technicon Industrial Systems: 1977, 'Industrial Methods No. 399-07, 98-70W, and 100-80W', Technicon Industrial Systems, Tarrytown, New York.
Thompson, G. B.: 1974, *History of Logging in Davis, West Virginia, 1884-1924*, McClain Printing Company, Parsons, WV.
Tiner, R. W., Jr.: 1987, *Mid-Atlantic Wetlands: A Disappering Natural Treasure*, U.S. Fish and Wildlife Service Region V, Newton Corner, MA, and U.S. Environmental Protection Agency Region III, Philadelphia, PA.
U. S. Department of Commerce, Weather Bureau.: 1980, *Climatic Summary of the U. S., Supplement for 1977, 1978, and 1979, West Virginia*, U. S. Government Printing Office, Washington, D.C.
Vitt, D. H., Achuff, P., and Andrus, R. E.: 1975, *Can. J. Bot.* **53**, 2776.
Vitt, D. H., and Slack, N. G.: 1975, *Can. J. Bot.* **53**, 332.
Walbridge, M. R.: 1982, 'Vegetation Patterning and Community Distribution in Four High-elevation Headwater Wetlands in West Virginia', M.S. Thesis, West Virginia University, Morgantown.
Walbridge, M. R., and Lang, G. E.: 1982, 'Major Plant Communities and Patterns of Community Distribution in Four Wetlands of the Unglaciated Appalachian Region', in B. R. McDonald (ed.), *Proceedings of the Symposium on Wetlands of the Unglaciated Appalachian Region*, West Virginia University, Morgantown.
Weakely, A. S. and Schafale, M. P.: 1994, *Wat. Air Soil Poll.* (in press).
Welch, W. H.: 1957, *Mosses of Indiana*, Bookwalter Company, Indianapolis, IN.
Wieder, R. K.: 1985, *Biogeochem.* **1**, 277.
Wieder, R. K., McCormick, A. M., and Lang, G.E.: 1981, *Castanea* **46**, 16.

CARBON DYNAMICS IN APPALACHIAN PEATLANDS OF WEST VIRGINIA AND WESTERN MARYLAND

J. B. YAVITT

Department of Natural Resources, Fernow Hall, Cornell University, Ithaca, NY 14853 USA

Abstract. Abundant production of organic matter that decomposes slowly under anaerobic conditions can result in substantial accumulation of soil organic matter in wetlands. Tedious means for estimating production and decomposition of plant material, especially roots, hampers our understanding of organic matter dynamics in such systems. In this paper, I describe a study that amended typical estimates for both production and decomposition of organic matter by measuring net flux of carbon dioxide (CO_2) over the peat surface within a conifer swamp, a sedge-dominated marsh, and a bog in the Appalachian Mountain region of West Virginia and western Maryland, USA. The sites are relatively productive, with net primary production (NPP) of 30 to 82.5 mol C m^{-2} yr^{-1}, but peat deposits are shallow with an average depth of about 1 m. In summer, all three sites showed net CO_2 flux from the atmosphere to the peat during the daytime (-20.0 to -30.5 mmol m^{-2} d^{-1}), supported by net photosynthesis, which was less than net CO_2 flux from the peat into the atmosphere at nighttime (39.2 to 84.5 mmol m^{-2} d^{-1}), supported by ecosystem respiration. The imbalance between these estimates suggests a net loss of carbon (C) from these ecosystems. The positive net CO_2 flux seems to be so high because organic matter decomposition occurs throughout the peat deposit -- and as a result concentrations of dissolved inorganic carbon (DIC) in peat pore waters reached 4,000 μmol L^{-1} by late November, and concentrations of dissolved organic carbon (DOC) in peat pore waters reached 12,000 μmol L^{-1}. Comparing different approaches revealed several features of organic matter dynamics: (i) peat accretion in the top 30 cm of the peat deposit results in a C accumulation rate of about 15 mmol m^{-2} d^{-1}; however, (ii) the entire peat deposit has a negative C balance losing about 20 mmol m^{-2} d^{-1}.

1. Introduction

In peatland ecosystems, the accumulation of partially decomposed organic matter (née peat) is the foundation of the structure and function of the system. However, describing the dynamics of this material presents several problems. One traditional approach tries to measure the balance between the production and decomposition of organic matter, resulting in peat accumulation (Brinson *et al.*, 1981 and references cited therein). A different approach tries to measure the accretion rate of peat by dating deposition of successive layers in the profile (cf., Craft and Richardson, 1993). I suggest in this paper that one can significantly improve their understanding of organic matter dynamics in peatlands by studying net flux of CO_2 over the peat surface which integrates independent production measurements (assimilation of CO_2 by net photosynthesis) with decomposition measurements (remineralization of organic matter to CO_2 by respiration).

There are inherent limitations in any measurement of organic matter dynamics. For example, measurements of organic matter production that rely on allometric relationships derived from destructive harvests have an annual time scale at best and are difficult to relate to estimates of organic matter decomposition that rely on mass loss of litter samples confined in mesh bags for a much shorter period of time (Brock and Bregman, 1989;

Water, Air and Soil Pollution **77**: 271–290, 1994.

Moore, 1989; Rochefort *et al.*, 1990; Vitt, 1990). Another serious limitation involves the role played by plant roots in organic matter dynamics. Production estimates require frequent sampling and arduous work counting very small roots (Wallen, 1986; Finer, 1989; Conlin and Lieffers, 1993), while one could argue that confining roots in litter bags is so unrealistic that an estimate of decomposition is meaningless (Fahey, 1992). In contrast, measurements of peat accretion require accurate dates for different depth intervals and a thorough understanding of peat compaction with increasing depth in the peat (Johnson *et al.*, 1990; Malmer and Wallen, 1993). These considerations are not trivial, and misconceptions and short-cuts help fuel controversy about organic matter dynamics in peatlands (cf., Gorham, 1991).

One can amend some of these limitations by measuring net flux of CO_2 over the peat surface. Such a measurement integrates CO_2 flux between the atmosphere and vegetation (both aboveground and belowground) plus soil. Daytime measurements represent the following equation:

$$\text{Net flux of } CO_2 = \phi_{peat} + (R_{plant} - A)$$

where

ϕ_{peat} = CO_2 flux between peat and the atmosphere (supported by root and microbial respiration),

R_{plant} = plant respiration of aboveground components

A = assimilation of CO_2 by photosynthesis.

Positive values indicate that respiration exceeds assimilation of CO_2, and negative values indicate that assimilation of CO_2 exceeds respiration. Nighttime measurements represent the following equation:

$$\text{Net flux of } CO_2 = \phi_{peat} + R_{plant}$$

and estimate net ecosystem respiration for both aboveground and belowground components of the ecosystem.

In addition, one also should include the change (d) in storage of peat CO_2. One actually measures the change in dissolved inorganic carbon (dDIC) that accounts for the pH-dependent reaction of dissolved CO_2 with water forming bicarbonate (HCO_3^-) or carbonate (CO_3^{2-}). The DIC represents a significant pool for CO_2 when respiration rates are high and water flux out of the wetland is slow.

Wieder *et al.* (1990) investigated organic matter dynamics in peatlands of the Appalachian Mountain region of West Virginia and western Maryland and suggested a net annual loss of C from the peat of these systems to the atmosphere despite seemingly high rates of organic matter production. Yavitt *et al.* (1993) measured net flux of CO_2 in a portion of one peatland, dominated by a weakly minerotrophic bog, that also suggested an annual loss of C to the atmosphere.

The purpose of the present study is to provide additional insight into organic matter dynamics by measuring net flux of CO_2 in a conifer swamp, a marsh, and an ombrotrophic bog that commonly occur within Appalachian peatlands. The results suggest that these Appalachian peatlands do release more CO_2 into the atmosphere than they assimilate on an annual basis.

2. Methods

2.1. STUDY AREA

The research was carried out in two Appalachian peatlands, Big Run Bog, West Virginia (39°07'N, 79°35'W), and Buckle's Bog, Maryland (39°35'N, 79°22'W). Peatlands in this region of Appalachia are similar to more northern (>45°N) counterparts, with ground cover of *Sphagnum* spp. mosses and ericaceous shrubs such as *Andromeda glaucophylla* and *Kalmia latifolia.*

Mean annual temperature of the region is 7.9°C, with a minimum monthly mean of -3.2°C in January and maximum monthly mean of 18.3°C in July. The average frost-free season (number of days between 0°C) is 97. Mean annual precipitation averages 133 cm, with a fairly even distribution throughout the year. Annual snowfall averages 305 cm, and at least 2.5 cm of snow is on the ground for an average of 70 days. The peat deposit rarely freezes in winter, but it will freeze when there is a marginal snowpack. On an annual basis, precipitation exceeds potential evapotranspiration. However, it is likely that any month between April and October might experience a water deficit (i.e., potential evapotranspiration exceeds precipitation). In such months, water table levels may drop as low as 20 cm below the peat surface, before returning to the surface by winter.

Big Run Bog is physiographically a minerotrophic fen (sensu Sjors, 1950), receiving water and nutrients in runoff from the 276-ha watershed dominated by both deciduous and coniferous species (Wieder, 1985). Some parts of Big Run Bog receive more minerotrophic water than other parts that result in a mosaic of plant communities. I sampled in two of the more minerotrophic sites (Table I). One site was a forested fen (referred to hereafter as the swamp site) in which *Sphagnum girgensohnii* covered 90% of the peat surface, with an open canopy of red spruce (*Picea rubens*) and a dense understory of *Rhododendron maximum*. The second site was a marsh (referred to hereafter as the marsh site) in which a sedge (*Carex canescens*) covered about 85% of the peat surface and mosses covered less than 5% of the surface.

Buckle's Bog is a reasonable example of a true ombrotrophic bog (sensu Sjors, 1950), with a domed central region isolated from water and nutrients in runoff from the surrounding forested watershed. This region of the bog also has microtopography typical of many northern wetlands, with raised hummocks covered by *Sphagnum magellanicum* and deeper hollows dominated by *Sphagnum fallax*. (This study site is referred to hereafter as the bog site.)

TABLE I
Some characteristics of the three study sites in Appalachian peatlands.

	Site		
	swamp	marsh	bog
Dominant plants			
vascular	red spruce, Rhododendron	*Carex canescens*	ericaceous shrubs
non vascular	*S. girgensohnii*	none	*S. fallax*, *S. magellanicum*
Peat depth (m)			
mean	0.35	0.35	1.0
maximum	0.70	0.55	2.5
Bulk density (g cm^{-3})			
0-5 cm depth	0.16	0.01	0.05
30-35 cm depth	0.50	0.35	0.15
Organic matter (%)			
0-5 cm depth	85	85	85
30-35 cm depth	20	75	90
pH			
0-5 cm depth	4.5	3.8	4.4
30-35 cm depth	6.8	6.3	4.6
Ca^{2+} (µmol L^{-1})			
0-5 cm depth	22.8	26.1	20.2
30-35 cm depth	16.9	25.4	18.1
K^+ (µmol L^{-1})			
0-5 cm depth	6.8	12.1	6.2
30-35 cm depth	6.4	13.1	15.5
SO_4^{2-} (µmol L^{-1})			
0-5 cm depth	167	242	137
30-35 cm depth	126	131	5

A more complete description of the study sites is given in Yavitt *et al.* (1988) and in Wieder *et al.* (1990).

2.2. METHODOLOGICAL CONSIDERATIONS

There is a fair amount of literature that deals with approaches for measuring net flux of CO_2 between terrestrial ecosystems and the atmosphere (Denmead, 1991, and references cited therein). Undoubtedly, micrometeorological techniques -- e.g., eddy correlation that measures vertical CO_2 transport in the atmosphere caused by eddy motion

-- are preferred because they are nondestructive and nonevasive. However, micrometeorological techniques require relatively large study areas, thereby precluding their use in such small peatlands studied here.

An alternative approach uses relatively small, open-bottom chambers made of clear plastic (cf., Vourlitis *et al.*, 1993). Net flux of CO_2 is estimated by measuring the change in CO_2 concentration within the chamber headspace upon placing the chamber on the ground and periodically thereafter. Because clear plastic is used, daytime measurements reflect the balance between assimilation of atmospheric CO_2 (by photosynthesis) and release of CO_2 to the atmosphere (supported by respiration). As mentioned in the introduction section, positive values indicate that respiration exceeds assimilation of CO_2 and negative values indicate that assimilation of CO_2 exceeds respiration.

Chambers do have their own host of problems, however. For example, chambers change the microclimate (temperature, wind, evapotranspiration rate) within the chamber headspace, leading to better -- or in some cases worse -- growing conditions. Moreover, small pressure differences inside versus outside the chamber can cause an extremely large error in the flux measurement. It is possible to minimize this error by venting the chamber headspace with outside air to maintain pressure equilibrium. Furthermore, daytime measurements must be as short as possible, since a rapid photosynthetic rate can reduce CO_2 concentration within the chamber to low enough levels that would limit further photosynthesis. Such short-term measurements inevitably present their own problem of scaling-up repeated measurements to an hourly or longer time scale. Nevertheless, chambers offer relatively inexpensive means for estimating net flux of CO_2 in the absence of any other estimate.

2.3. MEASUREMENT OF NET FLUX OF CO_2

Net flux of CO_2 was estimated at each of the three sites using open-bottom, clear plastic chambers (125-L volume, covering 0.25 m^2). The chambers were tall enough to cover all of the vegetation in the marsh and in the bog sites. At the swamp site, however, the chambers were obviously too small to include the aboveground portions of trees and shrubs -- and thus the CO_2 flux measurements represent the balance between respiration in the peat, respiration by mosses on the peat surface, and photosynthesis by mosses.

Sampling consisted of four chambers per site, placed about 3 m from each other, with measurements made at midday (1000 to 1300 EST) at least monthly from August 1987 to March 1989. The procedure in the field involved placing a chamber over the peat surface as carefully as possible without disturbing the peat surface, turning on a radial fan to mix the chamber headspace, then collecting five gas samples from the chamber headspace consecutively at 3-min intervals, allowing ambient air to replace the volume of air sampled to equalize pressure. The CO_2 concentration of ambient air also was sampled to correct for the addition to the headspace of the chamber. On three sampling dates in 1988 measurements were carried out at 2 hr intervals from 0800 to 1800 (EST) to estimate net flux of CO_2 throughout the course of a day.

I also monitored air temperature, relative humidity and solar irradiance both inside and outside each chamber as well as wind speed and peat temperature at 2 and 15 cm depths for each measurement period.

2.4. MEASUREMENT OF DIC AND DOC

Two (replicate) equlibrators (sensu Hesslein, 1976) were used per site to sample concentrations of DIC and DOC in peat pore water seven times in 1988. Each equilibrator had nine rows of wells at 5-cm vertical intervals. The wells were filled with degassed, deionized water and covered with a sheet of 0.45 μm-pore-size membrane filter. Each equilibrator placed in the peat remained for a 10-day period. Within 10 min after retrieval, the water from each well was removed using a 30-mL syringe, acidified, and DIC was stripped into 10 mL of air by shaking vigorously for 2 min. The acidified water sample was saved for analysis of DOC.

2.5. ANALYTICAL MEASUREMENTS

The concentration of CO_2 in air samples was determined by gas chromatography using a thermal conductivity detector and a Poropak column (2 m X 3 mm) maintained at 35°C with He carrier gas (30 mL min^{-1}). I quantified gas concentrations by comparing peak area for samples and standards. Certified standards (286 to 10,000 μL L^{-1} of CO_2 in N_2) bracketed every 10-15 samples. Analytical precision was <0.2% and accuracy was within 2% of each standard.

The concentration of DOC in water samples was determined using a dedicated total organic carbon analyzer.

3. Results

3.1. DAYTIME MEASUREMENTS OF NET CO_2 FLUX

Daytime measurements of net CO_2 flux varied considerably among the three sites sampled (Figure 1). The bog site had a mean net CO_2 flux of -2.60 mmol m^{-2} hr^{-1} or about 2-times greater than the mean CO_2 flux of -1.26 mmol m^{-2} hr^{-1} at the swamp site and 3-times greater the mean CO_2 flux of -0.86 mmol m^{-2} hr^{-1} at the marsh site. These values represent mean fluxes made in the daytime across all sampling dates. The range of individual values was a maximal negative net CO_2 flux of -26 mmol m^{-2} hr^{-1} at the swamp site and a maximal positive net CO_2 flux of 16.5 mmol m^{-2} hr^{-1} at the marsh site.

Net CO_2 flux in daytime at all three sites was always positive for measurements made in December through February. Otherwise, negative values for net CO_2 flux occurred as early as March at all three sites, predominantly between May and September, and still

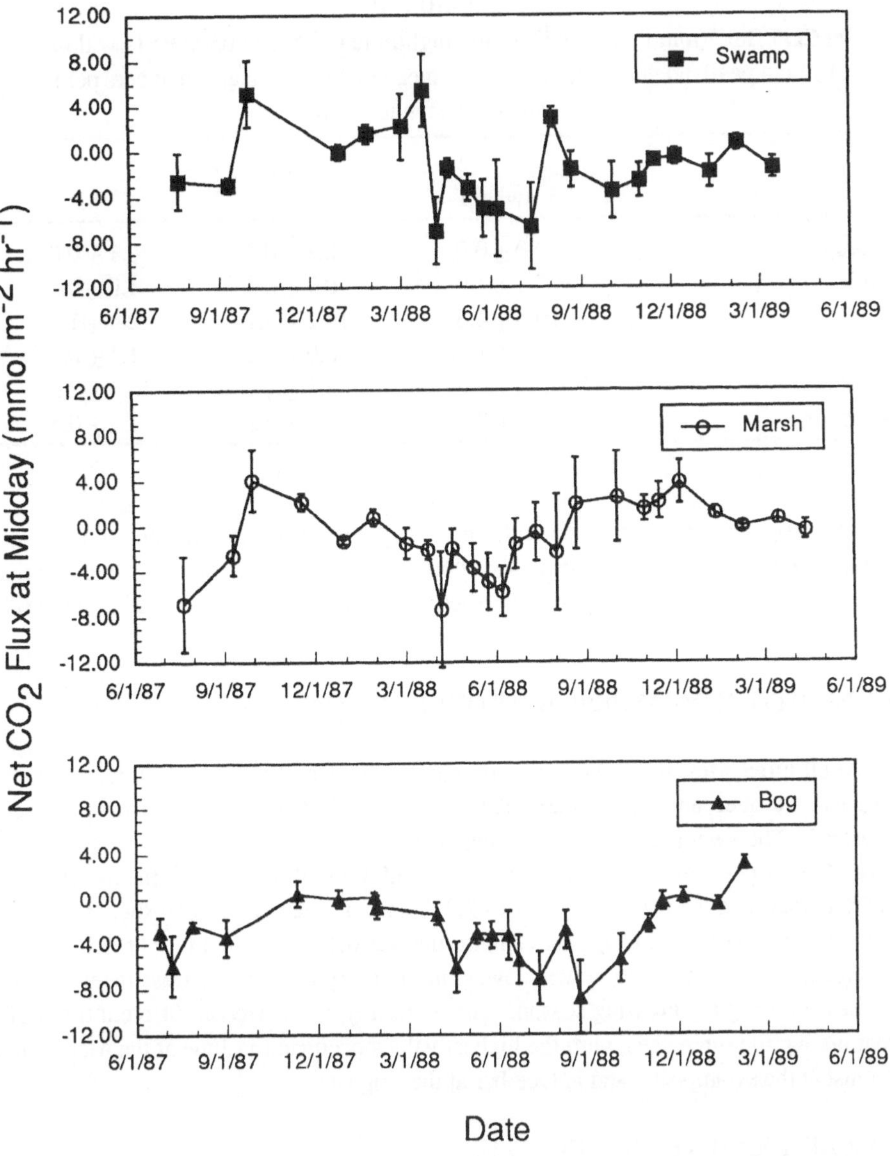

Fig. 1. Seasonal trend in midday measurements of net CO_2 flux in three different Appalachian peatland sites. Means for four replicate chambers ± 1 SE.

sporadically at all three sites in October and November. Positive values for net CO_2 flux in the daytime were <10% of the individual estimates made between May and September.

TABLE II

Net CO_2 flux (mmol m^{-2} hr^{-1}) during nighttime (2300 and 0300 EST) within three Appalachian peatland sites. Values are the mean of four replicate chambers per sampling date ± one standard error.

	Site		
	swamp	marsh	bog
February	0.0 ± 0.0	0.1 ± 0.0	0.1 ± 0.0
April	2.1 ± 1.0	0.7 ± 0.5	1.0 ± 0.3
June	10.6 ± 4.6	14.2 ± 8.1	2.4 ± 1.0
August	21.6 ± 10.0	6.0 ± 3.2	1.1 ± 0.2
October	2.1 ± 0.6	2.1 ± 1.1	10.4 ± 2.0
November	0.4 ± 0.2	0.4 ± 0.2	1.3 ± 0.5

The Bog site showed less variation among the four chamber measurements per sampling date, with a standard error about 40-60% of the mean, compared to the variation among measurements at the other two sites where the standard error was often 100% of the mean.

3.2. NIGHTTIME MEASUREMENTS OF NET CO_2 FLUX

At all three sites, net CO_2 flux at nighttime was measured in February, April, June, August, October and November (Table II) and always showed positive values as expected. The swamp site had the highest mean net CO_2 flux of 6.13 mmol m^{-2} hr^{-1} or about 56% higher than the mean CO_2 flux of 3.92 mmol m^{-2} hr^{-1} at the marsh site and 125% higher than the mean CO_2 flux of 2.72 mmol m^{-2} hr^{-1} at the bog site. The range of individual values for net CO_2 flux at nighttime was 0.1 to 32 mmol m^{-2} hr^{-1}.

Net CO_2 flux at nighttime was lowest in February at each site (less respiration) and increased during the growing season. The highest flux occurred at different times of the year among the three sites, with the highest values occurring in June at the Marsh site, in August at the swamp site, and in October at the Bog site.

3.3. DAILY RATES OF NET CO_2 FLUX

Measurements of net CO_2 flux were made repeatedly throughout the day at all three sites on three dates in 1988. At all three sites, net CO_2 flux was positive at nighttime and negative soon after sunrise, with a maximal value being reached at noon before declining steadily after that (Figure 2). The set of measurements made on 14 June for a mostly clear day (photosynthetic photon flux [PPFD] > 1500 μmol m^{-2} s^{-1} throughout midday) showed a relatively similar daily pattern as that for a partly cloudy day on August 13 (PPFD as low as 500 μmol m^{-2} s^{-1} during cloudcover of 10 to 30 min), except that

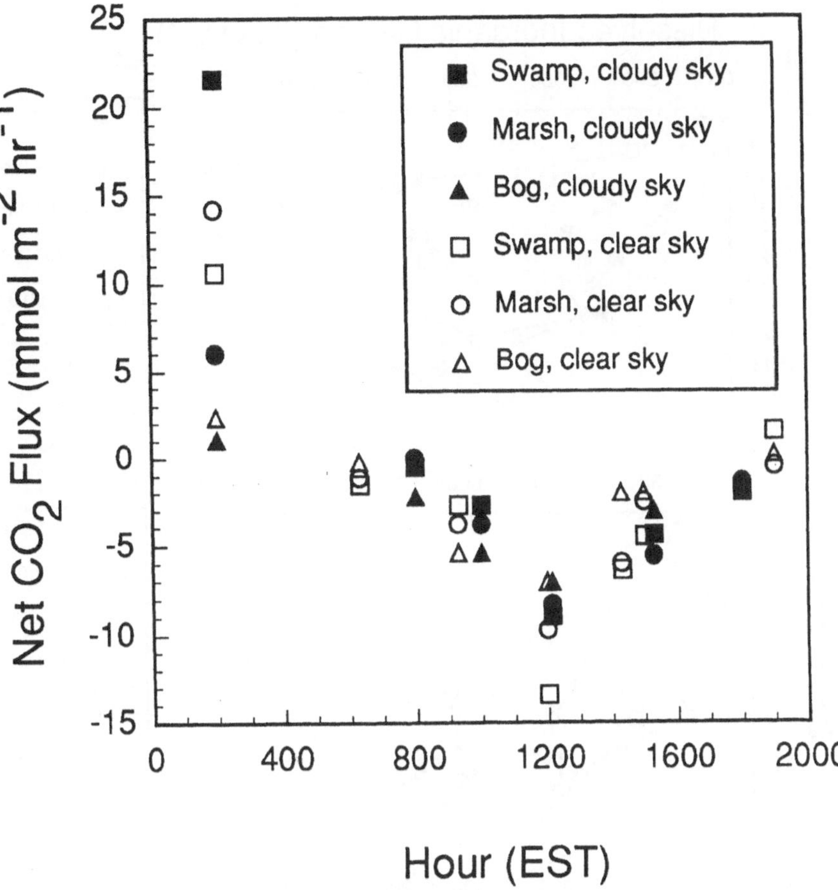

Fig. 2. Diurnal trend of net CO_2 flux for a clear day in June and for a mostly cloudy day in August in three different Appalachian peatland sites. Means for four replicate chambers.

negative net CO_2 flux at midday was not as low (less assimilation of CO_2) as that on the sunny day.

3.4. DIC AND DOC IN PEAT PORE WATER

Vertical profiles of DIC concentrations in peat pore water showed considerable variation within site per sampling date (Figure 3). The swamp site developed the highest DIC concentrations of nearly 3,000 μmol L^{-1} at the 15 to 25 cm depths, whereas concentrations at the marsh and bog sites seemed to increase with increasing depth but did not exceed 2,500 μmol L^{-1}. At the swamp site concentrations of DIC were lowest early in the season and increased significantly ($p < .05$ repeated measures analysis of

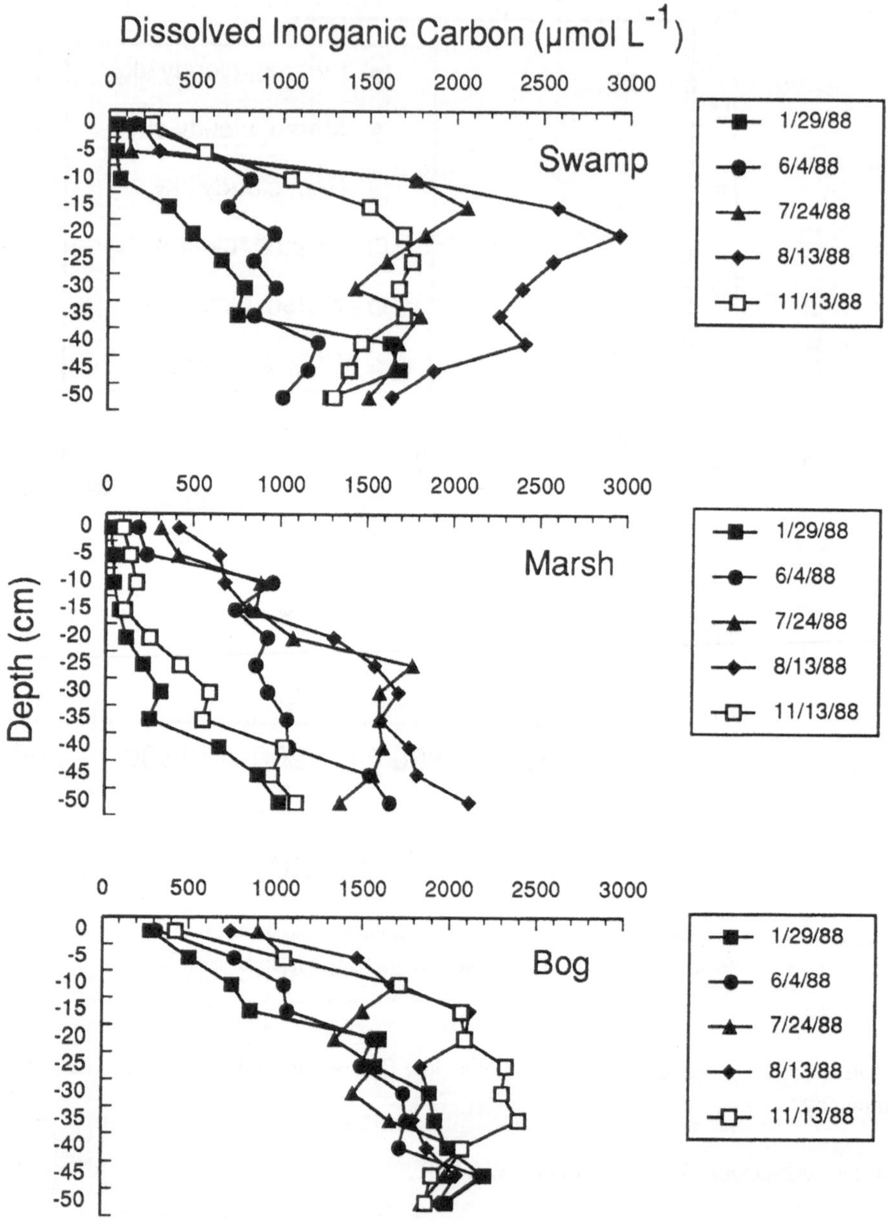

Fig.3. Seasonal trends in dissolved inorganic carbon concentrations (DIC) in peat beneath three different Appalachian peatland sites.

variance [ANOVA]) throughout the depth profile to maximal concentrations in August, before declining to still moderate DIC concentrations of 500 μmol L^{-1} in November. At

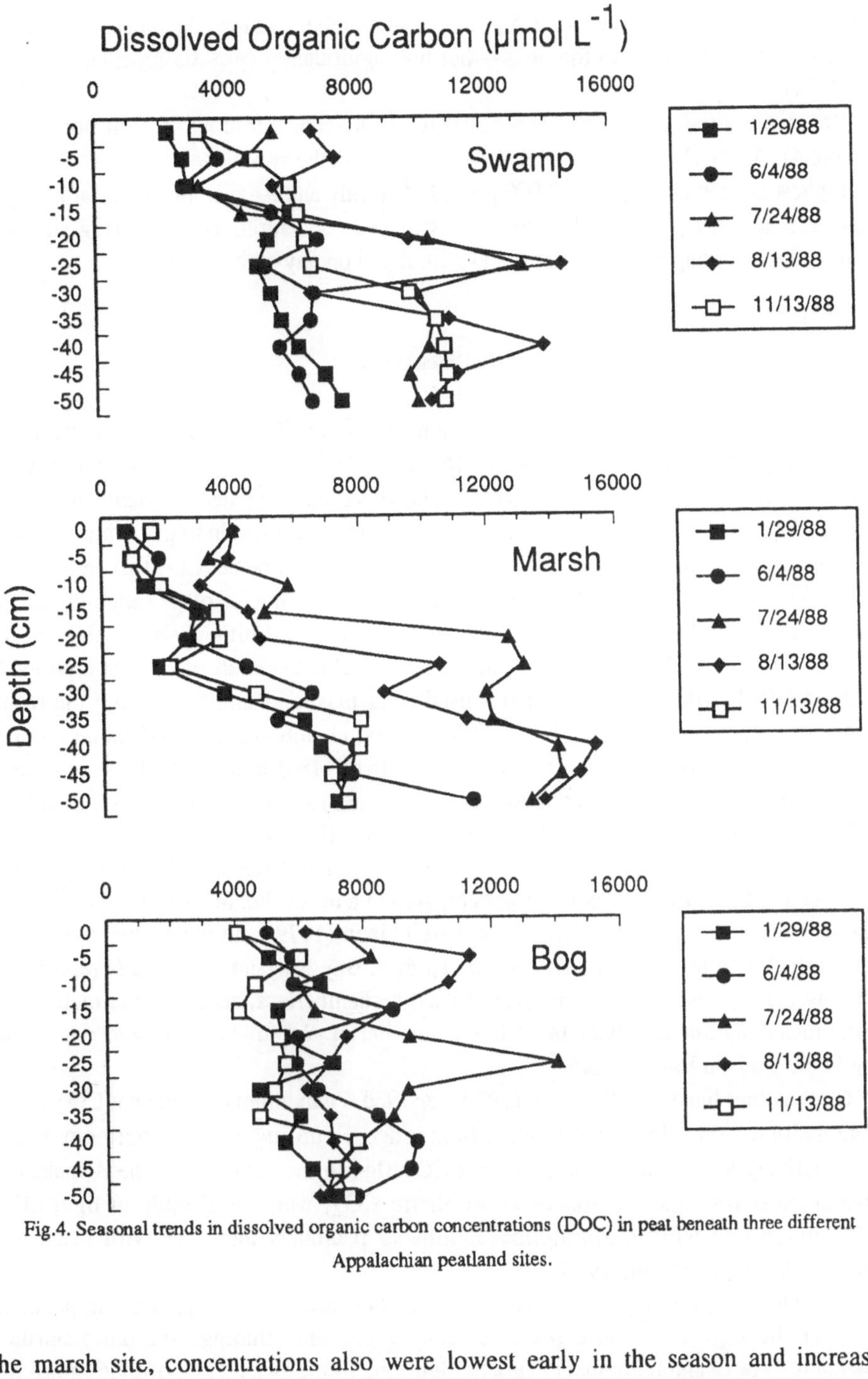

Fig.4. Seasonal trends in dissolved organic carbon concentrations (DOC) in peat beneath three different Appalachian peatland sites.

the marsh site, concentrations also were lowest early in the season and increased significantly ($p < .05$ repeated measures ANOVA) to maximal concentrations in July and

August but decreased to low concentrations by November. At the bog site, concentrations fluctuated from one date to the next -- but not significantly ($p > .05$ repeated measures ANOVA).

Relatively similar patterns were noted for DOC concentrations in peat pore waters (Figure 4). Vertical profiles of DOC concentrations at the swamp and marsh sites showed the highest concentrations of >12,000 μmol L^{-1} in July and August, in particular at depths >20 cm below the peat surface. In contrast, DOC concentrations at the Bog site were quite variable among sampling dates and with depth on any given sampling.

4. Discussion

I am not aware of any daytime measurements of net CO_2 flux in conifer-dominated swamps that consider only the peat and low-growing vegetation to compare the results reported here. For the other two study sites, however, daytime measurements of net CO_2 flux are relatively low (i.e., less negative) compared to previous findings in more northern peatlands. For example, midsummer measurements of daytime net CO_2 flux in the marsh site of 0 to -7 mmol m^{-2} hr^{-1} are less negative than findings for other sedge-dominated sites of -11.4 mmol m^{-2} hr^{-1} in western Alaska reported by Whiting *et al.* (1992) and -8.2 and -13.1 mmol m^{-2} hr^{-1} in two sites in the Hudson Bay lowlands of Canada reported by Whiting (1994). Both of those studies used clear plastic chambers, making the results comparable to results reported here. Others using chamberless, aerodynamic methods have reported daytime net CO_2 fluxes of -9.0 to -18.0 mmol m^{-2} hr^{-1} for sedge-dominated peatlands near Barrow, Alaska (Coyne and Kelley, 1975), and -8.2 mmol m^{-2} hr^{-1} for sedge-dominated peatlands in western Alaska (Fan *et al.*, 1992).

Likewise, the bog site with daytime net CO_2 flux in midsummer of -3 to -8 mmol m^{-2} hr^{-1} had a relatively low negative flux compared to a mean flux of -6.8 mmol m^{-2} hr^{-1} in southern Finland reported by Silvola and Heikkinen (1979) and mean fluxes of -8.8 to -28.6 mmol m^{-2} hr^{-1} for two sites in the Hudson Bay lowlands of Canada reported by Whiting (1994). Neumann *et al.* (1994) used a chamberless, aerodynamic method and found mean daytime net CO_2 fluxes in midsummer of -2.4 to -14.7 mmol m^{-2} hr^{-1} for a bog in the Hudson Bay lowlands.

On the other hand, Grulke *et al.* (1990) reported a maximum daytime net CO_2 flux of -0.42 mmol m^{-2} hr^{-1} for tussock tundra (mix sedge, shrub, and moss) in northern Alaska. The relatively low negative values of net CO_2 flux in the daytime for the Appalachian sites suggest less assimilation of atmospheric CO_2, which is notable in light of the presumption that NPP in *Sphagnum*-dominated peatlands increases with decreasing latitude (Wieder and Lang, 1983).

In contrast, nighttime measurements of net CO_2 flux in these Appalachian peatlands are near the high (i.e., more positive) end of previous findings for more northern peatlands. For example, net CO_2 flux at nighttime in the swamp site of 21.6 mmol m^{-2} hr^{-1} in August is at least 73% higher than the nighttime flux of 4.7 mmol m^{-2} hr^{-1} in a forested bog in Alaska reported by Luken and Billings (1985) and 12.5 mmol m^{-2} hr^{-1} in a black spruce stand established on peat in Alaska reported by Schlentner and Van Cleve (1985).

Likewise, net CO_2 flux at nighttime in the marsh site of 14.2 mmol m^{-2} hr^{-1} in June is at least 80% higher than nighttime flux of 1.6 to 7.9 mmol m^{-2} hr^{-1} measured in several sedge-dominated sites in Alaska (Billings *et al.*, 1977; Luken and Billings, 1985; Giblin *et al.*, 1991) and 2.8 mmol m^{-2} hr^{-1} in a sedge-dominated fen in subarctic Canada (Moore, 1986). Net CO_2 flux at nighttime in the bog site of 10.4 mmol m^{-2} hr^{-1} in October is at least 37% higher than nighttime flux of 2.4 mmol m^{-2} hr^{-1} measured in a bog in Sweden (Svennson, 1980) and 7.6 mmol m^{-2} hr^{-1} measured in a bog in Alaska (Luken and Billings, 1985). Kim and Verma (1992) did report net CO_2 flux at nighttime of 2.8 to 17 mmol m^{-2} hr^{-1} for a *Sphagnum*-dominated peatland in Minnesota. It also is notable that the Appalachian sites have positive net CO_2 flux during the winter which agrees with net CO_2 fluxes of 0.2 to 2.0 mmol m^{-2} hr^{-1} reported Zimov *et al.* (1993) for unfrozen peat above the permafrost in Siberian tundra in winter.

In addition to the biosphere-atmosphere exchange of CO_2, some CO_2 accumulates as DIC in peat pore water of peatlands. The DIC concentrations reported here are somewhat lower than CO_2 concentrations of peat pore water reported by Nilsson and Bohlin (1993) for several peatlands in Sweden (970 to 6,500 μmol L^{-1}) and by Benstead and Lloyd (1994) for a peatland in Scotland (up to 4,500 μmol L^{-1}). The lower concentrations in Appalachian peatlands might result from preferential flux of CO_2 to the atmosphere rather than accumulation in peat pore water. For example, the data in Table II and in Figure 4 suggest that seasonal patterns in nighttime measurements of net CO_2 flux matched subtle differences in the vertical profiles of DIC concentration in peat pore water. Accordingly, the highest values for net CO_2 flux in the swamp site occurred in August at the same time DIC concentration in peat pore water reached a maximum. Likewise, the marsh site showed maximal net CO_2 flux and DIC concentration in June, whereas the bog site had the highest net CO_2 flux and DIC concentrations in November.

I did find a lower mean DIC concentration in the marsh site (844 μmol L^{-1}), dominated by *Carex*, than in the swamp (1,226 μmol L^{-1}) and bog (1,602 μmol L^{-1}) sites, dominated by *Sphagnum*, as confirmed by Nilsson and Bohlin (1993). The reason for this difference in DIC concentrations among peat types is not clear and could relate to different rates of organic matter decomposition or hydrologic regimes (Nilsson and Bohlin, 1993).

The DOC concentrations in peat pore water found here are much higher than mean concentrations of 2,000 to 4,900 μmol L^{-1} commonly reported for other peatlands (McKnight *et al.*, 1985; Moore, 1987; Marin *et al.*, 1990; Dalva and Moore, 1991). Such high DOC concentrations in these Appalachian peatlands probably result from relatively incomplete decomposition of organic matter, producing soluble organic matter, compared to that in the other northern peatlands. Moreover, lower mean concentration of DOC in the marsh site (6,346 μmol L^{-1}) than in the swamp (7,228 μmol L^{-1}) and bog (6,936

μmol L^{-1}) sites seems to agree with findings in other peatlands (cf., Moore, 1987; Dalva and Moore, 1991).

5. Synthesis

The measurements reported here provide the basis for ecosystem C budgets. Such budgets indicate whether each site is presently accumulating or losing C (i.e., net sink or source for atmospheric CO_2). This is a pertinent question for Appalachian peatlands such as Big Run Bog and Buckle's Bog that formed under a much cooler climate 13,000 years ago than their present climate (Maxwell and Davis, 1972; Watts, 1979). While persistence of these peatlands in the present-day temperate climate is obvious, persistence does not mean that they still accumulate C. They could be losing C because assimilation of C by photosynthesis and release of C by respiration have different responses to increasing temperature (cf., Townsend *et al.*, 1992). Respiration rates increase with increasing temperature, whereas photosynthetic rates have an asymptotic relationship with increasing temperature. Thus, it is possible that peat deposited in Appalachian peatlands in the past is no longer accumulating C but rather is being mobilized to CO_2 (and CH_4; Yavitt *et al.*, 1993).

For example, these Appalachian peatlands release CO_2 to the atmosphere from December to early April despite being snow covered. To estimate the amount of CO_2 lost from each site during the winter (1 December to 15 April) I developed linear relationships between peat temperature (2 cm depth) and positive net flux of CO_2, then used daily peat temperatures for the region (National Oceanic and Atmospheric Administration, 1988) to estimate total CO_2 flux to the atmosphere. It is notable that such equations (not shown) described only 40 to 50% of the variation in nighttime net CO_2 flux; including data on peat moisture content (i.e., water table depth; cf., Kim and Verma, 1992) did not improve their predictive capability. Nevertheless, I estimated CO_2 release to the atmosphere of 2.9 mol m^{-2} from the swamp site, 1.4 mol m^{-2} from the marsh site, and 3.1 mol m^{-2} from the bog site during the winter, representing roughly 3.5% of the C in annual NPP in the swamp site and 5% of the C in annual NPP in the marsh and bog sites (Wieder *et al.*, 1989).

During the snow-free season (15 April to 30 November), the swamp site was the only site where the highest nighttime net CO_2 flux coincided with the highest peat temperature (19°C at 2 cm depth) in August. In contrast, the nighttime net CO_2 flux reached its maximal value before the warmest time of the year in June in the Marsh site (12°C at 2 cm depth) and in autumn in the Bog site (8°C at 2 cm depth). This disparity between maximal nighttime net CO_2 flux and peat temperature among the different sites makes scaling-up the limited number of measurements reported here to seasonal -- or annual -- estimates unrealistic.

Scaling-up the daytime measurements of net CO_2 flux reported here to estimate C balance is even more problematic because it requires more than just temperature as the

TABLE III

Summary of C fluxes for three Appalachian peatland sites. Negative values indicate assimilation of atmospheric CO_2, and positive numbers indicate release of CO_2 to the atmosphere. See text for assumptions and derivations of estimated values.

	Site		
	swamp	marsh	bog
Net CO_2 flux (mmol m^{-2} d^{-1})			
daytime	-30	-20	-27
nighttime	83	53	39
daily-integrated	53	33	12
dDIC (mmol m^{-2} d^{-1}) (June to August)	11	6	3
dDOC (mmol m^{-2} d^{-1}) (June to August)	-66	-116	-23
NPP (mol m^{-2} yr^{-1})			
aboveground	-61	-15	-39
belowground	-21	-15	-22
total	-82	-30	-61
Peat accretion (mol m^{-2} yr^{-1})	-5	-5	-5

independent variable in a simple model; i.e., photosynthetic rates vary in response to changes in irradiance, temperature, humidity, CO_2 concentration (Nobel, 1983), as well as water table level in peatlands. In this regard, I previously found poor correlations ($r < .40$) for net CO_2 flux and a suite of environmental factors for other sites within Big Run Bog (Yavitt *et al.*, 1993). Therefore, I did not repeat the correlations with the data set presented here because I expected equally poor results. Rather, I realize that more studies, possibly experimental in nature, are required to explanation the patterns in daytime net CO_2 flux reported here.

For three sampling dates, I did have several daytime measurements of net CO_2 flux to estimate separate daytime- and nighttime-integrated net CO_2 fluxes. A surprising outcome of these measurements was that negative CO_2 flux in the daytime did not account for the positive net CO_2 flux at nighttime at all three sites (Table III). As a result, the daily-integrated net CO_2 flux was positive, suggesting a net loss of CO_2 to the atmosphere from each site. Oechel *et al.* (1993) also reported that arctic tundra in Alaska was a net CO_2 source at rates of 2.8 to 18 mmol m^{-2} d^{-1}.

A partial explanation for this outcome is the relatively low negative values for net CO_2 flux in the daytime. It is notable that measurements made both in June (clear sky and PPFD of about 1500 μmol m^{-2} s^{-1}) as well as those made in August (partly cloudy sky and PPFD of about 500 μmol m^{-2} s^{-1}) showed the maximal value at noon, with the assimilation of CO_2 declining steadily after that. Others (Grulke *et al.*, 1990; Whiting, 1994) have found much more negative values for net CO_2 flux after noon on clear days than on partly cloudy days, suggesting that assimilation of CO_2 closely tracks levels of PPFD. The daytime patterns in net CO_2 flux in these Appalachian peatlands are more like those in upland situations on well-drained soil, where midday water stress lowers net CO_2 flux in late afternoon on clear days by reducing stomatal conductance and limiting CO_2 assimilation.

A different -- and certainly plausible -- explanation for the relatively low negative values for net CO_2 flux in the daytime at these sites is that the plants are taking up CO_2 emitted from the peat surface in addition to the uptake of atmospheric CO_2 (cf., Sternberg, 1989). This phenomenon would lower the amount of atmospheric CO_2 taken up to support photosynthesis, without limiting production of organic matter. Certainly more study is need to document this explanation -- but it seems likely nonetheless.

A net C balance close to -- and possibly less than -- zero is a reasonable assumption for these Appalachian peatlands. While these systems seem to have relatively high NPP, they maintain shallow peat deposits, suggesting rapid decomposition of the annual production. For example, data on NPP for the three study sites are incomplete but relevant. Wieder *et al.* (1989) reported that aboveground production (on a C basis) in the bog site of 39 mol m^{-2} yr^{-1}. That study did not specifically measure belowground production but speculated that it could be 22 mol m^{-2} yr^{-1} on the basis of equal production aboveground and belowground by the vascular plants at the site. The same approach for the marsh site yields 30 mol m^{-2} yr^{-1}, with half of that aboveground and the other half belowground.

An estimate of NPP at the swamp site is more problematic. I estimated shrub, herb and moss production on the basis of the same approach as that for the bog site, except correcting for differences in species composition between the sites (Walbridge, 1982). No data are presently available for the tree component. However, Grigal *et al.* (1985) estimated annual tree production (aboveground plus belowground) of 27 mol m^{-2} yr^{-1} for a similar spruce-dominated peatland in Minnesota. Using this value for the swamp site results in annual NPP of 82.5 mol m^{-2} yr^{-1}.

On the other hand, Wieder *et al.* (199x) measured the rate of peat accretion in the top 35 cm of the peat deposit at Big Run Bog -- and at four other peatlands in the eastern U.S.A. -- but unfortunately those studies did not consider the sites in this study. Nevertheless, the results suggest that peat accretion as measured by ^{210}Pb dates of peat at 2-intervals is equal to a C accumulation rate of only 5.5 mol m^{-2} yr^{-1} among the study sites ranging from northern Minnesota to the Pocono Mountains of Pennsylvania to Big Run Bog and Cranesville Swamp in West Virginia. Assuming this accumulation rate applies to the three sites I studied, then C accretion in the surface peat is only 6 to 18% of the NPP (Table III).

The results of the studies reported here show that organic matter decomposition occurs throughout the year in the peat deposit at each site, resulting in (i) CO_2 that largely escapes to the atmosphere even in winter, and (ii) partial decomposition products that accumulate as DOC. Through the combination of approaches, it appears the three sites are net sources of atmospheric CO_2. Certainly, more rigorous measurements are necessary to prove this conclusion. Nevertheless, C cycling is quite active in these Appalachian ecosystems, relying on large inputs and loss of CO_2. Consequently environmental changes including improper management that affect one -- or more -- aspect of C cycling could alter the overall dynamics of these ecosystems. I encourage others to use these methods to help understand the biogeochemistry of C in a wide range of Appalachian peatlands.

Acknowledgments

Several undergraduate students at West Virginia University provided invaluable assistance in the field and laboratory. Research was supported by grants from the National Science Foundation.

References

Benstead, J., and Lloyd, D.: 1994, Direct mass spectrometric measurement of gases in peat cores. *FEMS Microbiol. Ecol.* 13, 233-240.

Billings, W.D., Peterson, K.M., Shaver, G.R., and Trent, A.W.: 1977, Root growth, respiration, and carbon dioxide evolution in an arctic tundra soil. *Arct. Alp. Res.* 9, 129-137.

Brinson, M.M., Lugo, A.E., and Brown, S.: 1981, Primary productivity, decomposition and consumer activity in freshwater wetlands. *Ann. Rev. Ecol. Syst.* 12, 123-161.

Brock, T.C.M., and Bregman, R.: 1989, Periodicity in growth, productivity, nutrient content and decomposition of *Sphagnum recurvum*-var-*mucronatum* in a fen woodland. *Oecologia* 80, 44-52.

Conlin, T.S.S., and Lieffers, V.J.: 1993, Seasonal growth of black spruce and tamarack roots in an Alberta peatland. *Can. J. Bot.* 71, 359-360.

Coyne, P.I., and Kelley, J.J.: 1975, CO_2 exchange over the Alaskan arctic tundra: meteorological assessment by an aerodynamic method. *J. Appl. Ecol.* 12, 587-611.

Craft, C.B., and Richardson, C. J.: 1993, Peat accretion and nitrogen, phosphorus and organic carbon accumulation in nutrient-enriched and unenriched Everglades peatlands. *Ecol. Appl.* 3, 446-458.

Dalva, M., and Moore, T.R.: 1991, Sources and sinks of dissolved organic carbon in a forested swamp catchment. *Biogeochemistry* 15, 1-19.

Denmead, O.T.: 1991, Sources and sinks of greenhouse gases in the soil-plant environment. *Vegetatio* 91, 73-86.

Fahey, T.J.: 1992, Mycorrhizae and forest ecosystems. *Mycorrhiza* 1, 83-89.

Fan, S.M., Wofsy, S.C., Bakwin, P.S., Jacob, D.J., Anderson, S.M., Kebabian, P.L., McManus, J.B., Kolb, C.E., and Fitzjarrald, D.R.: 1992, Micrometeorological measurements of CH_4 and CO_2 exchange between the atmosphere and subarctic tundra. *J. Geophys. Res.* 97, 16,627-16,643.

Finer, L: 1989, Fine root length and biomass in a pine mixed birch-pine and spruce stand on a drained peatland. *Suo (Helsinki)* 40, 155-161.

Giblin, A.E., Nadelhoffer, K.J., Shaver, G.R., Laundre, J.A., and Mckerrow, A.J.: 1991, Biogeochemical diversity along a riverside toposequence in arctic Alaska. *Ecol. Monogr.* 61, 415-435.

Gorham, E.: 1991, Northern peatlands: role in the carbon cycle and probable responses to climatic warming. *Ecol. Appl.* 1, 182-195.

Grigal, D.F., Buttleman, C.G., and Kernik, L.K.: 1985, Biomass and productivity of the woody strata of forested bogs in northern Minnesota. *Can. J. Bot.* 63, 2416-2424.

Grulke, N.E., Riechers, G.H., Oechel, W.C., Hjelm, U., and Jaeger, C.: 1990, Carbon balance in tussock tundra under ambient and elevated atmospheric CO_2. *Oecologia* 83, 485-494.

Hesslein, R.H.: 1976, An in situ sampler for close-interval interstitial water studies. *Limnol. Oceanogr.* 21, 912-914.

Johnson, L.C., Damman, A.W.H., and Malmer, N.: 1990, *Sphagnum* macrostructure as an indicator of decay and compaction in peat cores from an ombrotrophic south Swedish peat-bog. *J. Ecol.* 78, 633-647.

Kim, J., and Verma, S.B.: 1992, Soil surface CO_2 flux in a Minnesota peatland. *Biogeochemistry* 18, 37-51.

Luken, J.O., and Billings, W.D.: 1985, The influence of microtopographic heterogeneity on carbon dioxide efflux from a subarctic bog. *Holarc. Ecol.* 8, 306-312.

Malmer, N., and Wallen, B.: 1993, Accumulation and release of organic matter in ombrotrophic bog hummocks processes and regional variation. *Ecography* 16, 193-211.

Marin, L.E., Kratz, T.K., and Bowser, C.J.: 1990, Spatial and temporal patterns in the hydrogeochemistry of a poor fen in northern Wisconsin. *Biogeochemistry* 11, 63-76.

Maxwell, J.A., and Davis, M.B.: 1972, Pollen evidence of pleistocene and holocene vegetation on the Allegheny Plateau, Maryland. *Quat. Res.* 2, 506-530.

McKnight, D.M., Thurman, E.M., Wershaw, R.L., and Hemond, H.F.: 1985, Biogeochemistry of aquatic humic substances in Thoreau's Bog, Concord, Massachusetts. *Ecology* 66, 1339-1352.

Moore, T.R.: 1986, Carbon dioxide evolution from subarctic peatlands in eastern Canada. *Arct. Alp. Res.* 18, 189-193.

Moore, T.R.: 1987, Patterns of dissolved organic matter in subarctic peatlands. *Earth Surf. Processes Landforms* 12, 387-397.

Moore, T.R.: 1989, Plant production, decomposition, and carbon efflux in a subarctic patterned fen. *Arct. Alp. Res.* 21, 156-162.

National Oceanic and Atmospheric Administration: 1988, *Local climatological data, West Virginia report.* Natl. Clim. Data Cent., Asheville, N.C.

Neumann, H.H., den Hartog, G., King, K.M., Chipanshi, A.C.: 1994, Carbon dioxide fluxes over a raised bog at the Kinosheo Lake tower site during the northern wetlands study (NOWES). *J. Geophys. Res.* 99, 1529-1538.

Nilsson, M., and Bohlin, E.: 1993, Methane and carbon dioxide concentrations in bogs and fens - with special reference to the effects of the botanical composition of the peat. *J. Ecol.* 81, 615-625.

Nobel, P.S.: 1983, *Biophysical plant physiology and ecology*, W.H. Freeman, San Francisco. 608 pp.

Oechel, W.C., Hastings, S.J., Vourlitis, G., Jenkins, M., Riechers, G., and Grulke, N.: 1993, Recent change of arctic tundra ecosystems from a net carbon dioxide sink to a source. *Nature* 361, 520-523.

Rochefort, L., Vitt, D.H., and Bayley, S.E.: 1990, Growth, production and decomposition dynamics of *Sphagnum* under natural and experimentally acidified conditions. *Ecology* 71, 1986-2000.

Schlentner, R.E., and Van Cleve, K.: 1985, Relationships between CO_2 evolution from soil, substrate temperature, and substrate moisture in four mature forest types in interior Alaska. *Can. J. For. Res.* 15, 97-106.

Silvola, J., and Heikkinen, S.: 1979, CO_2 exchange in the *Empetrum nigrum-Sphagnum fuscum* community. *Oecologia* 37, 273-283.

Sjors, H.: 1950, On the relation between vegetation and electrolytes in north Swedish mire waters. *Oikos* 2, 241-258.

Sternberg, L.D.S.L.O.: 1989, A model to estimate carbon dioxide recycling in forests using carbon-13 carbon-12 ratios and concentrations of ambient carbon dioxide. *Agric. For. Meteorol.* 48, 163-174.

Svennson, B.H.: 1980, Carbon dioxide and methane fluxes from the ombrotrophic parts of a subarctic mire. *Ecol. Bull.* 30, 235-250.

Townsend, A.R., Vitousek, P.M., and Holland, E.A.: 1992, Tropical soils could dominate the short-term carbon cycle feedbacks to increased global temperatures. *Clim. Change* 22, 293-303.

Vitt, D.H.: 1990, Growth and production dynamics of boreal mosses over climatic, chemical and topographic gradients. *Bot. J. Linn. Soc.* 104, 35-59.

Vourlitis, G., Oechel, W.C., Hastings, S.J., and Jenkins, MA.: 1993, A system for measuring *in situ* CO_2 and CH_4 flux in unmanaged ecosystems: an arctic example. *Funct. Ecol.* 7, 369-379.

Walbridge, M.R.: 1982, Vegetation patterning and community distribution in four high-elevation headwater wetlands in West Virginia. M.S. Thesis, West Virginia University, Morgantown.

Wallen, B.: 1986, Above and below ground dry mass of the three main vascular plants on hummocks on a subarctic peat bog. *Oikos* 46, 51-56.

Watts, W.A.: 1979, Late quaternary vegetation of central Appalachia and the New Jersey coastal plain. *Ecol. Monogr.* 49, 427-468.

Whiting, G.J.: 1994, CO_2 exchange in the Hudson Bay lowlands: Community characteristics and multispectral reflectance properties. *J. Geophys. Res.* 99, 1519-1528.

Whiting, G.J., Bartlett, D.S., Fan, S., Bakwin, P.S., and Wofsy, S.C.: 1992, Biosphere/atmosphere CO_2 exchange in tundra ecosystems: community characteristics and relationships with multispectral surface reflectance. *J. Geophys. Res.* 97, 16,671-16,680.

Wieder, R.K.: 1985, Peat and water chemistry at Big Run Bog, a peatland in the Appalachian Mountains of West Virginia. *Biogeochemistry* 1, 277-302.

Wieder, R.K., and Lang, G.E.: 1983, Net primary production of the dominant bryophytes in a *Sphagnum*-dominated wetland in West Virginia. *Bryologist* 86, 280-286.

Wieder, R.K., Yavitt, J.B., and Lang, G.E.: 1990, Methane production and sulfate reduction in two Appalachian peatlands. *Biogeochemistry* 10, 81-104.

Wieder, R.K., Novak, M., Schell, W.R., and Rhodes, T.: 199x, Rates of peat accumulation over the past 200 years in five *Sphagnum*-dominated peatlands in the United States. *J. Paleolimnol.* in press.

Wieder, R.K., Yavitt, J.B., Lang, G.E., and Bennett, C.A.: 1989, Aboveground net primary production at Big Run Bog, West Virginia. *Castanea* 45, 209-216.

Yavitt, J.B., Lang, G.E., and Downey, D.M.: 1988, Potential methane production and methane oxidation rates in peatland ecosystems of the Appalachian Mountains, United States. *Global Biogeochem. Cycles* 2, 253-268.

Yavitt, J.B., Wieder, R.K., and Lang, G.E.: 1993, CO_2 and CH_4 dynamics of a *Sphagnum*-dominated peatland in West Virginia. *Global Biogeochem. Cycles* 7, 259-274.

Zimov, S.A., Zimova, G.M., Daviodova, S.P., Daviodova, A.I., Voropaev, Y.V., Voropaev, Z.V., Prosiannikov, S.F., Prosiannikov, O.V., Semiletova, I.V., Semiletova, I.P.: 1993, Winter biotic activity and production of CO_2 in Siberian soils: a factor in the greenhouse effect. *J. Geophys. Res.* 98, 5015-5023.

PART IV

VEGETATION DYNAMICS AND ECOLOGY

HYDROLOGIC AND WETLAND CHARACTERISTICS OF A PIEDMONT BOTTOM IN SOUTH CAROLINA

D.D. HOOK[1], W.H. MCKEE, JR.[3], T.M. WILLIAMS[1], S. JONES[4], D. VAN BLARICOM[2], AND J. PARSONS[5]

[1]Department of Forest Resources, [2]The Strom Thurmond Institute, Clemson University, Clemson, SC. [3]U. S. Forest Service, Charleston, SC. [4]Environmental Services, Stone Mountain, GA, and Department of Biological and Agricultural Engineering, North Carolina State University, Raleigh, NC.

Abstract. A four hectare mixed bottomland hardwood site on Ninety Six Creek in the Piedmont of South Carolina near Ninety Six, SC was studied for two years to characterize wetland traits. The soils were thermic Fluventic or Fluvaquentic Dystrochrepts predominantly Shellbluff series and well drained. Overbank flooding occurred on the average of 4 times per year and 1.5 times during the growing season for a 13 year period. High water table levels during the early growing season were related to rainfall events. A hydrologic model (WATRCOM-2D), soils, water table levels, and GIS techniques were used to estimate the portion of the bottom that met wetland criteria similar to those defined in the 1987 and 1989 federal wetland delineation manuals. Less than one hectare met these criteria. The wetland "status" of the vegetation within the bottom and adjacent slope was not correlated with water table levels, predicted wetland areas, or landforms. Wetland traits of the site were closely related to hydric soil traits within the upper 25 cm of the Chewacla and Chenneby soils and landform characteristics. Wetlands in this bottom were primarily driven by local precipitation and not by overbank flooding as originally suspected. Songbirds and small mammals were relatively abundant in the small bottom during the spring and summer of 1992. Protection of only the jurisdictional wetlands in this bottom would not be adequate to sustain riverine functions (conveyor) and to provide wildlife travel corridors between adjacent forested areas.

Keywords: Overbank flooding, redox potentials, wetland criteria, hydrologic modeling, hydrogeomorphology.

1. Introduction

In the Piedmont region of South Carolina there are approximately 80,000 hectares (ha) of floodplain adjacent to streams which receive frequent overbank flooding and support vegetation typically adapted to live in a saturated soil. They may not be classified as "wetlands" because the soils are entisols with poorly defined horizons, alluvial in origin, and relatively recent in their formation; hence they do not exhibit well defined hydric indicators (U. S. Army Corps Engineers 1991). Riparian areas typically exhibit functional characteristics similar to wetlands (Brinson 1993 a,b), but may not receive protection under the Clean Water Act unless the areas meet federal jurisdictional wetland criteria. Since delineation of wetlands in small riparian areas are hampered by lack of precise wetland indicators, a project** was initiated to quantify the hydrology, soils, and selected bird and small mammal habitat characteristics of a small Piedmont floodplain on Ninety Six Creek near Ninety Six, South Carolina.

**This project was partially supported by the U. S. Army Corps of Engineers, Charleston District, Charleston, SC.

The purpose of the study was to determine the relationships among long-term hydrological conditions (flooding frequency, duration, soils, water table levels, soil redox potentials, vegetation, and timing), rainfall data, and selected wildlife habitats. It was hypothesized that these relationships would provide accurate and defensible delineation techniques in the field. Small mammals and songbirds were inventoried during the summer of 1992 as indicators of habitat quality in this riparian area.

2. Approach

2.1. Study Area

A four ha mature bottomland-hardwood stand below Enoree Road (SC 288) on Ninety Six Creek in Greenwood County, SC was selected as the study site (Figure 1). A stream gage station has been maintained on the bridge of Ninety Six Creek since September 1980 by the U. S. Geologic Survey. The site is occupied by a mature bottomland forest that has been relatively undisturbed within the past 50 years. The watershed above the gaging station encompasses 17.4 square miles (45.1 km^2) and is a rural area with a mixture of residential property, farmland, pastures, and small woodlands.

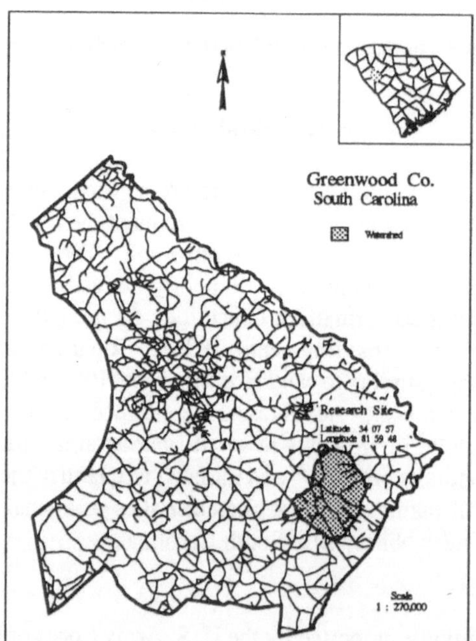

Figure 1. Location of research site within Greenwood County, South Carolina. Watershed area is shaded and other patterns on the map are roads.

2.2. WATER TABLE LEVELS, SOIL REDOX POTENTIALS AND SOILS

The study area was surveyed to 15 cm topographic intervals in the spring of 1992 and elevations were related to the datum of the gage station at the bridge on Enoree Road. In March 1991, five water table level Wells were installed to a depth of about 1.4 m or to the restrictive layer, whichever occurred first. In January 1992, four additional Wells were installed. Four Wells were located along each of two transects that ran perpendicular from the creek to where the slope increased sharply to about 20%, a distance of about 90 m. Well 5 was offset from hydrologic transect 1 (HT1) about 30 m south of Well 4 (Figure 2). A PVC pipe, 1.5 m long and 7 cm in diameter, was perforated with holes except for 25 cm on the top end. The perforated pipe was covered with a screen to reduce sedimentation through the bored holes and installed in each Well with about 25 cm exposed above the soil surface. The exposed portion of the pipe was covered with an aluminum can to reduce filling with debris. Two redox potential electrodes were installed within 1.5 m of each Well site, one at 15 cm and one at 25 cm soil depths.

The first five Wells and accompanying electrodes were measured once a week or twice a month from Julian Day (JD) 74 through JD 294 in 1991. In 1992, the five original Wells and four new Wells were measured twice a week from JD 3 through JD 363 in 1992. All Wells were read again from JD 64 through JD 151 in 1993. On JD 285 in 1992, the meter for measuring redox potential was found to be malfunctioning; thereafter redox readings were terminated. Redox potential measurements for JD 140 (May 20) and greater for 1992 were not used because there were questions about the proper function of the meter during this period. All but three of the electrodes had become unreliable by the spring of 1993, so no redox potential measurements were taken the third year. Consequently, redox potential data were available only from JD 45 to JD 139 in 1992. Soil pH was measured at 15 and 25 cm depths in 1993.

Redox electrodes were constructed according to the specifications of Letey and Stolzy (1964). Calibration of electrodes was done using a standard recommended by Light (1972), and a calomel electrode was used as a reference. Readings were converted to Eh by adding 240 mv to the meter readings and correcting to pH 7 (Gambrell and Patrick, 1978).

2.3. SOILS

The soils at each Well site were classified by two soil scientists of the U. S. Soil Conservation Service, Columbia, SC. A soils map was constructed for the site based on the soil classifications at each Well, examination of soil profiles within the study area, and topographic and vegetative features using GIS techniques.

2.4. HYDROLOGIC MODEL

A hydrologic model, WATRCOM-2D (Parsons et al., 1991a,b), was used to predict long-term hydrologic conditions on the site. Observed water table data from the two hydrologic transects at the research site were used to calibrate the model. U. S. Geologic

Survey stream gaging data records from September 1991 to August 1992 from Ninety Six Creek were used as boundary conditions for model calibration and long-term simulations. Measured soils data along with the USDA-SCS soils information were used to estimate the model's required soil information. Daily historic precipitation data from Columbia, SC were used for the long-term simulations.

After calibrating the model, four possible wetland criteria were simulated. These were: 1) water table depth within 15 cm of the soil surface (SS) for 7 days; 2) water table depth within 15 cm of SS for 14 days; 3) water table depth within 30 cm of SS for 7 days; and 4) water table depth within 30 cm of SS for 14 days. For each simulation, WATRCOM-2D predicted the water table depth along each transect perpendicular to Ninety Six Creek. WATRCOM-2D used the predicted water tables to delineate areas along the transects that met the wetland criteria under consideration. The hydrology of the remaining areas of the bottom was estimated using the WATRCOM-2D results and a GIS system. This enabled the mapping of areas in the bottom that met selected wetland criteria.

2.5. Vegetation

Vegetation surveys were conducted along three parallel transects from the creek to the base of the slope (see Figure 7, vegetation plot diameters are to scale on the map). Nineteen circular plots (0.016 hectares) were systematically located at 22 m intervals along the transects. Within each plot, all vegetation greater than 1.37 m tall (T) was inventoried by diameter and species. Percent cover was ocularly estimated, and density was determined by species for woody and herbaceous vegetation less than 1.37 m tall (S) on three randomly located 0.0004 ha plots within the larger plot. The data for T vegetation were summarized by three diameter classes at 1.37 m height (i. e. 0-5.0 cm, 5.1-25.0 cm, and > 25.0 cm) and S vegetation by density and percent cover. All vegetation strata (i. e. annuals, perennial, shrubs, and trees) were grouped for analysis as they are for delineation computations in the Federal Manual for Identifying and Delineation of Jurisdictional Wetlands (An Interagency Cooperative Publication 1989).

For each plot, the vegetation for each stratum (S vegetation and T diameter classes; 0-5 cm, 5.1-25.0 cm, and >25 cm) of each plot was classified as hydric vegetation (obligate or facultative wet) or non-hydric vegetation (facultative, facultative upland, and upland) and coded (0 = non wetland vegetation and 1 = wetland vegetation based on the National Plant List of Species; Reed, 1988). The sample plots were located on the map via GIS for each classification category. A hydric vegetation rating was assigned to each plot by summarizing the coded values for each stratum, dividing by the number of strata (4), and converting to percent. For instance, if three strata had wetland vegetation and one did not, the plot was designated 75 % wetland vegetation. Prevalence Indices (PI) were also computed for each plot and vegetation strata per the Federal Manual for Identifying and Delineating Jurisdictional Wetlands.

Maps with PI for each plot were constructed for dominant S and T vegetation. PI were computed for all S vegetation as S1 all species and S2 dominant vegetation, all T vegetation as (T1), dominant vegetation (T2). PI were also computed independently for each of the three diameter classes of T vegetation.

2.6. Avifauna

Data were collected on avifauna between astronomical sunrise and 0900 hours once a week unless it rained from JD 135 (May) through JD 190 (July) of 1992. The average temperature during the bird census was 19.2 degrees Celsius. Time of visit to the each plot was varied to minimize bias, with counts beginning immediately upon arrival at plots.

A 65 m buffer zone from the exterior (road) to the interior of the 4 hectare study was excluded and two interior plots were located at 175 m apart. Distance (x) of species from plot center was recorded at x < 25 m, x > 25 but < 50 m, and x > 50 m. Observations were made at each plot for 15 minutes. Records were kept of birds heard for first 3 minutes, next 2 minutes, next 5 minutes, and last 5 minutes and estimated distance from observer. The time between plots was 15 minutes or longer. This method is comparable to the Point Count System employed by the U. S. Forest Service (Ralph et al 1993). Species were identified primarily by songs and calls; visual observations were rare due to the tall and closed canopy.

2.7. Small Mammal Inventory

The research site is joined on one side by housing, a church yard, and open farm land; thus it serves as a natural travel corridor between adjacent forests above and below the site. Three transect lines were established 25 m apart near the center of the study area. Due to the irregular shape of the site, transect lines were unequal in length and contained 15, 18, and 9 traps, respectively. Sherman live traps were placed at 15 m intervals along each transect. The traps were baited with rolled oats and peanut butter. Cotton balls were placed in the traps to help the mammals maintain their body temperatures. Traps were checked every morning and rebaited. Plot number, species, sex, age, weight, and reproductive condition were recorded for each capture. Captured animals were released as soon as measurements were completed. Traps were set for 6 consecutive nights during the waxing moon.

3. Data Interpretation

GIS techniques were used to map predicted hydrologic boundaries and soil series boundaries. GIS was also used to map the wetland vegetation status of each plot. The redox potentials for 1992 were corrected by soil pH measurements taken at 15 and 25 cm depths for JD 64 through 139 in 1993. These corrections lowered the redox potentials from 0 to -177 mv. A comparison of the curves before and after corrections showed a small decrease in redox potential for all location but no significant change in pattern over time. Water table levels and redox potentials at 15 and 25 cm depths for 1992 were compared to creek flood stages for the same period to determine relationship between soil saturation and flood stages.

A matrix was computed for all Wells and redox potential measurements to determine correlations among the water table levels and redox potentials. In addition, a matrix was computed for rainfall and Well readings and rainfall and redox potentials for each year. Daily rainfall data for 1991 and 1992 from the South Carolina Forestry Commission Epworth Fire Tower, located approximately 6 km SW of the research site, were used in the correlation matrices.

The observed water table data from hydrologic transect 1 (HT1; Wells 1-4) and hydrologic transect 2 (HT2; Wells 6-9) were used to calibrate the WATERCOM-2D model (Figure 6). Potential evapotranspiration was estimated from daily temperature records from Columbia, SC using the Thornthwaite method (Thornthwaite 1948). Since hourly rainfall data were not available for the Ninety Six site, hourly rainfall records from Columbia, SC were used for the calibration period, from November 1991-May 1993. U. S. Geologic Survey stream gaging data for Ninety Six Creek were obtained for the period September 1981-May 1992. These data along with the USDA-SCS soils (USDA, SCS 1980) information were used to set up WATRCOM-2D input data sets. Lateral saturated hydraulic conductivity was varied over the range of values given by USDA-SCS. Visual inspection of graphs of observed and simulated water tables versus time on each transect were used to select the saturated conductivity input values. The resulting input data sets yielded simulations that were in good agreement with the observed water table data. Well 5 had the highest correlation with water table levels of HT1 ($r = 0.84$ or better); therefore it was used to illustrate general water table relationships for the upper end of the site. Well 6 had the highest correlation with water table levels of HT2 ($r = 0.90$ or better); thus it was selected to represent the lower transect.

The GIS constructed maps were used to assess relationships among soils, hydrology, and vegetation. Avifauna and small mammal populations data were analyzed by tabular methods only because there were not enough samples and the area appeared to be too small to assess spatial relationships.

The vegetation data were also analyzed by multivariate classification procedures (DECORANA and TWINSPAN) to identify similarities and dissimilarities among sample plots.

4. Results

4.1. Soils and Hydrology

4.1.1. Soil classification and pH

The soils at Wells 1, 2, 3, 4, 5, and 6 were correlate to the Shellbluff series (fine-silty, mixed, thermic Fluventic Dystrochrepts). These series are classified as moderately well drained by the SCS (Figure 2). The soil at Well 7 was correlated to the Chewacla series (fine-loamy, mixed thermic Fluvaquentic Dystrochrepts) which is classified as somewhat poorly drained. Soils at Well 8 and 9 were correlated to the Chenneby Taxadjunct (fine-silty, mixed, thermic Fluvaquentic Dystrochrepts) which is classified as somewhat poorly drained. Soil pH values averaged 6.3 during the March to May period, but values varied by season and Well location within seasons (data not shown). The largest deviation from the mean occurred between Julian Days 64 and 90 when the soil pH dropped to an

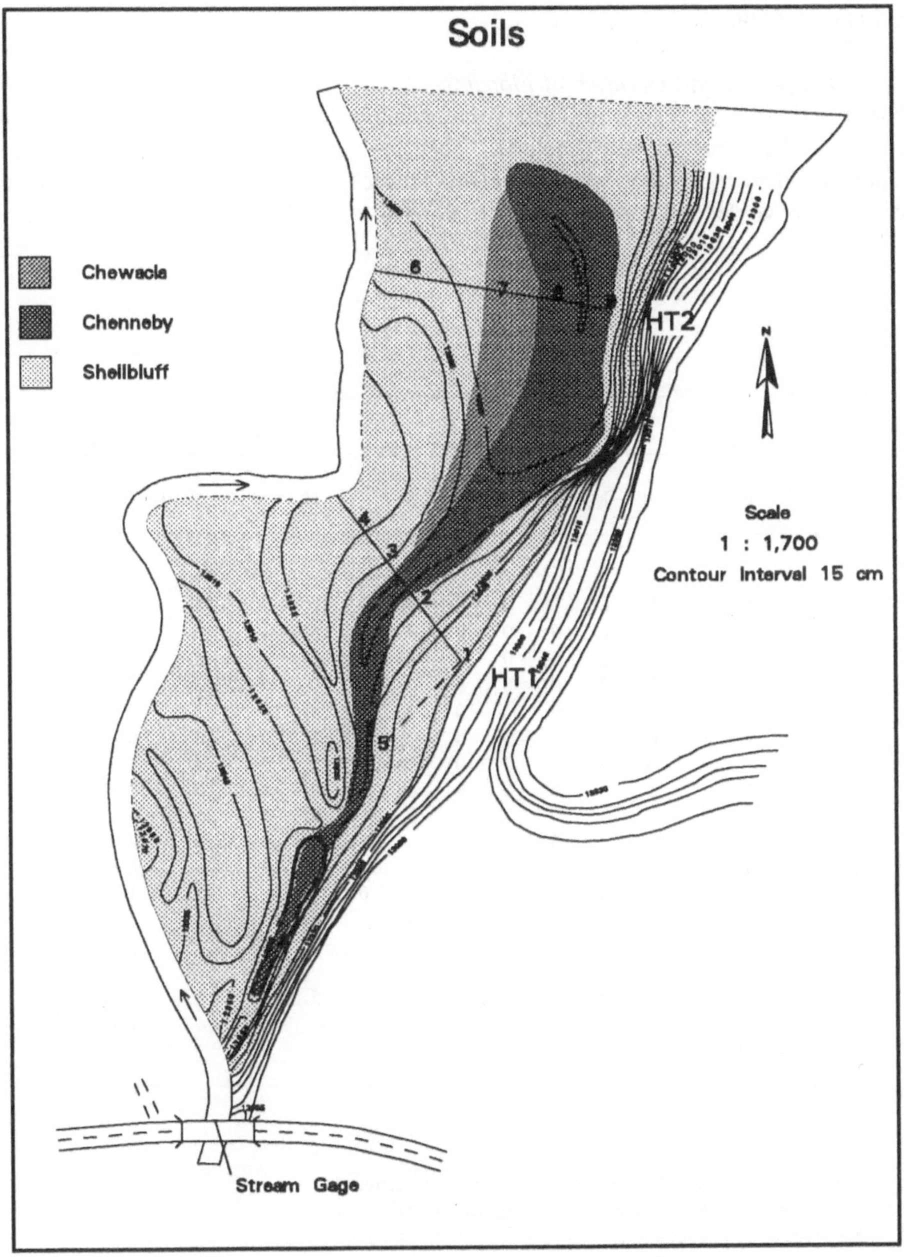

Figure 2. Soil series, hydrologic transects (HT1 and HT2), and topographic elevations of the Ninety Six Site research site.

average of 5.6. However, there was a large amount of variation in pH values with time and among Wells.

4.1.2. *Frequency and Duration of Flooding*

During a 13 year period, October 1980 to September 1993, the research site flooded an average of 4.1 times per year (Table I). When the creek exceeded channel capacity, it generally did so with a deep overflow; debris lines could be found across the bottom up to 0.5 m above the soil surface. The site flooded for two consecutive days only five

TABLE I

Number of days the stream gage exceeded flood stage (290 cfs) at Ninety Six Creek by year and month.

Year	J	F	M	A	M	J	J	A	S	O	N	D	Total	GS
80	-	-	-	-	-	-	-	-	-	1	0	0	1	1
81	0	1	0	0	0	0	0	0	0	0	0	2	3	0
82	2	2	0	0	0	0	0	0	0	0	0	1	5	0
83	2*	1	1	1	0	0	0	0	0	0	0	1	6	2
84	2	3	0	0	2*	0	0	0	0	0	0	0	7	2
85	0	2	0	0	0	0	0	0	0	0	3	0	5	0
86	0	0	0	0	0	0	0	0	0	0	0	0	0	0
87	2	1	1	0	0	0	0	0	0	0	0	0	4	1
88	0	0	0	0	0	0	0	0	0	0	0	0	0	0
89	0	0	1	1	0	0	1	0	0	1	0	0	4	4
90	0	1	0	0	0	0	0	0	0	2	0	0	3	2
91	2*	0	4**	0	0	0	0	0	0	0	0	0	6	4
92	0	0	1	0	0	0	0	0	0	0	2	1	4	1
93	1	1	3	0	0	0	0	0	0	-	-	-	5	3
Sum	11	12	11	2	2	0	1	0	0	4	5	5	53	20
Mean													4.1	1.5
***	46	54	46	15	8	0	8	0	0	23	15	31		

GS = Growing Season (March through October).

* Periods when flood stage occurred for two consecutive days.
** Flooded two consecutive days twice in March 1991.
*** Probability of flooding by month (number years flooding occurred/13 years X 100 = %).

times during the 13 year observation period (January 22-23, JD 22-23 1983; February 28-March 1, JD 59-60 1987; May 3-4, JD 123-124, 1984 and March 2-3 and 29-30, JD 61-62 and 88-89 1991). Thus, the duration of flooding was generally less than 24 hours per event and frequency of flooding was 1.5 +/- 1.5 times per year during the growing season (March 1-October 31; JD 60-304). Probability of flooding was highest for February (54%) and January and March were next highest each with probability of 46% (Table I). Of the eleven times the site flooded in March, seven occurred before March 15. Therefore, the probability of flooding to occur after the average date of last freeze (March 25, USDA, SCS 1980) was very small, usually less than 20%.

4.1.3. Water Table Levels

In general, depth to the water table declined throughout the measurement period in 1991, except for small rises associated with flooding on JD 88 and 89 and periodic rises due to rainfall during the measurement period (Figure 3). The water table responses to the two consecutive days of flooding may have been more dramatic than Well readings indicated. The Wells were read on JD 86 and 92 in 1991; thus the main effect of overbank flooding on JD 88 and 89 in 1991 may have been missed in the six day interval.

In 1992, the depth to the water table in all Wells began to rise on JD 66 before the flooding event on JD 68 (Figure 3b & c). Water table levels on the lower transect, especially Wells 8 & 9, began to rise again in early fall (about JD 280 to 310; Figure 3c). Water table levels in all Wells rose to near the surface following overbank flooding on JD 326 and 330.

Changes in water table levels in response to precipitation and overbank flooding are shown in Figure 4. Until JD 50 the water table levels in all Wells were very low and unresponsive to precipitation. From JD 50 until JD 180 the water table rose after each precipitation event. After JD 180 the water tables fell and the Wells remained dry or very low until day 326 despite frequent summer precipitation events.

The lack of response of the water table levels in all Wells to precipitation during June, July, and August (JD 151-243) was probably due to climatic conditions. High solar radiation loads and high temperatures typically combine during these months in the South to create high rates of evapotranspiration that exceed the rate of rainfall (Thornthwaite 1948). As forests withdraw water to meet high transpiration demands, the water table characteristically drops and the soil profile becomes unsaturated and may dry to near the wilting point (Pritchett and Fisher 1987). Precipitation will not percolate to the water table until the unsaturated profile reaches field capacity (Hanks and Ashcroft 1980). The water table does not rise again until percolating precipitation exceeds drainage. The water table on the research site did not begin to rise after the 1991 dry period until late February 1992 (Figure 3b). By JD 180 of 1992, the water table had fallen so that all Wells except Well 8 were dry. After this date, even large precipitation events did not bring the unsaturated soil to field capacity until JD 320.

4.1.4. Soil Redox Potentials

The soil redox potential data were limited but were sufficient to show that when the water table levels were near the surface during the early spring, redox potentials tended

a

b

c

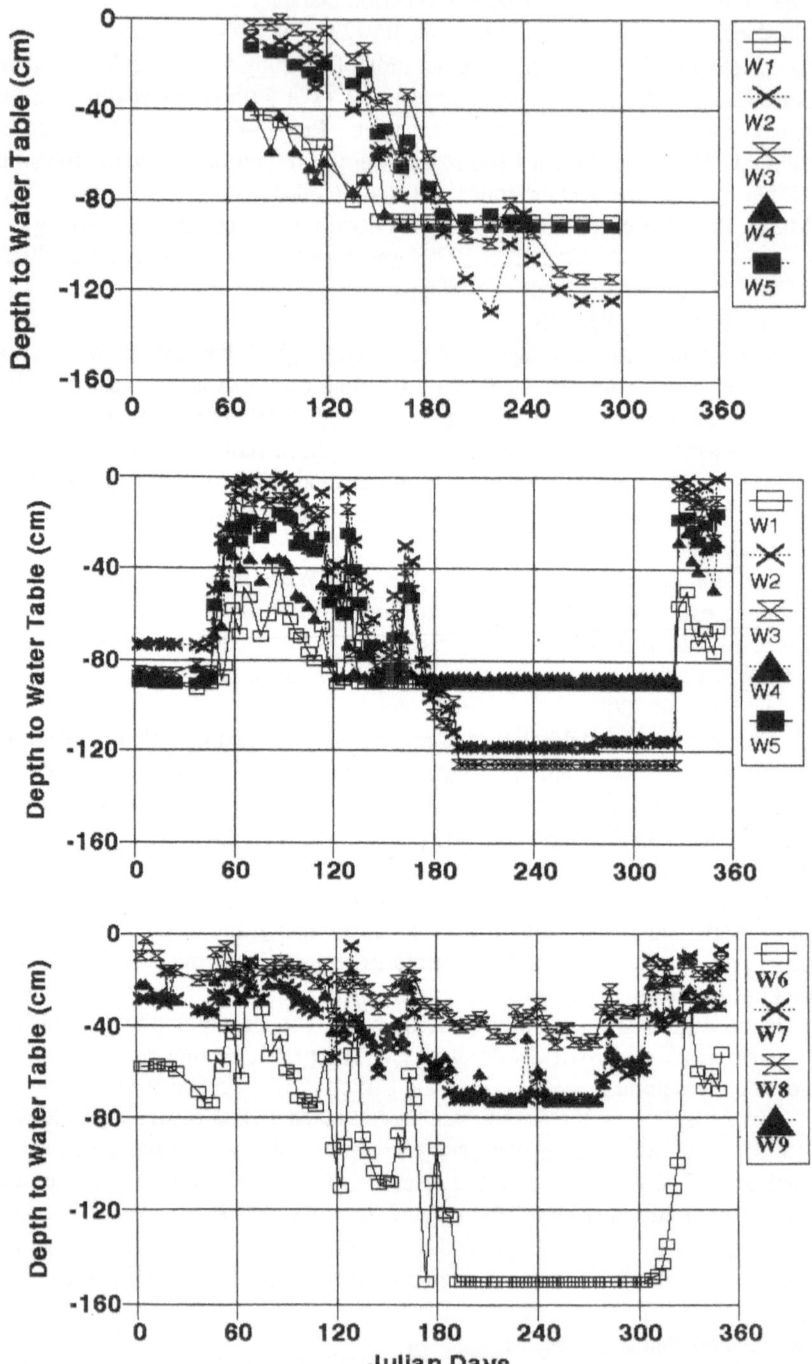

Figure 3. Depth to water table level (cm) by Well for 1991 (a) and 1992 (b & c) by Julian Day.

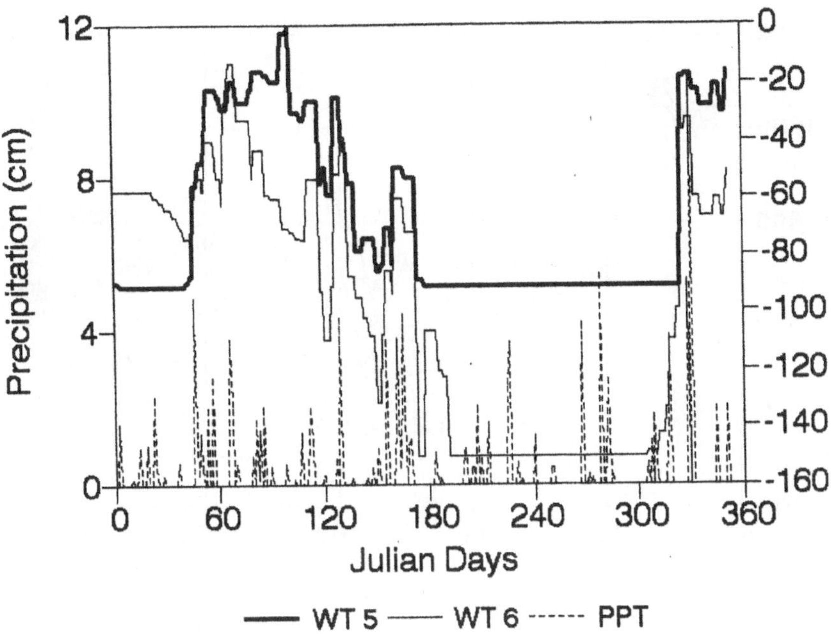

Figure 4. Relationship of precipitation to water table level for HT1 (Wells 5) and HT2 (Well 6) for 1992.

to decrease. Also, the data showed that reducing conditions existed for several days to several weeks among Well locations during the spring wet period (Figure 5a & b).

Soil reduction begins around +340 mv at pH 7 and potentials less than +100 mv are considered highly reduced (Gambrell and Patrick, 1978). Early in 1992, the soils at all Well sites were well aerated as indicated by redox potentials between 990 and 1400 mv (Figure 5a and b). Well 2 generally displayed highly reduced conditions from JD 63 through JD 108 in 1992. The other Well sites on this transect generally remained above + 400 mv except for occasional reducing conditions. Water table levels in Well 3 were similar to those in Well 2 (Figure 3a & b), but Well 3 seldom had redox potentials less than +200 mv (Figure 5a). On the lower transect, Wells 7, 8, and 9 generally had high reducing conditions from JD 59 through JD 139 in 1992 (Figure 5b). In contrast, the redox potential of Well 6 did not read less than +400 mv during the same period.

There were no direct correlations between precipitation and water table levels or between precipitation and soil redox potentials.

4.1.5. Longterm Hydrologic Simulations

The model, WATRCOM-2D, predicted four closely related areas as potential wetlands using four distinct wetland criteria (Figure 6). The smallest predicted wetland area resulted from the criterion of the water table within 15 cm of the soil surface for 14 days and the largest area fit the criterion of the water table within 30 cm of the soil surface for 7 days (Figure 6). Accuracy of predicted wetland boundaries was within one meter

a

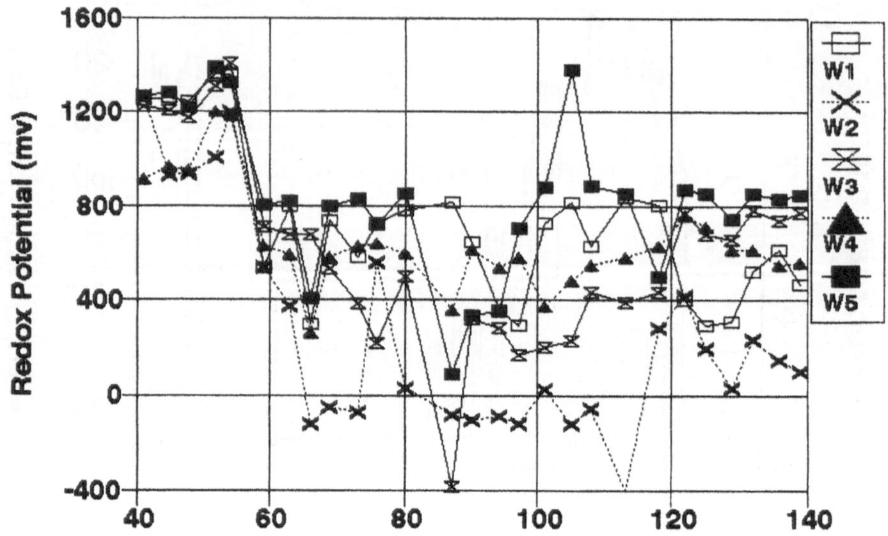

b

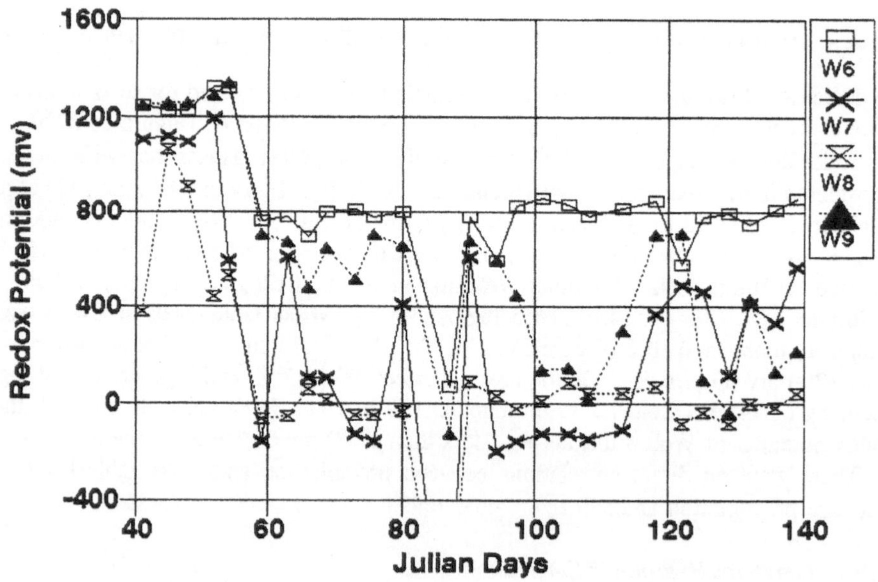

Figure 5. Redox potentials (Eh in mv) for 15 cm soil depths at all Well locations for Julian Days 41 through 139 for 1992. Values have been correct to pH 7.

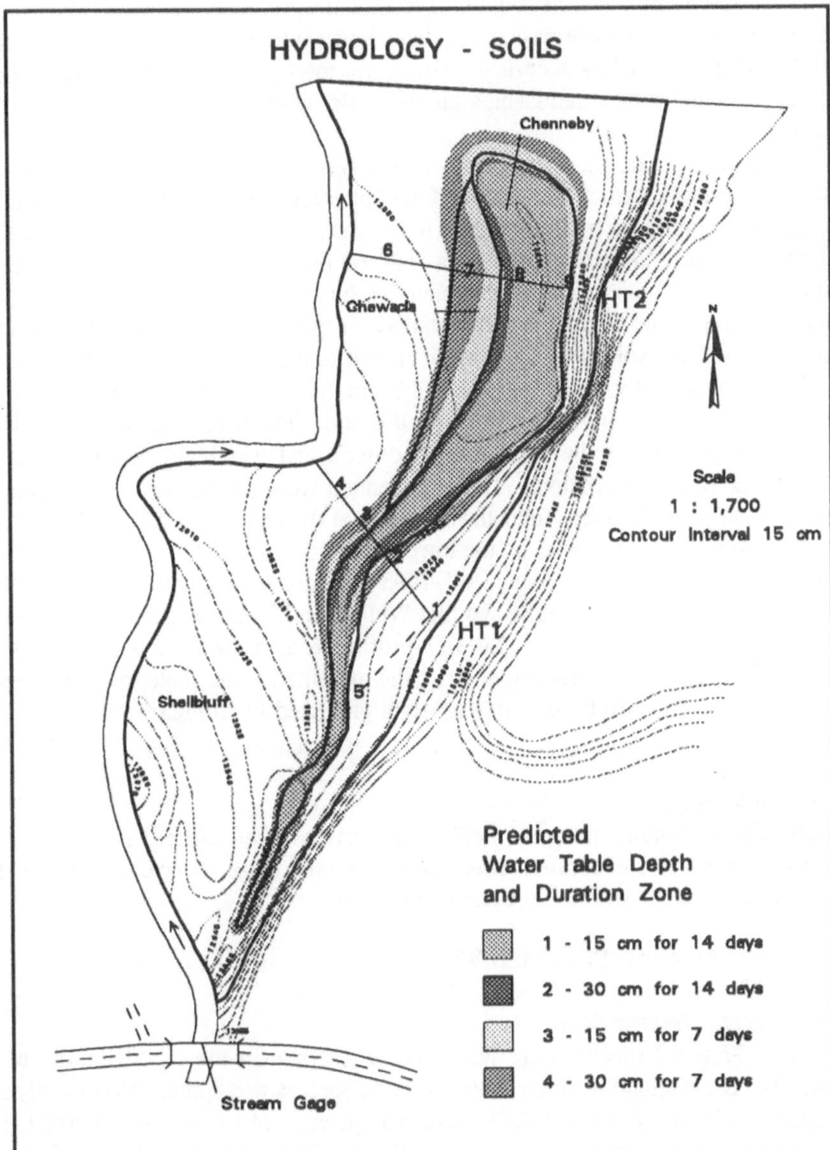

Figure 6. Predicted wetland boundaries for four wetland criteria for the Ninety Six Site. Data was based on long-term precipitation data and 13 year flooding history of the site and were calibrated with Soils and water table data from the site.

where the hydrology transects (Figure 6) crossed the predicted wetlands. The hydrologic boundaries away from the transects may have larger interpretive errors but were probably within two meters or less accuracy. Therefore, the hydrologic maps were sufficiently accurate for analyzing relationships among soils, vegetation, and long-term predicted hydrology.

The location of Chewacla and Chenneby soil series boundaries were closely associated with the long-term predicted wetland criteria boundaries (Figure 6). All of the Chenneby series met the predicted hydrology criterion of the water table being within 30 cm of the soil surface for 7 days, but less than half of the Chewacla series was within the hydrology criterion of the water table being within 15 cm of the soil surface for 7 days. Only a portion of the Chewacla series showed positive hydric soil indicators within the upper 30 cm of the soil, (i. e. the portion closest to the Chenneby series between Wells 7 and 8). Apparently, waterlogging at 30 cm for 7 days is not sufficient to cause positive hydric indicators above 30 cm in this soil. The hydrologic criterion of 15 cm for 14 days closely fits the 1987 manual criteria and was highly correlated with the Chenneby series in this bottom. The most lenient wetland criterion tested (water table within 30 cm of soil surface for 7 days) exceeded the area of the two hydric soil series (i. e. Chenneby and Chewacla) on the lower end of the wetland and near Wells 2 and 3.

Water table level records from 1991-1993 (all data are not shown but 1991-92 data are shown in Figure 3a & b) reflected that Wells 2 and 3 had water tables that generally stayed in the upper 25 cm of the soil profile from early March until early May. Consequently, for three consecutive years the measured water table levels of these two Wells were closely correlated with long-term predicted hydrologic wetland boundaries (Figure 6).

4.1.6. Landform

The areas in this bottom that were predicted as wetlands were closely associated with the concave portions of the bottom. Most of the wetland areas occurred below the 12908 cm elevation contour within the bottom (Figure 6).

4.2. Vegetation and Predicted Wetlands

4.2.1. Species Wetland Status

A total of 77 plant species were identified on the site in the early summer of 1992 (Table II). The wetland indicator status of these species was spread fairly evenly across the spectrum from upland to obligate wetland species. Most were facultative (31) and about equal amounts were facultative wet (19) and facultative upland (20). Only a few species were found in the obligate (3) and upland (4) categories.

When the vegetation strata were assessed with regard to wetland, non-wetland bottom, and upland slope areas, there was no clear segregation of wetland status of the vegetation among the three landform types (Figures 7, 8 and 9). The composite of all canopies (S vegetation and three diameter class of T vegetation) showed that 75 % of the canopies on four plots within the predicted wetlands area had predominantly hydric vegetation (Figure 7). However, one plot within the predicted wetlands area had 50 % of the

TABLE II

List of plant species found on the Ninety Six site in the early summer of 1992 and their wetland indicator status.

Scientific Name	Common Name	Wetland Indicator Status
Acer negundo	Box-elder	FACW
Acer rubrum	Red maple	FAC
Acer saccharum	Sugar maple	FACW
Arisaema dracontium	Green dragon	FACW
Arthraxon hispidus	Joint-head Anthraxon	FACU
Arundinaria gigantea	Switchcane	FACW
Asarum canadense	Wild ginger	FACU?
Asimina triloba	Pawpaw	FAC
Aster paludosus	Southern swamp aster	FACW
Bazzania trilobata	Liverwort	UPL?
Bignonia capreolata	Crossvine	FAC
Campsis radicans	Trumpet creeper	FAC
Carpinus caroliniana	American hornbeam	FAC
Carya cordiformis	Bitternut hickory	FAC
Celtis laevigata	Sugarberry	FACW
Chimaphila spp	Pipsissewa	UPL
Cimicifufa racemosa	Black snakeroot	FACU
Cornus florida	Dogwood	FACU
Crataegus marshallii	Parsley hawthorn	FAC
Elephantopus carolinianus	Carolina elephant-foot	FAC
Epilobium angustifolium	Fireweed	FACU
Eulalia viminea	Nepal microstegium	FAC
Euonymus americanus	Strawberry bush	FAC
Eupotrium spp	Eupotrium spp	FAC?
Fagus grandifolia	American beech	FACU
Fragaria virginiana	Wild strawberry	FAC
Fraxinus pennsylvanica	Green ash	FACW
Galium asprellum	Bedstraw	FACW
Geranium maculatum	Wild geranium	FACU
Geum canadense	White avens	FAC
Hydrangia aborescens	Wild hydrangia	FACU
Hydrocotyle umbellata	Penny-wort	OBL
Hypericum apocynifolium	St. John's-wort	FACW
Impatiens pallida	Impatiens	FACW
Ipomoea purpurea	Common morning glory	FACU
Juglans nigra	Black walnut	FACU
Juncus effusus	Soft rush	FACW
Juniperus virginiana	Eastern red cedar	FACU
Lactuca serriola	Prickley lettuce	FAC
Lespedeza striata	Japanese clover	FACU
Ligustrum sinense	Chinese privet	FAC
Liquidambar styracifula	Sweetgum	FAC

TABLE II (continued)

Scientific Name	Common Name	Wetland Indicator Status
Lirodendron tulipifera	Yellow poplar	FAC
Lonicera japonica	Japanese honeysuckle	FAC
Lyonia ligustrina	Maleberry	FACW
Mitchella repens	Partridgeberry	FACU
Morus rubra	Red mulberry	FAC
Parthenocissus quinquefolia	Virginia creeper	FAC
Passiflora lutea	Yellow passionflower	FACU or UPL
Peltandra luteospadix	Green arrow-leaf	OBL
Persea borbonia	Redbay	FACW
Pinus taeda	Loblolly pine	FAC
Polygonum virginianum	Virginia knotweed	FAC
Polystichum acrostichoides	Christmas fern	FAC
Prunus serotina	Black cherry	FACU
Quercus alba	White oak	FACU
Quercus coccinea	Scarlet oak	UPL
Quercus falcata var. pagodifolia	Cherrybark oak	FAC
Quercus michauxii	Swamp chestnut oak	FACW
Quercus nigra	Water oak	FAC
Quercus phellos	Willow oak	FACW
Quercus rubra	Northern red oak	FACU
Quercus Shumardii	Shumard oak	FACW
Rubus Argutus	Blackberry	FACU
Rudbeckia laciniata	Cone flower	FACW
Sambucus pubens	Elderberry	FACU or UPL
Saururus cernuus	Lizard's tail	OBL
Smilacina racemosa	False Solomon's-seal	FACU
Smilax glauca	Green catbrier	FAC
Smilax rotundifolia	White catbrier	FAC
Solidago mirabilis	Goldenrod	FACW
Toxicodendron radicans	Poison-ivy	FAC
Ulmus americana	American elm	FACW
Ulmus rubra	Slippery elm	FAC
Uniola latifolia	Inland sea oats	UPL
Viola spp	Violet spp	FAC?
Vitis rotundifolia	Muscadine	FAC

Number Species by Wetland Status

UPL	FACU	FAC	FACW	OBL	Total
4	20	31	19	3	77

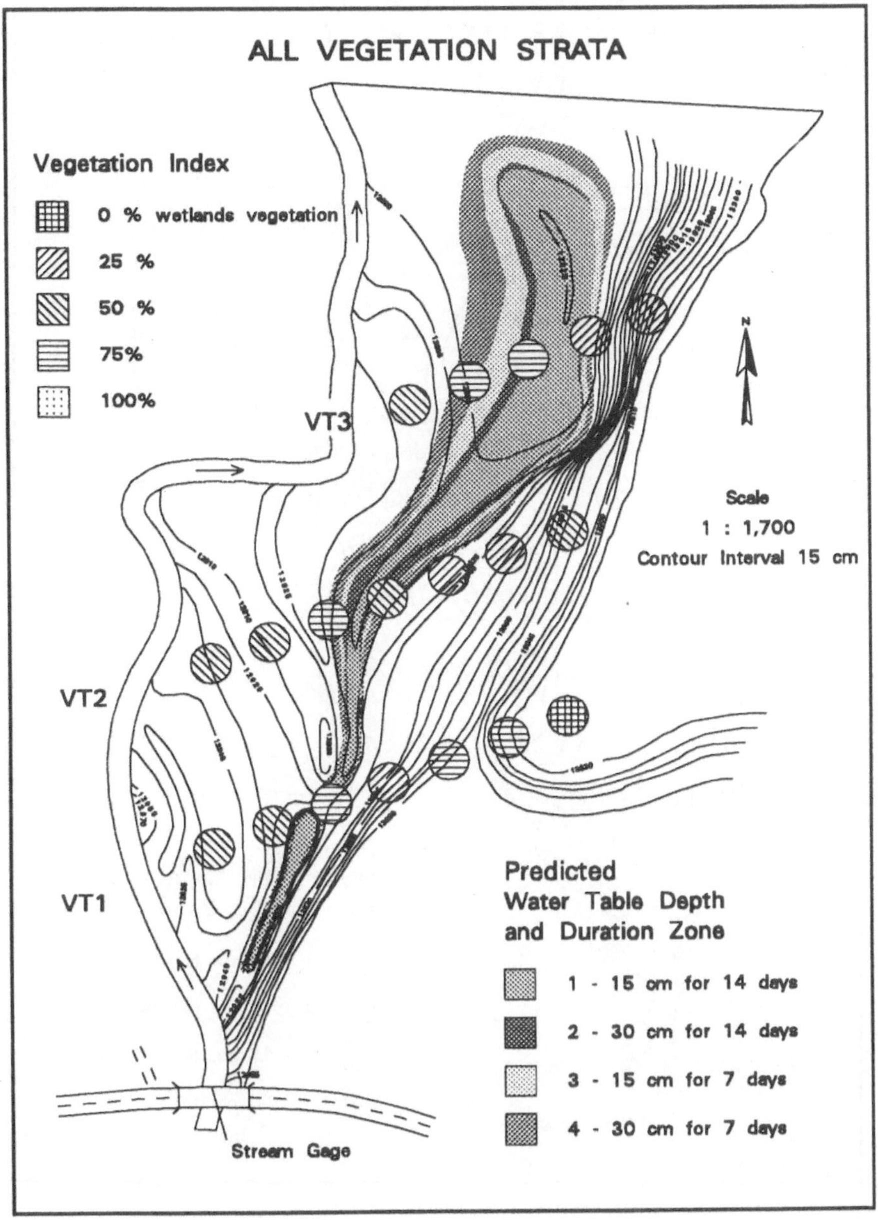

Figure 7. Vegetation indices for all strata on all plots based on percent of species that were rated as facultative wet (FACW) and obligate (OBL). The circlar areas are the plot locations along each transect drawn to scale.

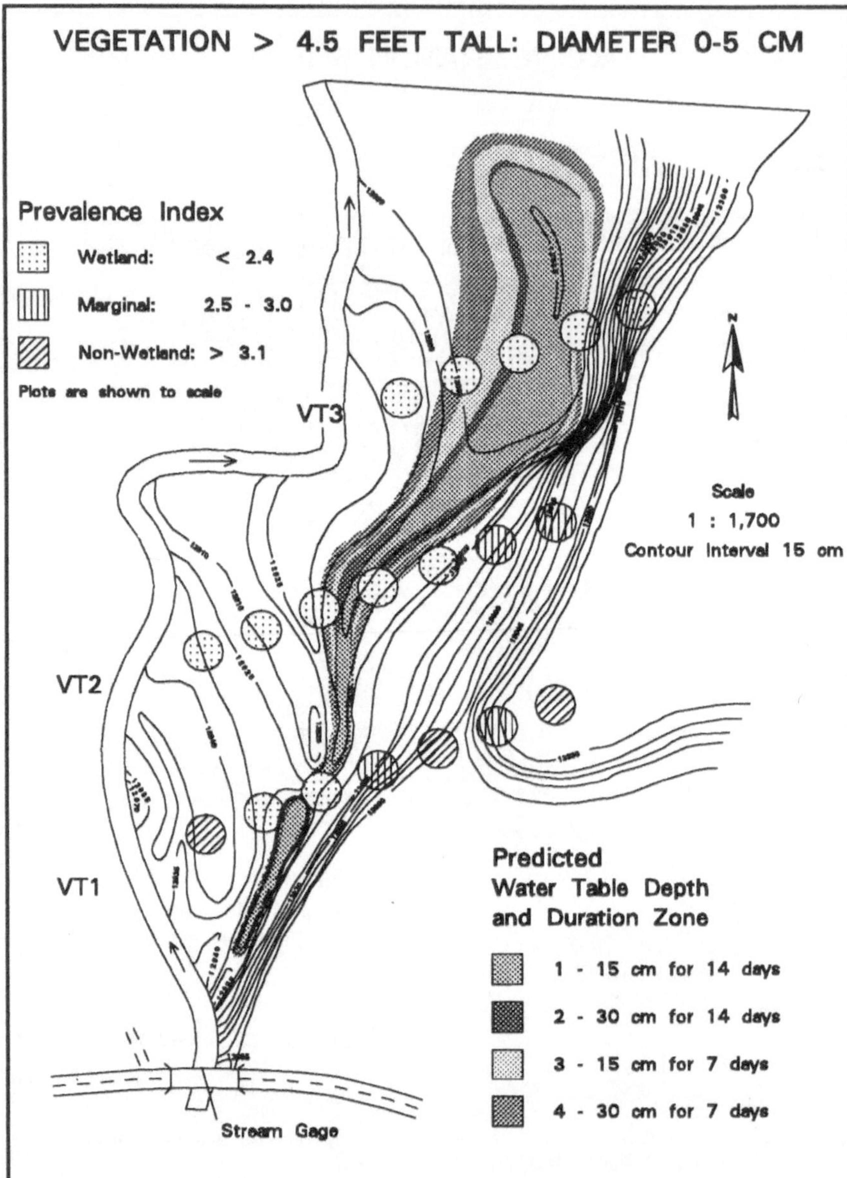

Figure 8. Prevalence indices for nineteen plots for the 0-5 cm diameter class of vegetation.

canopy as predominantly hydric vegetation and one plot had only 25 % hydric vegetation. Two plots with 75 % predominance of hydric vegetation in the canopy occurred on non-wetland areas and three plots with only 25 % hydric vegetation in the canopy occurred in the non-wetland bottom area.

4.2.2. Prevalence Index

Prevalence indices (PI) for all canopies and various combinations of vegetation categories showed a lack of correlation between PI and the three landform types (Figures 8 and 9). One would not expect a large difference in wetland vegetation status between the wetland and non-wetland area of the bottom, but one would expect the vegetation characteristics of the adjacent steep slope to be much different. However, the PI for the slope areas were more closely related to the wetland areas than the non-wetland bottom (Figure 9).

DECORANA and TWINSPAN analyses showed that several plots segregated out from the others but segregation was not correlated with soil classification, wetland and non-wetland characteristics, or predicted hydrology of the site.

4.2.3. Avifauna

Thirty-two species of bird species were identified during the spring-summer breeding season (May-July; Table III). Several species were closely associated with riparian woodlands (Dickson 1978) and time of activity varied among the species (Table IV). These species, Eastern Phoebe, Downy Woodpecker, Common Yellowthroat, Yellow Warbler, Acadian Flycatcher, Song Sparrow, White-breasted Nuthatch, Hooded Warbler, and Barred Owl, were found within a 50 m radius of plot centers, indicating a preference for forest interiors. Species identified at greater than 50 m included Rufous-sided Towhee, American Crow, Mourning Dove, Bobwhite Quail, Mockingbird, Brown Thrasher, and Pine Warbler. The latter species prefer open brushlands or conifer stands (Dickson, in press), both of which are adjacent to the bottomland hardwood study site.

4.2.4. Small Mammals

Four species of small mammals were identified from 56 individuals captured in 252 trap nights (6 nights, 42 traps), a 22% success rate (Table V). The average number of captures per night was 9.3 (range 3-16, Table VI) which represents a very high success rate compared with studies in deciduous stands (Dueser and Shugart 1978 and Geier and Best 1980).

5. Discussion

The entire study site had the appearance of an active floodplain. Drift lines of debris (up to 0.5 m above the soil surface) from overbank flooding were common throughout the bottom and many wetland types of vegetation were present in the bottom. However, soils, water table level data, and long-term predicted hydrologic conditions demonstrated that only a small portion of the research site met wetland jurisdiction criteria (Figure 6).

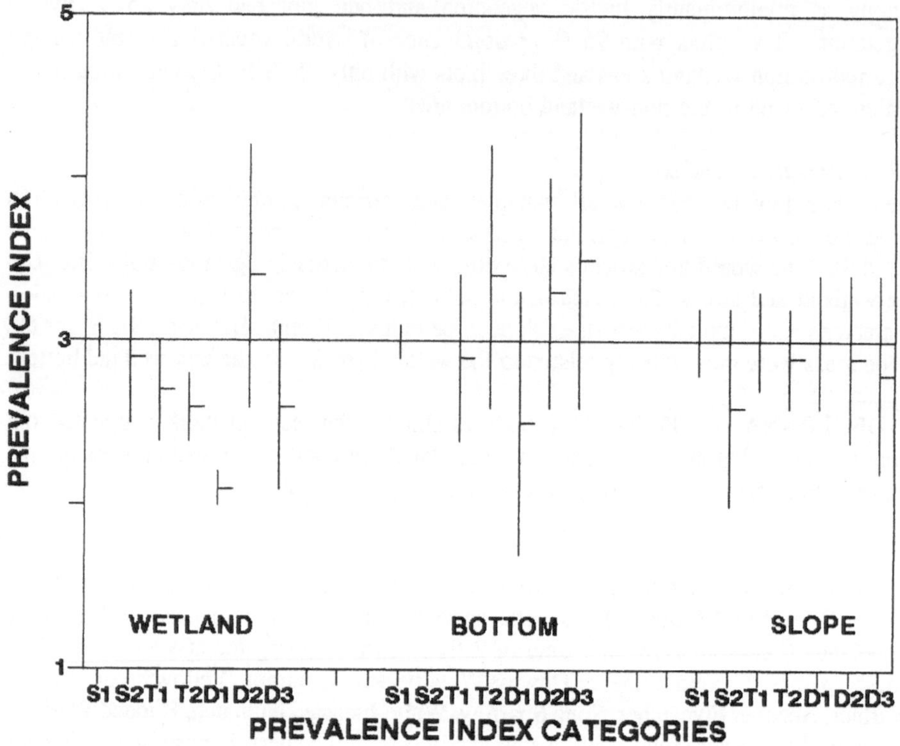

Figure 9. Prevalence indices computed for 7 categories of vegetation for the landform areas--wetland, bottom, and slope on the research site.

KEY: Vegetation less than 1.37 m tall; L1 = all species and L2 = dominant species based on percent cover. Vegetation greater than 1.37 m tall; T1 = all species; T2 = dominant species based on basal area; D1 = 0-5 cm diameter class; D2 = 5.1-25 cm diameter class; and D3 = > 25 cm diameter class.

TABLE III

Avifauna species identified during the period May 18-July 10, 1992 at Ninety Six Creek site near Ninety Six, SC.

Scientific Name	Abbreviation	Common Name
Caprimulgus vociferus	WHPW	Whip-poor-will
Cardinalis cardinalis	CARD	Northern Cardinal
Colinus virginianus	BWQU	Common Bobwhite
Contopus virens	EAPE	Eastern Pewee
Corvus brachyrhynchos	AMCR	American Crow
Cyanocitta cristata	BLJA	Blue Jay
**Dendroica petechia	YEWA	Yellow Warbler
Dendroica pinus	PIWA	Pine Warbler
**Empidonax virescens	ACFL	Acadian Flycatcher
**Geothlypis trichas	CYTH	Common Yellow-throat
Megaceryle alcyon	BEKF	Belted Kingfisher
Melanerpes erythrocephalus	RHWP	Red-headed Woodpecker **
Melospiza melodia	SOSP	Song Sparrow
Mimus polyglottos	MOBI	Mockingbird
Parus bicolor	TUTM	Tufted Titmouse
Parus carolinensus	CACH	Carolina Chickadee **
Pecoides pubescens	DOWP	Downy Woodpecker
Pipilo erythropthalmus	RSTOW	Rufous-sided Towhee
Piranga rubra	SUTA	Summer Tanager
Polioptila caerulea	BGNC	Blue-gray Gnatcatcher
**Sayornis phoebe	EAPH	Eastern Phoebe
**Sitta carolinensis	WBNH	White-breasted Nuthatch
Spizella passerina	CHSP	Chipping Sparrow
Spizella pusilla	FISP	Field Sparrow
**Strix varia	BAOW	Barred Owl
Thryothorus ludovicianus	CAWR	Carolina Wren
Toxostoma rufum	BRTHR	Brown Thrasher
Unknown	UN#1	Unidentified #1
Vireo flavifrons	YTVI	Yellow-throated Vireo
Vireo solitarius	SIVI	Solitary Vireo
Zenaida macroura	MODO	Mourning Dove
Zonotrichia albicollis	WTSP	White-throated Sparrow
Wilsonia citrina	HOWA	Hooded Warbler

**Species associated with riparian woodlands

TABLE IV

Avifaunal activity by date and species for Ninety Six Creek site near Ninety Six, SC during the spring-summer of 1992.

Abbreviation	Observation Dates									
	18May	24May	1Jun	14Jun	15Jun	17Jun	20Jun	30Jun	4Jul	10Jul
WHPW					A					
CARD	A	A	C	C	C	A	A			
BWBQ	A							A		
EAPE		A		A	B	B	B	A	A	
AMCR	C	C	C	C	C	A	C		B	C
BLJA	A	A	A	C	B	B	B	B	C	
YEWA**	B	B		C	B	B	B	B	B	B
PIWA			B		B					
ACFL**			A	A						
CYTH**	B	C	A		A					
BEKF	B									
RHWP		A								
SOSP**				B	B	B	A	A	A	A
MOBI							A			
TUTM	C	C	C	C	C	C	C	C	C	C
CACH	C	B	C	C	C	C	C	C	C	A
DOWP**	C	C	C	C	B	B				
RSTOW		A	A		A					
SUTA				B						
BGNC				A	A	A	A			
EAPH**	B	A	B	B	A	A	A			
WBNH**				B	A					
CHSP					B	B	B	B		
FISP				A						
BAOW**				B						
CAWR	C	B	B	C	C	C	C	B	B	
BRTHR								A		
UN#1								B		
YTVI	C	C	B	B	A	A	A	A	A	A
SOVI	B	B	B	A	C	A	B	B	B	
MODO	B	B	B		A		B	A	A	A
WTSP			A	C	A	C	B	B	A	
HOWA			A							

Key A = Observed Plot One
B = Observed Plot Two
C = Observed at Plots One and Two

** = Species associated with riparian woodlands.

TABLE V

Small mammal species captured and identified at Ninety Six Creek site near Ninety Six, SC.

Scientific Name	Common Name
Order: Insectivora	
Family: Soricidea	
Blarina carolinensis	Southern Shorttail Shrew
Order: Rodentria	
Family: Cricetidae	
Peromyscus leucopus	White-footed Mouse
P. gossypinus	Cotton Mouse
Family: Sciuridae	
Glaucomys volans	Southern Flying Squirrel

TABLE VI

Number of small mammal species captured per trap night.

Species	Night 1	2	3	4	5	6	Total Captures
Blarina carolinenesis	0	1	0	0	1	0	2
Peromyscus leucopus	1	4	4	8	11	6	34
P. gossypinus	1	1	2	7	4	4	19
Glaucomys volans	1	0	0	0	0	0	1
Totals (per night)	3	6	6	15	16	10	56

The wetland indicator status of the vegetation cover was not correlated with the areas that had hydric soil conditions and predicted wetland hydrologic conditions. Generally, the vegetation within the designated wetland areas had low PI (indicative of wetlands) but, many plots outside the wetland had low PI also. Therefore, the vegetation data were not helpful in determining jurisdictional wetland boundaries. This is not totally surprising, even though the area has been relatively undisturbed for about 50 years. Tiner (1991) states that "Plants did not evolve to become indicators species; this designation is a human attempt to use plants to designate wetlands."

The lack of correlation between the vegetation and hydric conditions in the floodplain suggests several points: 1) the reducing stress from soil saturation was not prolonged or frequent enough to significantly impact the composition of the plant community, 2) the soil saturation (waterlogging) events occurred so early in the season (early March to mid May) that they had little impact on the vegetation, 3) the wetland indicator status of several species may be in error for the Piedmont region, and/or 4) soil waterlogging stresses were offset by soil moisture stresses during dry periods of the year or in drought years. The long dry periods, typical of Piedmont summers, may be such that hydric species cannot compete on these sites. Only three obligate plant species were identified on the site and they were all perennial herbaceous species that can adapt to seasonal waterlogging changes more readily than woody species (Crawford 1992).

For about a two month period during the spring, the water tables stayed near the soil surface, and reducing conditions were prevalent in the upper 30 cm of the soil surface. The site flooded relatively infrequently and duration of inundation was usually less than 24 hours per event.

During the spring, water table levels generally rose after precipitation events. The sensitivity of the water tables to precipitation during this period was probably due, in part, to the mild spring temperatures and incomplete leafing out, which placed little stress on soil storage capacity (Pritchett and Fisher 1987). The soil was well aerated when the water table dropped below 25 cm (i. e. soil redox potentials were generally greater than +600 mv) but redox potentials typically declined after a rise in the water table.

The water table levels and hydric nature of the soils in this small bottom appeared to be influenced primarily by local precipitation and landform rather than overbank flooding and groundwater. During the three year observation period, water table levels were related to flooding events in the winter and occasionally in early March. Hence, overbank flooding did not appear to occur often enough or have sufficient duration to impact hydric nature of soils, hydrology, or vegetation. The position of the wetland at the toe of the steep slope, especially on the lower transect, suggests that it might receive groundwater input (Carter 1990; Brinson 1993a and b) but there was no evidence of such input. Well 9 was located between the steep slope and the wettest part of the study area, yet the water table level in Well 9 stayed about 25 cm below the water table level in Well 8 (nearest the wettest part of the bottom) for about 75 days during the summer in 1992. This suggests that the groundwater table associated with the upland slope may have dropped below the clay layer in the wettest area and left it as a perched water table. If this is correct, the wettest area received very little groundwater from the upland areas, especially during the long dry summer. Since the jurisdiction wetland area had no outlet during non-overbank flooding periods, it resembled an isolated wetland. Yet, its ecological processes probably include both donor and conveyor traits (Brinson 1993a) due to infrequent overbank flooding.

The areas of the study site that met various jurisdictional criteria tended to have soil saturation within 20 cm of the soil surface for about 60 days or about 15 percent of the year. Soil reduction of less than +200 mv within 15 cm of the soil surface occurred about 12 percent of the time at three Well locations that were followed by shorter periods of soil saturation and reduced soil conditions. These areas appear to be slightly drier than a mixed hardwood site in the Great Dismal Swamp which had soil saturation with 20 cm of the soil surface for more than 10 percent of the year (Day, West, and Tupacz 1988). Once the vegetation became fully leafed-out, evapotranspiration appeared to withdraw water from the soil faster than precipitation could replace it. Consequently, soil saturation seldom occurred within 25 cm of the soil surface after JD 150 except in the wettest portion of the area near Well 8. Apparently, the short duration of early season soil waterlogging was not sufficiently different across the site to have a detectable impact on the plant community composition.

Avifauna activities were relatively high in this bottom during the bird breeding season of 1992. Of the 32 bird species identified, eight were closely associated with riparian woodlands (Dickson 1978). During a 6 night period, four species of small mammals were caught and an average of 9.3 animals were caught per night.

The high capture rate of small mammals was probably related to the nature of the vegetation, riparian zone features, and habitat structure. Hodorff et al. (1988) found that closed canopy stands with dense shrub understory supported greater numbers of small mammals than open canopy stands with a shrub layer. Healy and Brooks (1988) found that while capture rates on 7 hardwood stands were not correlated with stand age or overstory structure, there was a positive correlation with shrub cover. In his study of 46 year-old, 66 year-old, and climax stage stands, Pearson (1959) noted the capture rates for white footed mice were greatest in stands with high shrub-tree cover. Mc'Closkey and LaJoie (1975) also found that the herb-shrub layer was an important component in habitat utilization. Dueser and Shugart (1978) found a higher density of Peromyscus at sites with a deciduous canopy, low density of trees, and high density of shrub understory. Geier and Best (1980) found a greater diversity of small mammal species on a wet floodplain than a dry floodplain or upland sites. In addition, they found significant positive correlations between abundance of woody plant debris (logs, brushpiles, or stumps) and small mammal numbers. Woody debris provides cover, perch sites, travel corridors, and protected sites for feeding, reproduction and food storage (Harmon et al. 1986; Thomas 1979). The study site at Ninety-Six Creek had a large amount of coarse woody debris as well as a dense shrub layer; therefore it provided excellent habitat for small mammals.

The bottom on the opposite side of the creek had been cleared and converted to residential sites, pasture, and a churchyard. Hence, the bottom provided a protective corridor for animals to traverse from woodlands above the site to those below the site as well as excellent on-site habitat.

If only the jurisdictional wetland area of this bottom were protected (Figure 6), the corridor would not be sufficient to connect the adjacent woodlands, and much of the value of this bottom as habitat and wetland conveyor functions would be lost or greatly reduced.

It appears that delineation of wetlands, in small Piedmont bottoms such as this, can best be achieved by examining soils for positive indicators (mottling and gleying) of waterlogging within 25 cm of the surface and combining soil information with landform (micro-topography) to determine wetland boundaries. Dependency on vegetation analysis for delineation may lead to the inclusion of areas as wetlands that are in fact not wetlands.

6. Conclusions

1) Only a small portion of this study site had hydric soil characteristics within the upper 25 cm of the soil profile. These hydric soils were closely associated with about one ha that was predicted (based on long-term hydrologic conditions) to meet wetland criteria similar to those of the 1987 and 1989 Federal Wetland Delineation Manuals.

2) Local precipitation appeared to influence soil saturation, soil reduction, and water table levels more than the infrequent floods of short durations that occurred primarily during the dormant season.

3) The wetland indicator status of the vegetation was not correlated with the wetland and non-wetland areas in the bottom or adjacent slope.

4) Analysis of vegetation composition by DECORANA and TWINSPAN segregated the vegetation into several groups, but the groups were not correlated with wetland and non-wetland areas. Thus if vegetation analysis were used to assess wetland boundaries on this site, the results would lead to the inclusion of areas that do not meet wetland criteria based on long-term hydrologic predictions, water table levels, and soil redox potential measurements.

5) Lack of correlation of vegetation to wetland areas within this bottomland forest may be related to the fact that hydric stress on this site occurs for such a short time early in the growing season that it has minimal impact on vegetation composition. Also, infrequent and short duration of flooding (usually early in growing season) and/or faulty wetland status of species may account for lack of close relationship between wetland areas and vegetation composition.

6) The hydric soils and predicted wetland areas occurred only in the concave landform portions of the bottomland hardwood forest. These areas existed near the toe of the upland slope.

7) The mixed bottomland hardwood stand in this small Piedmont bottoms provided excellent habitat for a number of songbird and small mammal species. It provided

on-site habitat, a travel corridor with adjacent forest stands, and appeared to function as receptor and conveyor riparian areas.

Acknowledgements

The U. S. Army Corps of Engineers, Charleston District, Charleston, SC partially supported this project. We appreciate the help of the USDA, Soil Conservation Service personnel (Edward Herren and Emory Holsonback) in classifying the soils on the site and in assisting with the topographic survey (Conner Franklin). We appreciate the help of Allyne J. Heiterer, Graduate Student, Department of Forest Resources and Dr. Susan Lobe, Ecologist, USDA, Forest Service, Southeastern Forest Experiment Station for conducting and analyzing the songbird and small mammal surveys. The South Carolina Forestry Commission for providing rainfall data from the Epworth Tower near the site and Bill Church of the U. S. Geologic Survey, Columbia, SC for providing up to-date data on stream flow.

References

An Interagency Cooperative Publication: 1989, Federal manual for identifying and delineating jurisdictional wetlands. U. S. Departments of Army, Fish and Wildlife Service, Environmental Protection Agency, and Soil Conservation Service. January 1989. Superintendent of Documents, U. S. Gov. Printing Office. Washington, DC 20402. 76 pp.

Brinson, M. M.: 1993a, Changes in the functioning of wetlands along environmental gradients. Wetlands 13 No. 2 Special Issue June 1993, pp. 65-74.

Brinson, M. M.: 1993b, A hydrogeomorphic classification for wetlands. US Army Corps of Engineers, Waterways Experiment Station, Vicksburg, MS. Wetlands Research Program Technical Report WRP-DE-4. 79 pp.

Carter, V.: 1990, The Great Dismal Swamp: An illustrated case study, in *Forested Wetlands*, A. E. Lugo, M. Brinson, and S. Brown (eds.). pp. 201-211. Ecosystems of the World, No. 15, Elservier, New York. p. 527.

Crawford, R. M. M.: 1992, Oxygen availability as an ecological limit to plant distribution. Advances in Ecological Research Vol 23. pp. 93-185. Academic Press Limited, London.

Day, F. P., West, S. K. and Tupacz, E. G.: 1988, The influence of ground-water dynamics in a periodically flooded ecosystem, the Great Dismal Swamp. Wetlands 8:1-13.

Dickson, J. G.: 1978, Forest bird communities of the bottomland hardwoods. pp. 66-73, in R. M. DeGraaf (tech coord). Proc of Workshop: Management of southern forests for nongame birds. USDA Forest Service, Gen. Tech. Rept. SE-14, Asheville, NC. 176 p.

Dickson, J. G. and Warren, M. V.: (in press), Wildlife and fish communities of eastern riparian forests, in Proceedings of Conference: Riparian ecosystems in the humid U. S. Functions, Values, and Management. March 15-18, 1993. Atlanta, GA. Sponsors U. S. Environmental Protection Agency, Fish and Wildlife Service, Department of Agriculture, Soil Conservation Service, and Forest Service; Tennessee Valley Authority and National Association of Conservation Districts.

Dueser, R. D., and Shugart, H. H.: 1978, Microhabitats in a forest-floor small mammal fauna. Ecol. 59:89-98.

Gambrell, R. P. and Patrick, W. H., Jr.: 1978, Chemical and microbiological properties of anaerobic soils and sediments, in Plant Life in Anaerobic Environments, D. D. Hook and R. M. M. Crawford (eds.). pp. 375-423. Ann Arbor Sci., Ann Arbor, MI. 564 p.

Geier, A. R. and Best, L. B.: 1980, Habitat selection by small mammals of riparian communities: evaluating effects of habitat alterations. J. Wild. Manage. 44(1):16-24.

Hanks, R. J. and Ashcroft, G. L.: 1980, Applied Soil Physics Advanced Series in Agricultural Sciences 8. Springer-Verlag, Berlin. 159 p.

Harmon, M. E., Franklin, J. F., Swanson, F. J., Sollius, P. S., Gregory, V., Sattin, J. D., Anderson, M. H., Cline, S. P., Aumeu, N. G., Sedell, J. R., Lieukaemper, G. W., Cromack, K. Jr., and Cummins, K. W.: 1986, Ecology of coarse woody debris in temperate ecosystems. Advances in Ecological Research.

Healy, W. M. and Brooks, R. T.: 1988, Small mammal abundance in Northern hardwood stands in West Virginia. J. Wildl. Manage. 52(3):491-96.

Hodorff, R. A., Sieg, C.H. and Linder, R. L.: 1988, Wildlife response to stand structure of deciduous woodlands. J. Wildl. Manage. 52(4):667-73.

Letey, J. and Stolzy, L. H.: 1964, Measurement of oxygen diffusion rates with the platinum microelectrode. I. theory and equipment. Hilgardia 35:545-554.

Light, T. S.: 1972, Standard solution for redox potential measurements. Analytical Chem. 44:1038-39.

Mc'Closkey, R. T. and LaJoie, D. T.: 1975, Determinants of local distribution and abundance in white footed mice. Ecol. 56:467-72.

Parsons, J. E., Skaggs, R. W. and Doty, C. W.: 1991a, Development and testing of a water management model (WATRCOM): Development. Trans. of the ASAE 34(1):120-128.

Parsons, J. E., Doty, C. W. and Skaggs, R. W.: 1991b, Development and testing of a water management model (WATRCOM): Field testing. Trans. of the ASAE 34(4):1674-1682.

Pearson, P. G.: 1959, Small mammals and old field succession on the Piedmont of New Jersey. Ecol. 40(2): 249-55.

Pritchett, W. L. and Fisher, R. F.: 1987, Properties and Management of Forest Soils. 2nd Edition. John Wiley and Sons. New York. 494 p.

Ralph, C. J., Geupel, G. R., Pyle, P., Martin, T. E. and DeSante, D. F.: 1993, Handbook of field methods for monitoring landbirds. USDA, Forest Service, Pacific Southwest Research Station. Gen. Tech. Rept. PSW-GTR-144, Albany, CA. 41 p.

Reed, P. B.: 1988, National list of plant species that occur in wetlands: Southeast (Reg.2). U. S. Fish and Wildlife Service. Biol. Rept. 88(26.2). 124 p.

Thomas, J. W. (tech. ed): 1979, Wildlife habitats in managed forests, the Blue Mountains of Oregon and Washington. USDA For. Ser. Handb. 553. Washington Off., Washington, DC.

Thornthwaite, C. W.: 1948, An approach toward a rational classification of climate. Geog. Rev. 38:55-94.

Tiner, R. W.: 1991, The concept of a hydrophyte for wetland identification. BioScience 41(4):236-247.

U. S. Army Corps of Engineers (CE): 1991, Charleston District, Regulatory Branch. Personal letter from Don Hill, May 21, 1991.

U. S. Department of Agriculture, Soil Conservation Service: 1980, Soil survey of Greenwood and McCormick Counties, SC. U. S. Government Printing Office 1980-232-310/81. 68 p.

LANDSCAPE-LEVEL PROCESSES AND WETLAND CONSERVATION IN THE SOUTHERN APPALACHIAN MOUNTAINS

SCOTT M. PEARSON[1]

Environmental Sciences Division, Oak Ridge National Laboratory[2],
P.O. Box 2008, MS: 6038, Oak Ridge, TN 37831-6038

Abstract. The function of wetland ecosystems is not independent of the landscapes in which they are embedded. They have strong physical and biotic linkages to the surrounding landscape. Therefore, incorporating a broad-scale perspective in our study of wetland ecology will promote our understanding of these habitats in the Southern Appalachians. Changes in the surrounding landscape will likely affect wetlands. Broad-scale changes that are likely to affect wetlands include: 1) climate change, 2) land use and land cover change, 3) water and air-borne pollution, 4) a shift in disturbance/recovery regimes, and 5) habitat loss and fragmentation. Changes in climate and land cover can affect the hydrology of the landscape and, therefore, the water balance of wetlands. Excessive nutrients and toxin transported by air and water to wetlands can disrupt natural patterns of nutrient cycling. Periodic disturbances, like flooding in riparian zones, is required to maintain some wetlands. A change in disturbance regimes, such as an increase in fire frequency, could alter species composition and nutrient cycles in certain wetlands. Many plant and animal species that found in small, isolated wetlands have populations that are dependent on complementary habitats found in the surrounding landscape. Loss or fragmentation of these complementary habitats could result in the collapse of wetland populations.

1. Introduction

Wetland habitats harbor some of the most unique and endangered plant and animal communities in the Southern Appalachian Mountains. To date, conservation efforts in this region aimed at protecting these rare and highly dispersed habitats have mainly focused on regulating wetland destruction through permits, acquiring wetlands outright, or contracting with landowners for wetland proteetion. While these measures prevent the immediate destruction of wetlands, they may not provide for the long term persistence of these unique communities.

The existence of wetland habitats in any landscape, including the Southern Appalachians, depends on the presence of a combination of conditions created and maintained by the broad scale conditions. These conditions include climate, which ultimately determines the availability of water, and landscape-level patterns of land covers, which affects hydrology and population dynamics of wetland species. Landscape characteristics such as topography and land cover determine water retention, discharge, and spatial patterns of flow (Gosselink and Turner, 1978). The combination of sufficient water and appropriate hydrology results in places where soils remain submerged or saturated on a regular and/or extended basis. These conditions favor the development of hydric soils suitable for the development of plant and animal communities adapted to wet conditions. Climate and land cover are broad scale characteristics that are affected by processes operating at regional and landscape levels. Thus, in order to understand and preserve wetland habitats, understanding the effects of broad scale characteristics on wetland ecology is necessary.

---

[1]Present address: Biology Department, Mars Hill College, Mars Hill, NC 28754 USA

[2]Managed by Martin Marietta Energy Systems, Inc. for the U.S. Department of Energy, under contract DE-AC05-840R21400.

Water, Air and Soil Pollution **77**: 321–332, 1994.

This paper will briefly describe the relationship between climate and landscape-level pattern on wetland hydrology, nutrient cycling, and population persistence. This paper will discuss the effect of changing climate on hydrology and disturbance, and its implication for Southern Appalachian wetlands. It will also discuss the influence of landscape-level patterning of habitats on hydrology and wetland populations.

Relationships between wetland ecology and broad-scale processes are illustrated in Figure 1. This figure is not an exhaustive list of all process and influences affecting wetlands, but it is useful for organizing this discussion. The three boxes denoting hydrology/water balance, nutrient cycling, and populations represent characteristics of wetlands. For example, wetlands are characterized by wet soils, populations of species adapted to wet conditions, and patterns of nutrient cycling (such as the formation of peat) that differ from more mesic and xeric habitats. Hydrology, nutrient cycling, and species composition interact by means of processes such as soil chemistry, evapotranspiration, and primary productivity within the wetland ecosystem. Climate and the landscape-level pattern of surrounding habitats are broad scale characteristics that may affect wetland ecology. The potential mechanisms of these broad scale effects will be explained below.

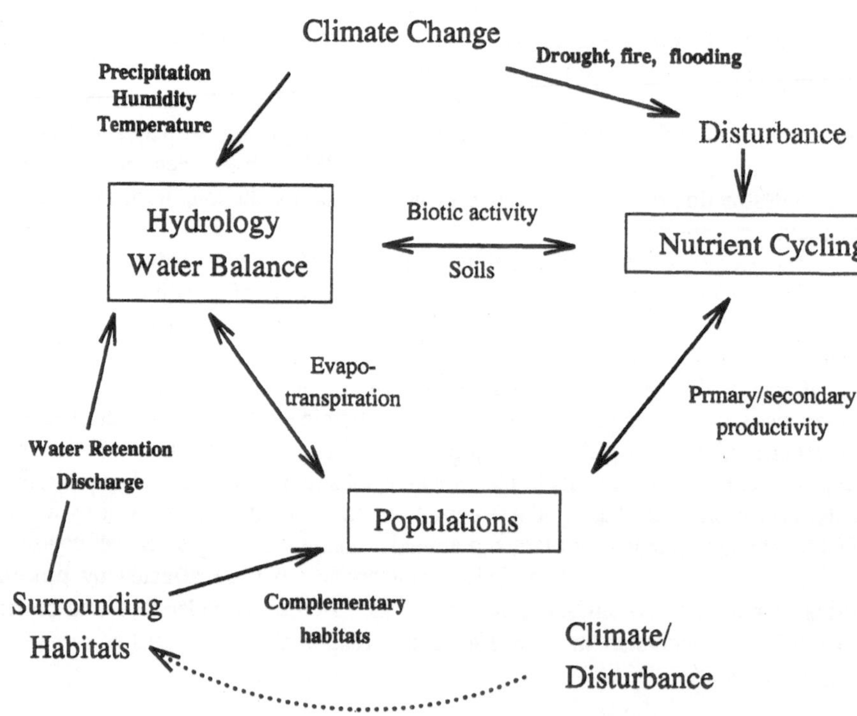

Figure 1. Conceptual relationships between wetland ecosystems and broad scale patterns of climate and the surrounding landscape. Boxes represent ecological processes operating within wetland habitats. Outside influences such as climate and landscape pattern of habitats may affect wetland hydrology, nutrient cycling, and population dynamics.

2. Climate and Hydrology

The hydrology of landscapes and wetland habitats is strongly related to climate. Precipitation provides the input of water for surface runoff and charging of ground water tables. Water loss occurs through surface and subsurface flows as well as through evapotranspiration. The resulting balance between water input and loss determines the location and seasonal wetness of potential wetland sites.

Evapotranspiration (ET) is a basic ecological phenomena related to primary productivity (Rosenzweig, 1968). For bogs (ombrotrophic wetlands), precipitation provides most or all of the water input. ET is the most important source of water loss. ET is also important for rheotrophic wetlands like fens and riparian zones because the stream flows and the amount of water available for subsurface flows can be reduced by high ET.

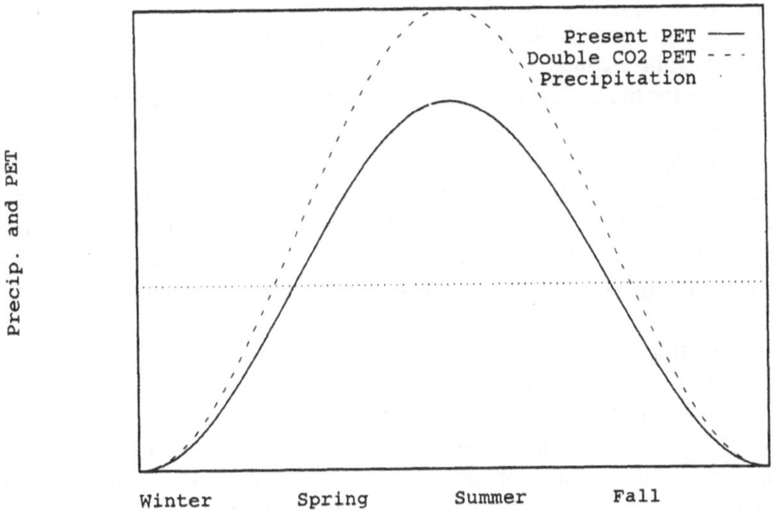

Figure 2. Annual patterns of precipitation and potential evapotranspiration (PET) in the Southern Appalachians. PET is less than precipitation but exceeds precipitation during the growing season. Warmer temperatures predicted to occur with increased atmospheric CO_2 would expand the growing season and increase PET.

Rates of ET depend on most strongly on the climatic variables of precipitation and temperature. Monthly precipitation totals are similar throughout the year in the Southern Appalachians. Though there is variation in the temporal pattern of rainfall from year to year, there are no regularly wet or dry seasons (NOAA, 1983; Swift *et al.*, 1988). The winters are characterized by high streamflows and wetter soils because precipitation (water input) exceeds water loss due to ET (Figure 2). ET is low during winter because of plant dormancy (reduced transpiration) and low temperatures (reduced evaporation).

During the growing season, soils begin to dry and streamflows drop as potential ET (PET) exceeds precipitation, resulting in a water deficit (Figure 2).

Global climate change models predict that, with double the current levels of CO2 in the atmosphere, the southeastern US could experience little change in precipitation but increased seasonal temperatures (Neilson *et al.*, 1989). The result would be greatly increased ET during the warm growing season months. The period to time where PET exceeds precipitation will increase, likely resulting in higher rates of water loss from ET (Figure 2). This could effect the water balance of both bogs and fens. Reduced periods of wetness could lead to local extinctions of species dependent on some minimal amount of submersion. Longer and more frequent droughts could promote the invasion of dryland species into wetland habitats.

3. Climate and Disturbance

Climate also effects the frequency and intensity of natural disturbances. Disturbance is an important factor for the maintenance of a diversity of plant communities in landscape (Pickett and White, 1985). These disturbances include flooding, drought, and fire.

Southern Appalachian forests are characterized by relatively small, highly scattered disturbances. Tree falls are the most common natural disturbance in this region, affecting approximately 1% of the deciduous-forested landscape annually (Trimble and Tryon, 1966; Runkle 1981, 1982). Recovery (canopy closure) from these small isolated disturbances is relatively quick. Periodic disturbances may also slow the transition of bogs into forest.

Both flooding and drought depend on seasonal pattern of water input to landscapes and water loss. Flooding is important for maintaining riparian wetlands like bottomland forest which depend on seasonal periods when soils are wet and anaerobic (Taylor *et al.*, 1990). A change in climate could change the seasonal patterns of streams flows. An increase in the severity of flood events could lead to the destruction of some alluvial habitats due to scouring by abnormally high stream flows. An increase in the frequency and length of droughts could result in the loss of long-lived species like trees that are intolerant to frequent drought stress. The most likely change due to climate for the Southern Appalachians would be a reduction in growing season stream flows due to higher ET driven by increased temperatures.

Fire depends indirectly on climate. Natural fire is a relatively rare source of disturbance for Southern Appalachian forests compared to forests in other regions of North America. In the past, fires have occurred most frequently on drier, south-facing slopes and lower elevation ridge tops (Barden and Woods, 1976; Harmon, 1982).

Southern Appalachian forests contain high biomass because of the great primary productivity of this region. However, the fire potential of these forests is seldom realized because of high rates of decomposition and prolonged periods when fuels are too wet to burn. At present, wetlands are seldom affected by natural fire. Wetter habitats such as forests on north-facing slopes and sheltered ravines burn infrequently (Harmon, 1982). Fires are common only during dry periods. If climate scenarios, such as those predicted by the climate change models, are realized, then periods of dryness will become much more common (Neilson *et al.*, 1990) raising the probability that wetter habitats will burn. As a result, the role of fire as source of disturbance will also increase.

In wetland systems, more frequent fires will effect nutrient cycling by affecting community composition of wetlands. That is, fire intolerant species may be reduce or extirpated. Also, fire has been attributed as the cause of reduced levels of N, S, and K in some wetland systems (Moore, 1990). The most important effect of fire on wetlands may be its ability to change the frequency and spatial configuration of habitats in the surrounding landscape.

4. Surrounding habitats and hydrology

Currently used strategies for protecting wetlands include regulating the outright destruction of wetlands within boundaries delineated around hydric habitats. Such a strategy could result in "protected" wetlands remaining as islands in a sea of highly impacted habitats (Figure 3). A highly altered landscape will surely differ significantly in its hydrology and its connectivity between wetlands and other habitats suitable for wetland populations.

While climate has an overriding and long term effect on the water balance of wetlands, the surrounding landscape also affects water balance. Vegetative cover affects water retention and discharge in landscapes (Swank *et al.*, 1988, see Richardson this volume). Studies such as those conducted at Coweeta Hydrological Lab demonstrate that changes in forest cover in Southern Appalachian watersheds affect patterns of stream flow (Swank and Crossley, 1988). Changes in stream flow consequently affect alluvial wetlands. Murdock (this volume) provides an example of how changes in plant cover surrounding a streamside wetland altered streamflow and channel morphology that resulted in a lowered water table.

Water retention and discharge is also important for rheotrophic mires like fens that receive water from the surrounding landscape. Land cover changes that divert more precipitation in runoff rather than to the ground water will impose a cycle of inundation and drought on a fen instead of a slower more constant release from ground water sources. Fens are also sensitive inputs of pollutants carried in water. The most common effect is that of eutrophication from leached fertilizers, particularly nitrates, from agricultural operations (Moore, 1990).

5. Surrounding habitats and wetland populations

Wetlands are often delineated by their characteristic plant and animal species. In the Southern Appalachians, wetlands tend to be small in areal extent and relatively rare in landscapes. These qualities imply that wetland populations are small and may not persist independent of populations in the surrounding landscape. While some of these species are confined solely to wetland habitats, others may be found in other less hydric habitats albeit at reduced densities. The "facultative" plant species listed in delineation guides belong to this group.

Small wetlands have small populations that are prone to stochastic extinction without an influx of individuals and genes from the outside. Indeed, small, isolated bogs have been found to have a larger component of plants with light, wind-dispersed seeds than larger, well-connected wetlands (Moore, 1990). This pattern implies that more vagile species are better able to recolonize small isolated wetlands after disturbance or stochastic

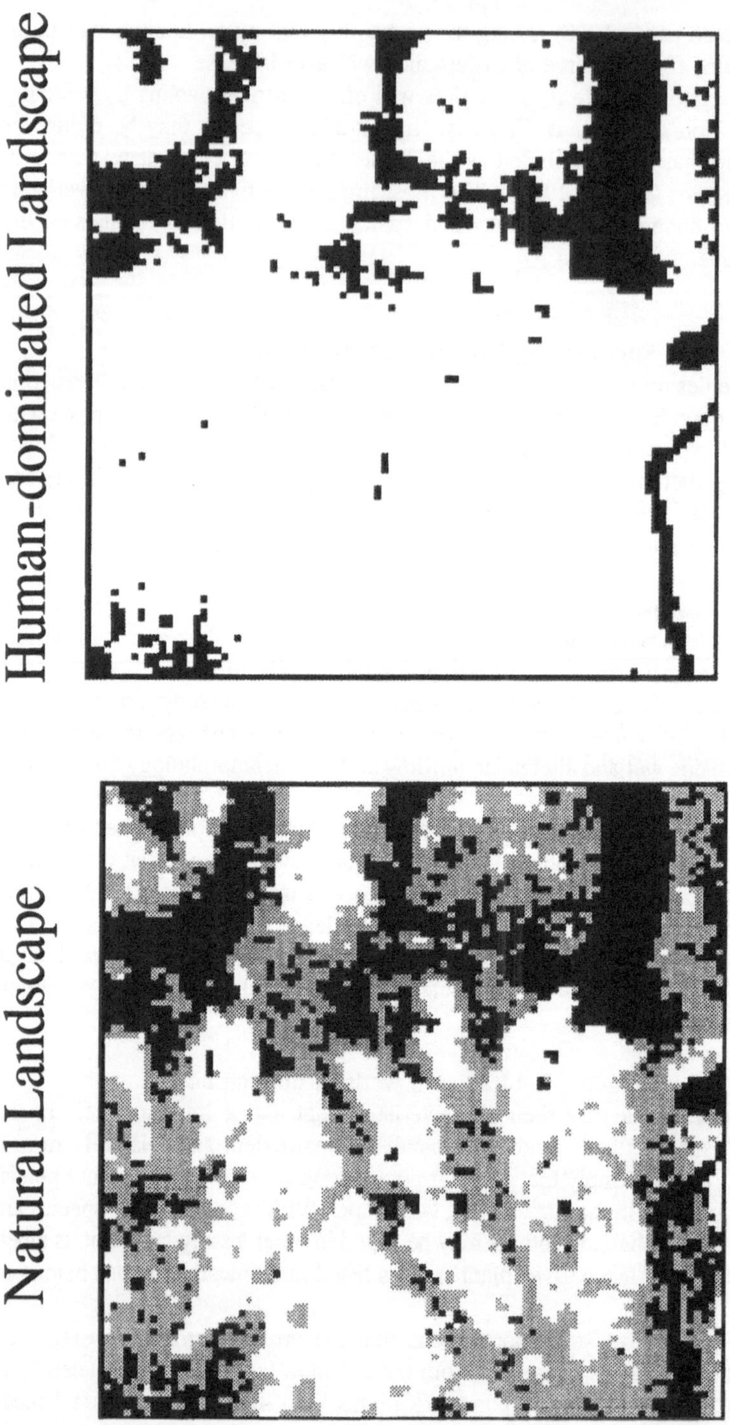

Figure 3. In natural landscapes, wetlands (black) may be surrounded by marginal habitats (shades of gray) that are suitable for dispersal and limited reproduction by wetland species and unsuitable habitats (white) that provide no resources for wetland species. If only wetlands are protected, a human-dominated landscape might develop where connectivity and additional resources provided by marginal habitats were lost through conversion to unsuitable habitat types.

extinctions (Figure 4). Therefore, populations in wetlands, particularly small ones, may not exist independent of populations inhabiting other wetlands and other habitats in the landscape.

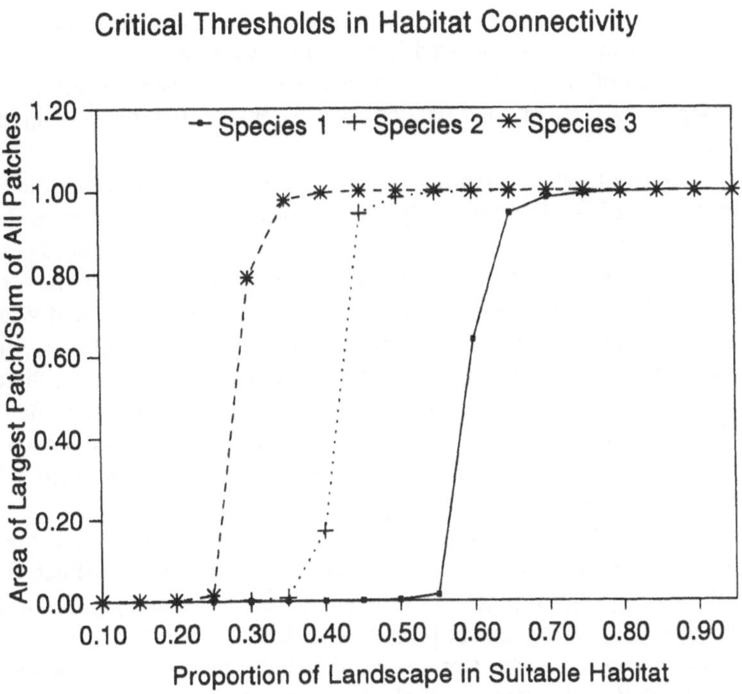

Figure 4. The ratio of the area of largest cluster (patch) of habitat to total area of habitat available (y-axis) provides an index of connectivity among the suitable habitat. Habitat is highly connected when this index is near 1 and highly fragmented when the index is near 0. This index is calculated for a simulated landscape where habitat is being lost in a random pattern. Species 1, 2, and 3 represent organism with low, medium, and high movement abilities, respectively. The index increases as the proportion of the landscape covered by suitable habitat increases. A rapid decline in index values associated with a small proportion of habitat loss indicates the presence of a critical threshold. These thresholds vary with movement ability of organisms. See Pearson *et al.*, (In press) for details of this analysis.

Less hydric habitats may contribute supplementary or complementary resources for wetland populations. Supplementary habitats may used by species in lieu of wetlands. These habitats provide a reservoir of dispersers for recolonization of wetlands after disturbance and of genetic diversity for wetland populations. Complementary habitats fulfill resource needs of a species that wetlands cannot. For example, wood frogs may spend most of the year foraging in surrounding forest but depend on the wetlands for breeding pools. Or, wetland plants may rely on pollinators that inhabit other habitats. Therefore, destruction of supplementary and complementary habitats in the surrounding landscape will have adverse consequences for populations in wetlands. Unfortunately, these linkages between wetlands and other habitats are not well known.

Populations in a particular wetlands may also be affected by connectivity or fragmentation in the surrounding landscape. Overall, the Southern Appalachians remain a forested landscape. Most of the intensive human land use in this region has occurred in the valley and lowland regions (Flamm and Turner, submitted). Unfortunately, a major portion of our wetlands are also located in these low-lying areas.

The impact of any land use on native ecosystems, including wetlands, depends on 1) the degree of habitat modification and 2) the spatial extent and pattern of modification.

The degree of habitat modification is the extent to which a modified land unit differs from the set of seral stages likely for that land unit under a natural disturbance regime (Pearson in press). For example, a paved parking lot in a valley does not resemble any of the set of possible land covers (seral stages) expected for that site. Moreover, heavily modified habitats show a high degree of contrast (sensu Kotliar and Wiens, 1990) to native habitats. The suitability of these modified habitats for wetland species is roughly related to the habitats' contrast to naturally occurring habitats. In nature, there are also sub-optimal habitat types used by many species. Land use by humans often does not totally destroy a patch of habitat. A particular land use may make the habitat only less than optimal for a period of time. Therefore, the contrast between impacted habitats and wetlands is important (Figure 3)

The spatial extent and spatial pattern of land use can affect the connectivity or fragmentation of landscapes. The loss of connectivity between native habitats will inhibit the ability of some ecological processes to move across the landscape. Connectivity can exert strong influences on ecological processes such as the movement and dispersal of organisms (Gardner *et al.*, 1989; Gardner *et al.*, 1991), the utilization of resources by animals (O'Neill *et al.*, 1988, Pearson *et al.*, in press), gene flow (Gilpin and Soule 1986), and the spread of disturbance (Turner *et al.*, 1991). Changing the patterns of connectivity can disrupt ecological processes that depend on movement within the landscape. For example, the persistence of a metapopulation of small mammals may depend on the ability of dispersing young to reach other populations. If the overall connectivity of the landscape is altered by the creation of habitats that act as barriers to dispersing young, the isolated populations of this small mammal could become extinct due to their own demographic instability (Brown and Kodric-Brown, 1977; Beuchner, 1989) or competitive interactions (Nee and May, 1992).

The spatial extent (area) of habitat loss reduces the total amount of habitat available. If habitat destruction is scattered about the landscape, habitat loss will affect the connectivity of remaining habitat. While suitable habitats are still common, scattered areas of habitat loss have little effect on connectivity. However, as more habitat is lost, landscapes can be transformed from well connected to highly fragmented very quickly (Figure 4). Crossing one of these critical thresholds of fragmentation can occur with only a small additional percentage of habitat lost. Moreover, the critical amount of habitat differs among species. Species with the ability to move or disperse over patches of unsuitable habitat are more tolerant of fragmentation than species with limited movement capabilities (Pearson *et al.*, in press).

The spatial pattern of habitat loss is also important. The spatial arrangement of habitats, within ecological constraints, is often determined by economics and social values in human-dominated landscape. As human populations increase, the amount of land converted to human use will increase, resulting in more habitat loss for many native species. Landscape connectivity can be maintained by organizing the spatial pattern of habitat loss and by protecting corridors for movement between wetlands.

The spatial pattern of habitat loss can be controlled by aggregating intensive land uses into one or a few locations. Aggregating habitat loss will protect the connectivity of the landscape even though the same spatial extent (amount of area) of the landscape is impacted (Figure 5). For example, if the same amount of acreage were to be allocated to each of a fixed number of homesites, aggregating the homesites into one or a few subdivisions would impact landscape connectivity less than scattering the homesites at low density across the landscape (e.g., Pearson *et al.*, in press).

Protecting corridors is another way of organizing habitat loss to protect landscape connectivity. Alluvial wetlands and riparian zones naturally serve as corridors because they provide connectivity between small bogs and fens. For wetland species that can also use more mesic habitats, the connectivity of non-wetland habitats in landscapes could affect the long term persistence of their populations. Protected corridors are intended to facilitate movement (dispersal and gene flow) between patches of suitable habitat (Diamond, 1975; Hudson, 1991). While satisfactory corridors permit animals to move between patches, they must be able to sustain plant populations if the distance between patch requires more than one generation to traverse. Corridors that satisfy the dispersal needs of the entire community must be short, wide, and free of environmental gradients that act as a barrier to wetland species. Corridors may not be effective for all species. Therefore, the best way to maintain connectivity is to limit the spatial extent of high intensity impacts keeping landscapes above the critical thresholds described above.

6. Conclusion

Wetlands of the Southern Appalachians cannot be preserved without considering the global and regional landscapes in which they are embedded. The hydrology of wetlands is affected by changes in climate and vegetative cover of these landscapes. The species that are characteristic of wetlands and give them value for conservation depend on the availability and spatial arrangement of suitable habitats in the landscape.

Climate and landscape patterns are being influenced by humans. Climatic patterns can be changed by carbon dioxide emissions and by changing global and regional vegetation. Using the land for food and fiber production alters disturbance regimes. Surface and ground water flows are affected by flood control, irrigation, municipal water uses, and hydroelectric power generation. New disturbances such as exotic species and fire are associated with human activities. The pattern of habitats on human-dominated landscapes is determined by social and economic values and ultimately on the numbers of humans using the land. Therefore, the ultimate fate of wetlands in the Southern Appalachians will depend on the stewardship of landscapes, not on the conservation of selected fragments of rare habitats. Long term, landscape-level processes must be considered to guarantee the long-term protection of these wetlands.

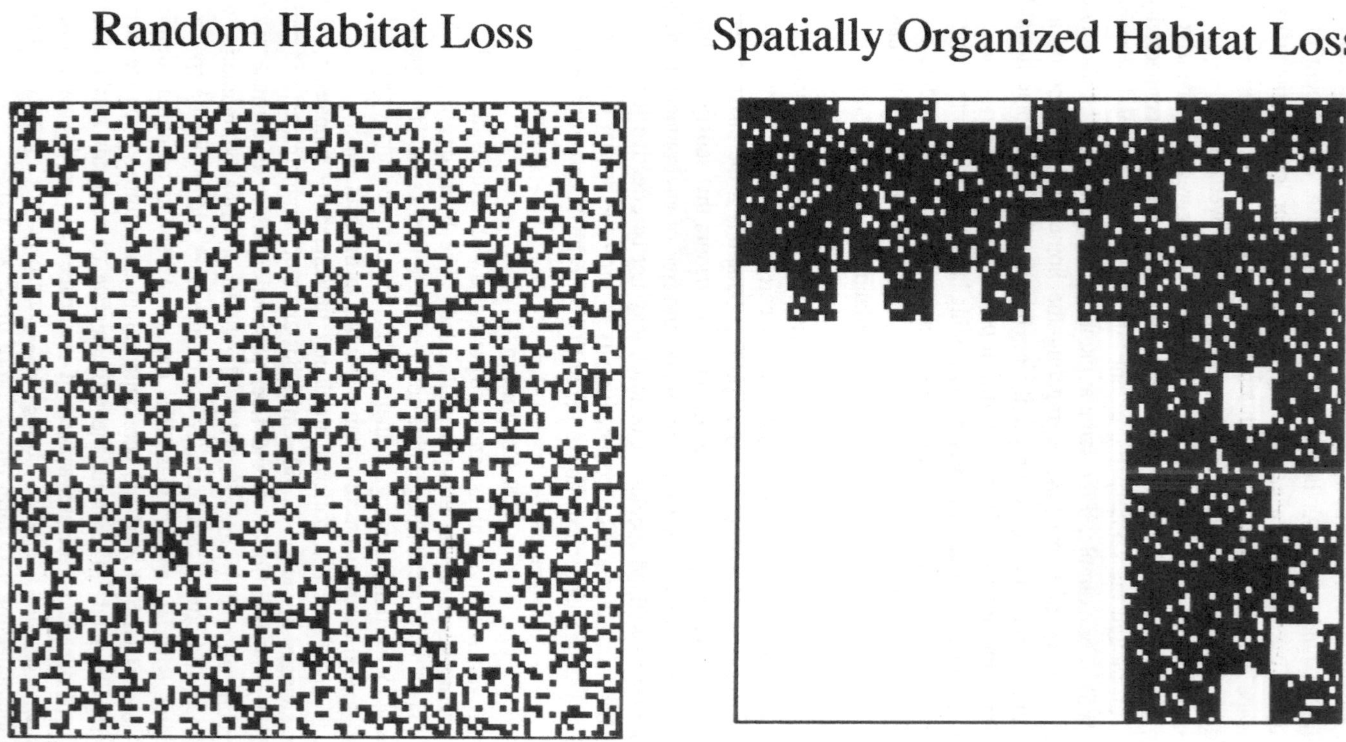

Figure 5. The spatial arrangement of habitat loss (open) can affect the connectivity of the remaining habitat (shaded). Both of these maps are 100 x 100 cells in extent and have 3280 cells (approximately 33%) of habitat remaining. Habitat loss that is random in space can result in a highly fragmented landscape consisting of many, small habitat patches. When habitat loss is spatially organized (aggregated), the remaining habitat is highly connected. (After Pearson *et al.*, in press).

Acknowledgements

This work received support from an Alexander Hollaender Distinguished Postdoctoral Fellowship sponsored by the U.S. Department of Energy administered through Oak Ridge Institute for Science and Education and from the Office of Health and Environmental Research, Department of Energy under contract No. DE-AC05-840R21400 with Martin Marietta Energy Systems, Inc. This is a publication of the Environmental Sciences Division of Oak Ridge National Laboratory.

References

Barden, L. S. and Woods, F. W. 1976. Effects of fire on pine and pine-hardwood forests in the southern Appalachians. *For. Science* **22**:399-403.

Beuchner, M. 1989. Are small-scale landscape features important factors for field studies of small mammal dispersal sinks? *Landscape Ecol.* **2**:191-199.

Brown, J. H. and Kodric-Brown, A. 1977. Turnover rates in insular biogeography: effects of immigration on extinction. *Ecology* **58**:445-449.

Diamond, J. M. 1975. The island dilemma: lessons of modern biogeographic studies for the design of natural reserves. *Biol. Conserv.* **7**:129-146.

Flamm, R. O. and Turner, M. G. Land cover changes in rural Western North Carolina, USA. Submitted to *Sistema Terra.*

Gardner, R. H., O'Neill, R. V., Turner, M. G., and Dale, V. H. 1989. Quantifying scale-dependent effects of animal movement with simple percolation models. *Landsc. Ecol.* **3**:217-227.

Gardner, R. H., Turner, M. G., O'Neill, R. V., and Lavorel, S. 1991. Simulation of the scale-dependent effects of landscape boundaries on species persistence and dispersal, in Holland, M. M., Risser, P. G., and Naiman, R. J. (eds) *Ecotones: the role of landscape boundaries in the management and restoration of changing environments*, Chapman and Hall, New York, pp. 76-89.

Gilpin, M. E. and Soule, M. E. 1986. Minimum viable populations: processes of species extinctions, in Soule, M. E. (ed) *Conservation biology: the science of scarcity and diversity*, Sunderland, Massachusetts: Sinauer Associates, Inc, pp. 19-34.

Gosselink, J. G. and Turner, R. E. 1978. The role of hydrology in freshwater wetland ecosystems, in Good, R. E., Whigham, D. F., and Simpson, R. L. (eds) *Freshwater wetlands: ecological processes and management potential.* Academic Press, New York, pp.63-78.

Harmon, M. 1982. Fire history of the westernmost portion of Great Smoky Mountains National Park. *Bull. Torreya Bot. Club* **109**:74-79.

Hudson, W., (ed). 1991. Landscape linkages and biodiversity. Island Press, Washington, D. C.

Kotliar, N. B. and Wiens, J. A. 1990. Multiple scales of patchiness and patch structure: a hierarchical framework for the study of heterogeneity. *Oikos* **59**:253-260.

Moore, P. D. 1990. Soils and ecology: temperate wetlands, in Willams, M. (ed.), *Wetlands: a threatened landscape*. Basil Blackwell, Inc., Cambridge, MA 02142, pp. 95-114.

Murdock, N. A. This volume. Rare and endangered plants and animals of southern Appalachian wetlands.

NOAA. 1983. Monthly precipitation probabilities for selected probability levels: supplement to climatology of the United States No. 81. National Oceanic and Atmospheric Administration, National Climatic Data Center, Federal Building, Asheville, NC 28801, November 1983.

Nee, S. and May, R. M. 1992. Dynamics of metapopulations: habitat destruction and competitive coexistence. *J. of Anim. Ecol.* **61**:37-40.

Neilson, R. P., King, G. A., DeVelice, R. L., Lenihan, J., Marks, D., Dolph, J., Campbell, B., and Glick, G. 1989. Sensitivity of ecological landscapes and regions to global climatic change. U.S. Environmental Protection Agency, Environmental Research Laboratory, 200 S.W. 35th Street, Corvallis, OR 97333. 103 pp.

O'Neill, R. V., Milne, B. T., Turner, M. G., and Gardner, R. H.. 1988. Resource utilization scales and landscape pattern. *Landsc. Ecol.* **2**:63-69.

Pearson, S. M. Understanding the impacts of forest fragmentation in the Southern Appalachian Mountains, in Sample, V. A. (ed) *Forest ecosystem management at the landscape level: the role of remote sensing and integrated GIS in resource management planning, analysis, and decision making.* Island Press, Washington, D.C. In Press.

Pearson, S. M., Turner, M. G., Gardner, R. H. and O'Neill, R. V. An organism-based prespective of habitat fragmentation, in Szaro, R. C. (ed) *Biodiversity in managed landscapes: theory and practice*, Oxford University Press: Cary, NC. In press.

Pickett, S. T. A. and White, P. S. (eds) 1985. *The ecology of natural disturbance and patch dynamics*. Academic Press, Inc.: New York.

Richardson, C. J. This volume. Wetland functions and values; state of our understanding.

Rosenzweig, M. L. 1968. Net primary productivity of terrestrial communities: prediction from climatological data. *Am. Nat.* **102**:67-74.

Runkle, J. R. 1981. Gap regeneration in some old-growth forests of the eastern United States. *Ecology* **62**:1041-1051.

Runkle, J. R. 1982. Patterns of disturbance in some old-growth mesic forests of the eastern North America. *Ecology* **63**:1533-1546.

Swift, L. W., Jr., Cunningham, G. B., and Douglass, J. E. 1988. Climatology and Hydrology, in Swank, W. T. and Crossley, D. A., Jr. (eds) *Forest hydrology and ecology at Coweeta*, Springer-Verlag, New York, pp. 35-55.

Swank, W. T. and Crossley, D. A., Jr. (eds) 1988. *Forest hydrology and ecology at Coweeta*. Springer-Verlag, New York.

Swank, W. T., Swift, L. W., and Douglass, J. E. 1988. Streamflow changes associated with forest cutting: species conversions, and natural disturbances, in Swank, W. T. and Crossley, D. A., Jr. (eds) *Forest hydrology and ecology at Coweeta*. Springer-Verlag: New York, pp. 297-312.

Taylor, J. R., Cardamone, M. A., and Mitsch, W. J. 1990. Bottomland hardwood forests: their functions and values, in Gosselink, J. G., Lee, L. C., and Muir, T. A. (eds) *Ecological processes and cumulative impacts: illustrated by bottomland hardwood wetland ecosystems*, Lewis Publishers, Inc., Chelseas, MI 48118, pp. 13-86.

Trimble, G. R., Jr. and Tryon, E. H. 1966. Crown encroachment into openings cut in Appalachian hardwood stands. *J. Forest.* **64**:104-108.

Turner, M. G., Gardner, R. H., and O'Neill, R. V. 1991. Potential responses of landscape boundaries to global climate change, in Holland, M. M., Risser, P. G., and Naiman, R. J. (eds) *Ecotones: the role of landscape boundaries in the management and restoration of changing environments*, Chapman and Hall, New York, pp. 52-75.

VEGETATION OF THREE HIGH ELEVATION SOUTHERN APPALACHIAN BOGS AND IMPLICATIONS OF THEIR VEGETATIONAL HISTORY

J. DAN PITTILLO

Department of Biology, Western Carolina University, Cullowhee, NC 28723

Abstract. In order to investigate the extent of vegetation change during the Pleistocene Period, a study of three high elevation southern Appalachian bogs was undertaken. Three North Carolina sites were chosen that were suitable for pollen and paleoecological analysis: Flat Laurel Gap bog near Pisgah Mountain south of Asheville (elevation 1500 m), Boone Fork bog near Blowing Rock (elevation 1024 m), and Long Hope Valley bog north of Boone (elevation 1450 m).

Existing vegetation was sampled by transects of 10 x 10 m plot relevés that crossed a section of forest and open bog areas. While each site included a characteristic southern Appalachian bog, each differed: Flat Laurel Gap is predominately a heath community with interfingerings of open grassy glades which grade into a mixture of northern hardwoods and spruce; Boone Fork bog is a disturbed mixture of northern hardwoods which grades into a mixture of scattered shrubs and open glades predominated by *Sphagnum*; and Long Hope *Menyanthes* bog is an open herbaceous and grassy glade with scattered shrubs which grades abruptly into northern hardwoods and old-growth spruce. As might be expected, the more northern Long Hope Valley site had more northern taxa, such as *Menyanthes trifoliata, Lonicera canadensis,* and *Vaccinium macrocarpon*. The southern site also had a few boreal taxa such as *Eriophorum virginicum*.

Perhaps the best interpretation for these long-established southern Appalachian bogs is that they have provided continuously suitable habitats for relict northern species since the peak of the glacial ice advance 18,000 years ago.

1. Introduction

Studies of the southern Appalachian bogs are infrequent and the presence of this type ecosystem is rare. While the Appalachian Plateaus have some relatively well developed sphagnum bogs (Rigg and Strausbaugh, 1949) their presence in the Appalachian Mountains to the south and east in the Blue Ridge Province is much less frequent (Gaddy, 1981). The definitive discussion of the types of wetlands that include what they term the "Southern Appalachian Bog" is found in Schafale and Weakley (1990, pp. 179-200). Described as a part of the "nonalluvial wetlands of the mountains and piedmont," Schafale and Weakley (1990, p. 183-186) characterize the soils, topography, hydrology, vegetation, dynamics, range, associations, distinguishing features, variation, rare plant species, and add some comments, synonyms, and examples. In

Water, Air and Soil Pollution **77**: 333–348, 1994.

this description they divide the Southern Appalachian Bogs into "Northern and Southern Subtypes," and list the three sites described here as the Northern Subtype. The distinguishing features of the Northern Subtype, besides occupying the higher elevations, is the presence of *Carex trisperma, C. buxbaumii, Rhynchospora alba, Vaccinium macrocarpon*, and *Saxifraga pensylvanica*. These bogs grade into forests of various types. The definition for "bog" followed here is that of Schafale and Weakley (1990). The most recent discussion of bogs and fens in the southern Appalachians is by Richardson and Gibbons (1994). Following the US Fish and Wildlife Service wetland classification system, they describe palustrine wetlands or "mountain bogs and fens" as separate entities from the Carolina bays, forested pocosins, and scrub-shrub pocossins (Richardson and Gibbons, 1994, p. 259). They noted bogs and fens occur generally from 356-1500 m elevation in the Blue Ridge, Ridge and Valley, and Piedmont provinces and on the Ozark Plateaus. They compiled a list of 78 known montane sites, 68 of them designated as bogs and 10 as fens.

Wetlands are infrequent ecosystems in the higher elevations of the southern Appalachians. Standing water, except for artificial impoundments, is found only in swamps or bogs in the Blue Ridge and the adjoining Ridge and Valley provinces. Even here these wetlands represent less than one per cent of the land area. With continued land-use conversions, there is much pressure to drain wetlands for other uses. Thus a project to describe three such sites was undertaken in the mid 1980's by the Center for Quaternary Research at the University of Tennessee while the sites remain in undisturbed condition.

The overall objective was to describe the vegetation present at each of these sites and to determine the degree that the pollen rain is reflected in the pollen deposition of the moss polsters and sediment of the bogs. The objective of this paper is to characterize the existing vegetation and discuss paleovegetation implications regarding the presence of certain species present. High elevations were chosen because these have not had paleoecological studies applied to them and them would be the preferred sites to test the hypothesis (P. Delcourt, 1985) that tundra occurred on our highest elevations during the full-glacial period 18,000 years ago.

Three sites were chosen which best met the objectives of the study. The southernmost site, Flat Laurel Gap, occurs at an elevation of 1500 m on the Pisgah Ridge of the Great Balsam Mountains in the Pisgah Recreational Area campground (82°45'25" W and 35°23'45" N). This site consists of heath covered terraces with interfingerings of grass-dominated wetlands or glades. The second site, Boone Fork Bog, is at an elevation of 1024 m in the Blue Ridge Mountains (81°46" W and 36°08'30" N). This site is a mixture of grass dominated glades with scattered trees and shrubs. The northernmost site is at an elevation of 1450 m in Long Hope Valley of the Stone Mountains along the headwater tributary of the New River (81°37'30" W and 36°19'37" N). This site occurs near the south end of the valley and is characterized by the presence of bogbean (*Menyanthes trifoliata*) and is a shrub/herb/ *Sphagnum* dominated bog.

2. Methods

Each site was explored and test samples of the sediment depth and characteristics were made to determine the location of the potential coring site, with the deepest available unoxidized sediment profile. Then transects were arranged in order to describe the vegetation in the bog and adjacent forest (Flat Laurel Gap had an established transect that was followed instead). Transects were laid out across the coring site at 10 m intervals with plots positioned alternatively left and right at each 10 m, the first being located by a flip of a coin. A minimum number of plots needed to obtain a less than 10% increase in new species on the species area curve was determined for each sampling location.

The Flat Laurel Gap transect followed line that was established by Horton and Hotaling (1981) for a study that assessed the state of the vegetation for a management recommendation. The plot locations in the Hotaling and Horton study were randomly positioned either to the left or right at distances of 0 to 10 m from the compass line. We relocated the plots and re-estimated the cover of the eight 10 x 10 m quadrats with the relevé method of Mueller-Dumbois and Ellenburg (1974).

At the Boone Fork site a transect was established near Terry Moore's (1972) study plots. A total of five 10 x 10 m plots was located in the glade (or grassy dominated portion of the bog) and an additional six in the adjoining woodlands.

At Long Hope Valley the transect began in the center of the bog and extended to the northwest (335°) and southwest (245°). This permitted the location of seven quadrats in the bog and six quadrats in the forest, three in the hardwoods to the southwest and four in the spruce-hardwoods to the northwest.

Relevé estimates were made for each sample point. At each point, quadrats included tree quadrats 10 x 10 m, shrub quadrats 4 x 4 m, and herb quadrats 0.5 x 4 m. The quadrats were nested with the shrub quadrat and herb quadrats located in the point-corner and the herb quadrat laid out with its long side along the compass line. The relevé included a visual estimate in per cent cover for the quadrat under consideration. In the Flat Laurel Gap site, no attempt was made to separate the heath and glade communities but at the Boone Fork and Long Hope Valley sites the community type(s) were sorted into either open bog and forest communities. Importance Values (Mueller-Dumbois and Ellenburg, 1974) were calculated from sums of the coverage relevés and frequency data. Nomenclature follows Radford, Ahles, and Bell (1968) and voucher specimens are deposited at WCUH.

3. Results

The major community dominants for each of the sites and communities within the sites are tabulated in Tables I-III. Flat Laurel Gap (Table I) is dominated by heath with the grassy areas dominated by a layer of sedges with sphagnum beneath or in the occasional areas devoid of herbs and grasses. It has only few trees; *Amelanchier arborea* var. *laevis* is most important. *Rhododendron catawbiense-Kalmia latifolia* dominates the shrub level; the main understory is cover is *Galax aphylla* (in the heath areas) and *Carex* spp. and *Sphagnum* spp. in the open areas.

Boone Fork Bog (TABLE II) is dominated by sphagnum with a scattering of *Scirpus* and *Osmunda cinnamomea*. An occasional hemlock or birch occurs scattered throughout and these alternate with occasional *Rhododendron maximum*, *Salix sericea*, or Vaccinium *macrocarpon* colony. *Tsuga canadensis* and *Betula lenta* are the most important trees in the open bog but *Rhododendron maximum* shares this importance with *Betula* in the forest. *Rhododendron maximum* also dominates the shrub level in both sites while *Sphagnum* spp. and secondarily *Scirpus-Osmunda cinnamomea* are most important in the open bog. The forest at Boone Fork was dominated by large rosebays, a somewhat unexpected occurrence considering shading by overstory trees. Understory dominants are *Rhododendron maximum*, *Dryopteris intermedia*, and *Trillium undulatum*.

TABLE I

Flat Laurel Gap. Importance Values of trees, shrubs, and herbs of the Flat Laurel Gap transect plots, Great Balsam Mountains. (Abbreviations: pc c = estimated percentage cover; rel c = relative cover; freq = frequency; rel fr = relative frequency; and I V = Importance Value -200).

TREES

Species	pc c	rel c	freq	rel fr	I V
Acer rubrum	33.0	29.4	0.25	22.2	51.7
Amelanchier arborea var. laevis	54.0	48.2	0.5	44.4	92.6
Ilex ambigua var. montana	25.1	22.4	0.4	33.3	55.7
SUMS	112.1	100.0	1.15	99.9	200.0

SHRUBS

Species	pc c	rel c	pc fr	rel fr	I V
Alnus serrulata	1.0	0.13	2.2	3.2	4.5
Kalmia latifolia	148.2	29.8	0.75	13.3	43.1
Leucothoe recurva	4.1	2.8	0.6	11.1	13.9
Lyonia ligustrina	1.1	0.2	0.25	4.4	4.7
Rhododendron catawbiense	187.0	37.6	0.9	15.6	53.1
Rhododendron maximum	105.1	21.1	0.75	13.3	34.4
Rhododendron viscosum	0.2	0.13	2.2	2.4	2.5
Sambucus canadensis	7.0	1.4	0.12	2.2	3.6
Sorbus americana	0.1	0.02	0.13	2.2	2.2
Sorbus arbutifolia	2.1	0.4	0.25	4.4	4.9
Sorbus melanocarpa	2.0	0.4	0.25	4.4	4.8
Vaccinium corymbosum	15.1	3.0	0.75	13.3	16.4
Viburnum cassinoides	9.1	1.8	0.63	11.1	12.9
SUMS	496.9	99.8	5.6	100.0	199.8

HERBS

Species	pc c	rel c	pc fr	rel fr	I V
Agrostis perennans	1.2	0.4	0.4	7.0	7.4
Calamagrostis cinnoides	0.1	0.03	0.13	2.3	2.4
Carex howei	0.1	0.03	0.13	2.3	2.4
Carex leptalea	0.1	1.1	0.4	0.25	4.7
Carex stricta	50.0	16.4	0.13	2.3	18.7
Chelone sp.	15.2	5.0	0.4	7.0	12.0
Cladonia sp.	0.1	0.03	0.13	2.3	2.4
Cuscuta rostrata	2.2	0.7	0.4	7.0	7.7
Danthonia compressa	0.1	0.03	0.13	2.3	2.4
Drosera rotundifolia	0.1	0.03	0.13	2.3	2.4
Eriophorum virginicum	0.1	0.03	0.13	2.3	2.4
Eupatorium rugosum	0.1	0.03	0.13	2.3	2.4
Festuca rubra	0.1	0.03	0.13	2.3	2.4

Table I, herbs, cont.

Species	pc c	rel c	pc fr	rel fr	I V
Galax aphylla	130.0	42.7	0.8	14.0	56.7
Holcus lanatus	0.1	0.03	0.13	2.3	2.4
Houstonia serpyllifolia	0.1	0.03	0.13	2.3	2.4
Hypericum punctatum	0.1	0.03	0.13	2.3	2.4
Juncus brevicaudatus	0.1	0.03	0.13	2.3	2.4
Lobelia sp.	0.1	0.03	0.13	2.3	2.4
Mitchella repens	0.1	0.03	0.13	2.3	2.4
Osmunda cinnamomea	3.0	1.0	0.5	9.3	10.3
Poa alsodes	0.1	0.03	0.13	2.4	2.4
Solidago sp.	5.2	1.7	0.4	7.0	8.7
Sphagnum sp.	95.0	31.2	0.25	4.7	35.8
Trillium undulatum	0.1	0.03	0.13	2.3	2.4
SUMS	305.4	100.3	5.4	100.0	200.3

TABLE II

Boone Fork Bog. Importance Values of trees, shrubs, and herbs in the open bog and adjacent tree-covered forests of Boone Fork Bog in the Blue Ridge Mountains. (Abbreviations: pc c = estimated percentage cover; rel c = relative cover; freq = frequency; rel fr = relative frequency; and I V = Importance Value - 200).

OPEN BOG TREES

Species	pc c	rel c	pc fr	rel fr	I V
Acer rubrum	1.1	4.3	0.6	25.0	29.3
Amelanchier arborea	2.1	8.2	0.4	16.7	24.9
Betula alleghaniensis	5.0	19.6	0.6	25.0	44.6
Oxydendrum arboreum	0.1	0.4	0.2	8.3	8.7
Pinus strobus	1.0	3.9	0.2	8.3	12.3
Tsuga canadensis	16.2	63.5	0.4	16.7	80.2
SUMS	25.5	100.0	2.4	100.0	200.0

OPEN BOG SHRUBS

Species	pc c	rel c	pc fr	rel fr	I V
Kalmia latifolia	1.1	0.56	0.2	3.8	4.4
Lyonia ligustrina	3.1	1.6	0.4	7.7	9.3
Rhododendron maximum	56.2	28.7	0.6	11.5	40.2
Rosa palustris	0.05	1.0	0.2	3.8	3.9
Rubus alleghaniensis	14.1	7.2	0.8	15.4	22.6
Rubus argutus	0.1	0.05	0.2	3.8	3.9
Rubus flabellaris	2.1	1.07	0.2	3.8	4.9
Salix cinerea	4.0	2.04	0.2	3.8	5.9

Table 2, open bog shrubs, cont.

Species	pc c	rel c	pc fr	rel fr	IV
Salix sericea	41.0	20.9	0.6	11.5	32.5
Vaccinium corymbosum	4.0	2.04	0.6	11.5	13.6
Vaccinium macrocarpon	60.0	30.6	0.2	3.8	34.5
Viburnum cassinoides	10.2	5.2	1.0	19.2	24.4
SUMS	196.0	100.0	5.2	100.0	200.0

OPEN BOG HERBS

Species	pc c	rel c	pc fr	rel fr	I V
Carex leptalea	22.0	3.4	0.6	6.8	10.2
Carex muricata var. angustata	22.2	3.4	0.8	9.1	12.5
Dryopteris cristata	0.2	0.03	0.4	4.5	4.6
Epilobium leptophyllum	0.1	0.02	0.2	2.3	2.3
Eupatorium perfoliatum	3.0	0.5	0.2	2.3	2.7
Galium tinctorium	1.3	0.2	0.8	9.1	9.3
Habenaria lacera	0.1	0.02	0.2	2.3	2.3
Impatiens capensis	0.1	0.02	0.2	2.3	2.3
Juncus effusus	1.0	0.15	0.2	2.3	2.4
Lycopus virginicus	0.2	0.03	0.4	4.5	4.6
Osmunda cinnamomea	55.1	8.4	0.6	6.8	15.3
Polygonum sagitattum	1.4	0.21	1.0	11.4	11.6
Scirpus spp.	56.0	8.6	1.0	11.4	19.9
Solidago patula	10.1	1.5	0.4	4.5	6.1
Sphagnum sp.	480.0	73.5	1.0	11.4	84.8
Viola macloskeyi	0.4	0.06	0.8	9.1	9.2
SUMS	653.0	100.0	8.8	100.0	200.0

FOREST TREES

Species	pc c	rel c	pc fr	rel fr	I V
Betula alleghaniensis	50.0	3.3	0.5	7.9	11.2
Betula lenta	260.0	16.9	0.8	13.2	30.1
Hamamelis virginiana	70.1	4.6	0.7	10.5	15.1
Ilex ambigua var. montana	56.1	3.7	0.5	7.9	11.5
Liriodendron tulipifera	95.0	6.2	0.5	7.9	14.1
Magnolia fraseri	101.0	6.6	0.7	10.5	17.1
Prunus serotina	20.0	1.3	0.17	2.6	3.9
Quercus alba	30.0	2.0	0.17	2.6	4.6
Quercus rubra	85.0	5.5	0.3	5.3	10.8
Rhododendron maximum	625.0	40.7	1.0	15.8	56.5
SUMS	1535.4	100.0	6.33	100.0	200.0

Table II, cont.

FOREST SHRUBS

Species	pc c	rel c	pc fr	rel fr	I V
Hamamelis virginiana	5.1	3.9	0.3	40.0	43.9
Ilex sp.	1.1	0.8	0.3	40.0	40.8
Rhododendron maximum	125.0	95.3	0.17	20.0	115.3
SUMS	131.2	100.0	0.83	100.0	200.0

FOREST HERBS

Species	pc c	rel c	pc fr	rel fr	I V
Aster acuminatus	0.2	0.8	0.3	11.8	12.6
Dennstadtia punctilobula	5.1	20.3	0.3	11.8	32.1
Dryopteris intermedia	10.1	40.2	0.3	11.8	52.0
Goodyera pubescens	0.1	0.4	0.17	5.9	6.3
Lycopodium obscurum	0.1	0.4	0.17	5.9	6.3
Maianthemum canadense	0.2	0.8	0.3	11.8	12.6
Medeola virginiana	1.0	4.0	0.17	5.9	9.9
Mitchella repens	0.1	0.4	0.17	5.9	6.3
moss polsters	3.0	12.0	0.17	5.9	17.8
Trillium undulatum	5.2	20.7	0.5	17.7	38.4
unknown seedling	0.01	0.04	0.17	5.9	5.9
SUMS	25.1	100.0	2.83	100.0	200.0

Long Hope Valley bog (TABLE III) has the occasional *Picea rubens* and *Tsuga canadensis* with rather extensive patches of *Vaccinium macrocarpon* and scattered *Lyonia ligustrina* overtopping the *Sphagnum* and *Osmunda cinnamomea* herb layer. In the adjacent forest *Rhododendron maximum* and *Viburnum cassinoides* dominates the shrub layer and the herb level is dominated by *Solidago patula*-mosses and with a lesser importance by the two ferns, *Thelypteris novaboracensis* and *Dennstadtia punctilobula.*

4. DISCUSSION

Findings from this study generally corroborate those of Hotaling and Horton (1981). One of the differences was in the amount of the two rhododendrons, their ranking of *Rhododendron maximum* ahead of *R. catawbiense.* Certain differences may have been variance in cover estimates due to leaf characteristic difference between low and high elevations and their addition of density values to the I.V. In any case, I concur

TABLE III

Long Hope Valley. Importance Values of trees, shrubs, and herbs in the open bog and adjacent canopy-covered forests of the *Menyanthes* bog in the Stone Mountains. (Abbreviations: pc c = estimated percentage cover; rel c = relative cover; freq = frequency; rel fr = relative frequency; and I V = Importance Value - 200).

OPEN BOG TREES					
Species	pc c	rel c	pc fr	rel fr	I V
Acer rubrum	0.2	0.3	0.3	28.6	28.9
Amelanchier sp.	0.1	0.2	0.14	14.3	14.4
Betula sp.	0.1	0.16	0.14	14.3	14.4
Picea rubens	41.0	66.8	0.3	28.6	95.3
Tsuga canadensis	20.0	32.6	0.14	14.3	46.9
SUMS	61.4	100.0	1.0	100.0	200.0
OPEN BOG SHRUBS					
Species	pc c	rel c	pc fr	rel fr	I V
Ilex ambigua var. montana	10.3	2.1	0.14	3.7	5.8
Ilex collina	10.3	2.1	0.14	3.7	5.8
Ilex verticillata	5.0	1.02	0.14	3.7	4.7
Lonicera canadensis	5.2	1.06	0.4	11.1	12.1
Lyonia ligustrina	92.3	18.8	0.7	18.5	37.3
Rosa palustris	36.2	7.4	0.9	22.2	29.6
Rubus canadensis	3.3	0.7	0.14	3.7	4.4
Salix sericea	115.2	23.5	0.6	14.8	38.3
Vaccinium macrocarpon	210.0	42.8	0.6	14.8	57.6
Viburnum cassinoides	3.0	0.6	0.14	3.7	4.3
SUMS	490.8	100.0	3.9	100.0	200.0
OPEN BOG HERBS					
SPECIES	pc c	rel c	pc fr	rel fr	I V
Andropogon virginicus	18.0	2.8	0.43	4.4	7.3
Anemone quinquefolia	0.1	0.02	0.14	1.5	1.5
Arisaema triphyllum	0.1	0.02	0.14	1.5	1.5
Aster puniceus	5.0	0.8	0.14	1.5	2.3
Betula sp. seedling	0.1	0.02	0.14	1.5	1.5
Carex buxbaumii	90.0	14.2	0.3	2.9	17.1
Carex lurida	0.1	0.02	0.14	1.5	1.5
Carex muricata	32.1	5.01	0.7	7.4	12.4
Chelone cf. glabra	55.0	8.7	0.3	2.9	11.6
Drosera rotundifolia	0.1	0.02	0.14	1.5	1.5
Epilobium leptophyllum	0.2	0.03	0.3	2.9	3.0
Galium asprellum	19.1	3.01	0.9	8.8	11.8

Table III, open bog herbs, cont.

SPECIES	pc c	rel c	pc fr	rel fr	I V
Houstonia serpyllifolia	51.1	8.1	0.6	5.9	13.9
Impatiens capensis	0.1	0.02	0.14	1.5	1.5
Juncus effusus	1.0	0.16	0.14	1.5	1.6
Luzula sp	0.4	0.06	0.63	5.9	5.9
Maianthemum canadensis	0.1	0.02	0.14	1.5	1.5
Menyanthes trifoliata	5.0	0.8	0.14	1.5	2.3
mosses	48.1	7.6	0.9	8.8	16.4
Osmunda cinnamomea	145.0	22.9	0.29	2.9	25.8
Saxifraga pensylvanica	10.0	1.6	0.14	1.5	3.0
Scirpus atrovirens	9.1	1.5	0.6	5.9	7.3
Selaginella apoda	3.0	0.5	0.14	1.5	1.9
Senecio aureus	3.0	0.5	0.3	2.9	3.4
Solidago patula	18.1	2.9	0.7	7.4	10.2
Sphagnum sp.	116.1	18.3	0.6	5.9	24.2
Symplocarpus foetidus	1.0	0.16	0.14	1.5	1.6
Thelypteris palustris	1.0	0.16	0.14	1.5	1.62
Viola blanda	0.1	0.02	0.1	1.5	1.5
Viola macloskeyi	0.2	0.03	0.3	2.9	3.0
Viola sp.	1.3	0.2	0.6	5.9	6.1
SUMS	633.6	100.0	9.7	100.0	200.0

FOREST TREES

Species	pc c	rel c	pc fr	rel fr	I V
Acer pensylvanicum	10.2	1.02	0.3	4.3	5.4
Acer rubrum	160.3	1 6.1	1.0	13.0	29.1
Acer saccharum	41.1	4.1	0.5	6.5	10.6
Amelanchier arborea	50.0	5.0	0.7	8.7	13.7
Betula alleghaniensis	70.2	7.0	0.8	10.9	17.9
Crataegus flabellaris	0.1	0.01	0.17	2.2	2.2
Fagus grandifolia	197.1	19.8	0.3	4.3	24.1
Fraxinus americana	0.1	0.01	0.17	2.2	2.2
Hamamelis virginiana	18.0	1.8	0.3	4.3	6.2
Ilex ambigua var. montana	31.1	3.1	0.5	6.5	9.6
Picea rubens	1.2	9.2	0.7	8.7	17.9
Quercus rubra	45.1	4.5	0.5	6.5	11.1
Rhododendron maximum	270.1	27.1	0.8	10.9	38.0
Tsuga canadensis	43.0	4.3	0.5	6.5	10.8
SUMS	995.8	100.0	7.7	100.0	200.0

Table III, cont.
FOREST SHRUBS

Species	pc c	rel c	pc fr	rel fr	I V
Corylus cornuta	1.0	1.9	0.17	14.3	16.2
Hydrangea arborescens	0.1	0.19	0.17	14.3	14.5
Lonicera canadensis	4.0	7.5	0.17	14.3	21.8
Rubus canadensis	3.1	5.8	0.17	14.3	20.1
Sambucus pubens	10.0	18.8	0.17	14.3	33.0
Viburnum cassinoides	30.1	56.5	0.17	14.3	70.8
SUMS	48.3	100.0	1.0	100.0	200.0

FOREST HERBS

Species	pc c	rel c	pc fr	rel fr	I V
Arisaema triphyllum	0.3	0.7	0.5	9.4	10.1
Aster divaricata	0.1	0.23	0.17	3.1	3.4
Athyrium asplenioides	0.1	0.23	0.17	3.1	3.4
Carex pensylvanica	0.5	1.15	0.3	6.3	7.4
Circaea alpina	0.1	0.23	0.17	3.1	3.4
Conopholis americana	0.1	0.2	0.17	3.1	3.4
Dennstadtia punctilobula	10.0	23.0	0.17	3.1	26.1
Gentiana sp.	0.1	0.23	0.17	3.1	3.4
Glyceria melicaria	0.1	0.2	0.17	3.1	3.4
Houstonia serpyllifolia	0.1	0.23	0.17	3.1	3.4
Maianthemum canadense	5.0	11.5	0.17	3.1	14.6
mosses	7.1	16.3	0.8	15.6	31.9
Polygonatum biflorum	0.2	0.46	0.3	6.3	6.7
Prenanthes sp.	0.2	0.46	0.3	6.3	6.7
Solidago caesia/curtisii	8.2	18.9	0.67	12.5	31.4
Symplocarpus foetidus	0.1	0.23	0.17	3.1	3.4
Thelypteris novaboracensis	10.1	23.2	0.3	6.3	29.5
Veratrum parviflorum	1.0	2.3	0.17	3.1	5.4
Viola macloskeyi	0.1	0.23	0.17	3.1	3.4
Viola rotundifolia	0.1	0.23	0.17	3.2	3.4
SUMS	43.5	100.0	5.3	100.0	200.0

that the three most important shrubs are the two rhododendrons and *Kalmia latifolia*. I list 25 herb taxa to their listing of nine.

Moore (1972) found 116 vascular taxa compared to our 55. His study was not confined by the quadrats, the likely source of this difference in species numbers.

Each of the three sites have a different assemblage of species but also have species of northern origins. The elevations differ somewhat as is indicated by the more southern site at little higher in elevation but probably not enough to affect the microclimate very much. The isolation of each site and their probable different set of historical impacts would make it unlikely that the species composition of the three locations would remain unchanged over long periods. In a sense, these sites might be behaving as would islands, with new species invading and filling available niches and established populations becoming more depleted as their habitat warms and other influences reduce their ability to maintain themselves. We could, therefore, expect that the species composition would be primarily reflective of the surrounding vegetation and would secondarily represent relict sites for earlier established species. This pattern is suggested by the data for each of the sites.

Many of the species found in Southern Appalachian Bogs have been described as species with northern affinities while many others are considered southern Appalachian endemics. Gibson (1982) noted that many of the northern flora species left in the southern Appalachians after retreat of the continental glaciers and concurrent warming of the region have found suitable habitats near these seepage maintained wetlands. Some of these species, such as *Lilium grayi,* have probably evolved here as endemic species. See Richardson and Gibbons (1994) for a listing of many of these northern disjuncts and endemic taxa.

What are the implications of the vegetational history suggested by the types of species present at each of the sites? All sites have the presence of species more common to the north. Table IV lists some of the northern species present among the three sites. As one proceeds northward (Flat Laurel Gap to Boone Fork Bog to Long Hope Valley), there is an increase in the number of northern species. It is likely that these species have survived the most recent interglacial warming and I accept the basic premise (Delcourt and Delcourt, 1975; Shafer, 1988) that the tundra and boreal conditions which existed here during the past glacial period is the correct one. While it can be argued that there was not an even distribution of all northern species, their modern presence is directly linked to their distributions here during the glacial maximum 18,000 years before present (yr BP). Thus during the Pleistocene Period, the presence of tundra and southward extension of associated vegetation best accounts for the present northern disjuncts associated with these three southern Appalachian bogs.

TABLE IV

Northern affinity taxa present at one or more of the three sites. (FLG = Flat Laurel Gap, BFB = Boone Fork Bog, LHV = Long Hope Valley)

Species	FLG	BFB	LHV
Acer pensylvanicum			X
Carex buxbaumii			X
Carex leptalea	X	X	
Carex muricata var. angustata		X	X
Circaea alpina			X
Dryopteris cristata		X	
Epilobium leptophyllum		X	X
Eriophorum virginicum	X		
Lonicera canadensis			X
Maianthemum canadense		X	X
Menyanthes trifoliata			X
Picea rubens			X
Sambucus pubens		X	
Saxifraga pensylvanica			X
Symplocarpus foetidus			X
Trillium undulatum	X	X	X
Vaccinium macrocarpon		X	X

In support of this interpretation, Shafer (1984, 1986) and H. Delcourt (1985) found evidence for northern species in the sediment profiles at two of the sites. At Flat Laurel Gap, Shafer (1986, 1984) noted four assemblages with northern species as follows: The earliest period (3380-1280 yr BP) had plant macrofossils of *Picea rubens* and *Alnus crispa*, and pollen types of *Andromeda*, *Cornus canadensis*, and *Ledum/Chamaedaphne*. While *Picea* continued into the recent period, the other four dropped out. At Long Hope Valley H. Delcourt (1985) reported *Caltha palustris, Taxus canadensis, Menyanthes trifoliata*, and *Primula* pollen records. The *Menyanthes* and *Taxus* both remain in the valley, though *Taxus* occurs at a different site. Thus northern floristic element remains as a part of the northern flora that extended into this general region during the last glacial period.

From the foregoing discussions, it can be concluded that Southern Appalachian Bogs represent rare but valuable wetland ecosystems. Many of the species are relicts of the last glacial maximum 18,000 yr BP. While these wetland ecosystems differ from the more northern peat bogs and fens, their species composition and northern relationships are only occasionally documented in the literature (Gibson, 1982; Richardson and Gibbons, 1993; Schafele and Weakley, 1990). Protection of such sites for the interpretive information that may be gained is an important goal for these unusual ecosystems.

Acknowledgements

This project was supported by the Ecology Program of the National Science Foundation by funds through grant BSR-83-00345 to The University of Tennessee. Facilities and materials were provided by the Department of Biology at Western Carolina University. Dr. Hazel R. Delcourt and Dr. Paul Delcourt, along with their graduate student, David S. Shafer, assisted in the field sampling and are much appreciated for their help. I also thank Dr. Hazel Delcourt for her encouragement in this project. H.R. DeSelm's, Alan Weakley's and Carl Trettin's comments on the manuscript are much appreciated. The Southern Appalachian Man and Biosphere Conference and Conference Chairman Carl Trettin are also appreciated in bringing together this program and publication.

References

Cowardin, L. M., Carter, V., Golet, F. C. and LaRoe, E. T., 1979, *Classification of wetlands and deepwater habitats of the United States.* U. S. Fish and Wildlife Service, Office of Biological Services, Washington, DC. 136 p.

Delcourt, H. R., 1985, 'Holocene vegetational changes in the southern Appalachian mountains, U.S.A.,' *Ecol. Mediterr.* 11, 9-15.

Delcourt, P. A., 1985, 'The influence of Late-Quaternary climatic and vegetational change on paleohydrology in unglaciated eastern North America, 'Ecol. *Mediterr.* 11, 17-26.

Delcourt, H. R. and Delcourt, P. A., 1975., 'The Blufflands: Pleistocene pathway into the Tunica hills,' *Am. Mid. Nat.* 94, 385-400.

Delcourt, H. R. and Delcourt, P. A., 1985, 'Quaternary palynology and vegetational history of the southeastern United States,' in Bryant, V. M., Jr., and Holloway, R. G. (eds), *Pollen records of late-quaternary North American sediments.* American Assoc. of Stratigraphic Palynologists Foundation, Calgary, Alberta, Can., pp. 1-37.

Gaddy, L. L, 1981., 'The bogs of the southwestern mountains of North Carolina,' Report to the North Carolina Natural Heritage Program, Department of Natural Resources and Community Development, Raleigh. 6 p + appendix.

Gibson, J. R., 1982, 'Alder Run Bog, Tucker County, West Virginia: its history and vegetation,' in Clarkson, R. B. et al. (eds.), *Symposium on wetlands of the unglaciated Appalachian region.* Univ. of West Virginia, Morgantown, pp. 101-105.

Horton, J. H. and Hotaling, L. G., 1981, 'Floristics of selected heath communities along the southern section of the Blue Ridge Parkway,' U. S. Department of Interior, National Park Service, *NPS-SER Research/Resources Management Report No. 45,* Atlanta, GA, 36 p.

Moore, T., 1972, *The phytoecology of Boone Fork sphagnum bog,* M. S. Thesis, Appalachian State University, Boone, North Carolina, 45 p.

Mueller-Dumbois, D. and Ellenburg, H., 1974, *Aims and methods of vegetation ecology,* John Wiley & Sons, NY, 547 p.

Radford, A. E., Ahles, H. E., and C. R. Bell, 1968, *Manual of the vascular flora of the Carolinas,* UNC Press, Chapel Hill, NC, 1180 p.

Richardson, C. J. and Gibbons, J. W., 1994, 'Pocosins, Carolina bays, and mountain bogs,' p. 257-310, in Martin, W. H., Boyce, S. G. and Echternacht, A. C., *Biodiversity of the southeastern United States: lowland terrestrial communities,* John Wiley & Sons, Inc., New York.

Rigg, G. B. and Strausbaugh, P. D., 1949, 'Some states in the development of sphagnum bogs in West Virginia,' *Castanea* **14**, 129-148.

Schafale, M. P. and Weakley, A. S., 1990, *Classification of the natural communities of North Carolina (Third approximation),* NC Department of Environment, Health, and Natural Resources, Natural Heritage Program, Raleigh, NC. 325 p.

Shafer, D. S., 1984, *Late-Quaternary paleoecology, paleoclimatology, and geomorphology of Flat Laurel Gap, a high-elevation heath bog in the Blue Ridge Mountains of western North Carolina,* M. S. Thesis, The University of Tennessee, Knoxville, 148 p.

Shafer, D. S., 1986, 'Flat Laurel Gap bog, Pisgah Ridge, North Carolina: late Holocene development of a high-elevation heath bald,' *Castanea* **51**,1-9.

Shafer, D. S., 1988, 'Late Quaternary landscape evolution at Flat Laurel Gap, Blue Ridge Mountains, North Carolina,' *Quat. Res.* 30, 7-11.

WILDLIFE USE OF SOUTHERN APPALACHIAN WETLANDS

IN NORTH CAROLINA

A.C. BOYNTON

North Carolina Wildlife Resources Commission, Raleigh, North Carolina

Abstract. Wetlands provide structurally diverse habitats attractive to varied wildlife, both generalist and wetland specialist species. Wetlands in western North Carolina occupy a minor portion of the landscape, yet provide essential habitat for rare wildlife species. Structural features of western North Carolina wetlands that influence wildlife occurrence include meadows interspersed with shrub thickets, snags and hollow trees, fallen logs, deep mud and rivulets, and pools. Species lists keyed to structural features are presented.

1. Introduction

Western North Carolina encompasses about 1.6 million ha of mountainous terrain, drained by seven major river systems to both the Atlantic Ocean and Gulf of Mexico. Schafale and Weakley (1990) described several alluvial and non-alluvial wetland communities in the area using topographic, edaphic, hydrologic, and vegetative features (Table I). Western North Carolina wetlands are small and occur primarily on floodplains. Small wetlands also may occur on relatively steep slopes given suitable soil conditions. Wetlands are not evenly distributed across western North Carolina. Notable groups of wetlands occur in the region: the upper French Broad River valley, the New River valley just west of the Blue Ridge escarpment, the upper Nantahala River valley and the upper Linville River valley.

Human activity has caused a significant loss of mountain wetlands since the early part of this century. Some communities have diminished more than others (Weakley and Schafale, this volume), with wetlands in larger valleys receiving the brunt of impacts. Swamp Forest-Bog Complex was once the most commonly occurring wetland community in the region while other communities, like Spray Cliff and Upland Pool, were always rare (Weakley and Schafale, this volume).

The small size and scattered distribution of wetlands in the region today results in the rarity of wetland obligate species. Many generalist species, however, find food, cover and water in these communities. Abiotic and biotic features of wetland communities attract groups of species that use resources in a similar way. For example, beaver ponds with standing dead trees attract cavity nesting species. Almost all wetland communities in the region are important wildlife habitats, and several wetland communities stand out in their contribution to biodiversity. Certain species occur in several types of wetland habitats in the area, although other factors may make them rare (Table II).

In this paper, I discuss several structural attributes of wetlands and relate wildlife species to these attributes. Biologists have conducted few comprehensive faunal surveys in western North Carolina wetlands. Faunal lists for each wetland community type described by Shafale and Weakley (1990) are unavailable. Faunal surveys could provide crucial information to use in protecting and managing wetlands. I present information here on the expected occurrence of vertebrates in wetland communities based on North Carolina's Biological Characterization Database (The Nature Conservancy 1992), my

Water, Air and Soil Pollution **77**: 349–358, 1994.

TABLE I
Selected structural characteristics of wetland communities as described by Schafale and Weakley (1990) in western North Carolina

Structural Feature	Natural Community Type
Meadows and Shrubs	Southern Appalachian Bog Southern Appalachian Fen Piedmont/Mountain Semipermanent Impoundment
Forests	Swamp Forest-Bog Complex High Elevation Seep
Ponds	Piedmont/Mountain Semipermanent Impoundment
Snags and Hollow Trees	Swamp Forest-Bog Complex Piedmont/Mountain Semipermanent Impoundment
Fallen Logs	Swamp Forest-Bog Complex Piedmont/Mountain Semipermanent Impoundment
Deep Mud and Rivulets	Swamp Forest-Bog Complex Southern Appalachian Bog Southern Appalachian Fen Piedmont/Mountain Semipermanent Impoundment
Temporary Pools	Upland Pool Floodplain Pool

knowledge of the structural characteristics of area wetlands (Table I), and various studies of species.

I have organized additional species lists by wetland features. Due to elevational and other factors influencing distributions, not all species listed will occur in a particular wetland. Scientific names of species appear in the tables. North Carolina has listed several species as endangered, threatened, or special concern. These status categories are noted in the text where appropriate. I hope that this information assists biologists planning faunal surveys of wetland communities in western North Carolina.

2. Meadow and Brush Dominated Wetlands

Permanently saturated soils inhibit tree growth, producing meadows and brush, especially sedges, rushes, ferns, mosses, ericaceous shrubs, alder, and other low-growing plants. Deep mud may underlie these wetlands. Pools may occur, particularly where soils are disturbed by tree fall or vehicles. Snags and fallen logs are scarce, though sometimes present along edges. These wetlands can simulate early seral forests, attracting several terrestrial species that capitalize on disturbance. Luxuriant growth provides ample forage and nesting, escape, and thermal cover attractive to many vertebrate and invertebrate herbivores. Sedges, grasses, herbaceous plants, vines, shrubs, and small trees produce seeds and fruits that attract many species. Abundant vertebrate and invertebrate prey attract a variety of predators. Wetland openings provide a foraging area for birds and bats that specialize in catching flying insects.

Species of particular interest attracted to wetlands with meadows and brush include Bog Turtle (Threatened), Cooper's Hawk (Special Concern), Four-toed Salamander (Special Concern), Mountain Chorus Frog (Special Concern), Olive-sided Flycatcher (Special Concern), and Southern Pygmy Shrew (Special Concern) (Table III). The Bog Turtle has received considerable survey effort in western North Carolina over the last 20

TABLE II

Some vertebrate species potentially found using wetland habitats in western North Carolina

Scientific Name	Common Name
Cathartes aura	Turkey Vulture
Accipiter cooperi	Cooper's Hawk
Buteo jamaicensis	Red-tailed Hawk
Meleagris gallapavo	Wild Turkey
Bubo virginianus	Great Horned Owl
Sayornis phoebe	Eastern Phoebe
Corvus brachyrhynchos	American Crow
Corvus corax	Common Raven
Polioptila caerulea	Blue-gray Gnatcatcher
Turdus migratorius	American Robin
Vermivora pinus	Blue-Winged Warbler
Geothlypis trichas	Common Yellowthroat
Melospiza georgiana	Swamp Sparrow
Quiscalus quiscula	Common Grackle
Didelphis virginiana	Virginia Opossum
Sorex longirostris	Southeastern Shrew
Sorex hoyi winnemana	Southern Pygmy Shrew
Blarina brevicauda	Northern Short-Tailed Shrew
Condylura cristata	Star-nosed Mole
Pipistrellus subflavus	Eastern Pipistrelle
Lasiurus borealis	Red Bat
Sylvilagus floridanus	Eastern Cottontail
Sylvilagus obscurus	Allegheny Cottontail
Peromyscus maniculatus	Deer mouse
Peromyscus leucopus	White-footed Mouse
Microtus pennsylvanicus	Meadow Vole
Microtus pinetorum	Woodland Vole
Canis latrans	Coyote
Vulpes vulpes	Red Fox
Urocyon cinereoargenteus	Gray Fox
Ursus americanus	Black Bear
Procyon lotor	Raccoon
Mustela nivalis	Least Weasel
Mustela frenata	Long-Tailed Weasel
Spilogale putorius	Eastern Spotted Skunk
Lynx rufus	Bobcat
Odocoileus virginianus	White-tailed Deer
Fallen Log Associated Species	
Hemidactylium scutatum	Four-Toed Salamander
Chelydra serpentina	Snapping Turtle
Diadophis punctatus	Ringneck Snake
Regina septemvittata	Queen Snake
Storeria dekayi	Brown Snake
Storeria occipitomaculata	Redbelly Snake
Thamnophis sirtalis	Common Garter Snake
Agkistrodon contortrix	Copperhead
Mustela vison	Mink
Mud, Rivulet, and Pool Associated Species	
Pseudotriton montanus	Mud Salamander
Pseudotriton ruber	Red Salamander
Notophthalmus viridescens	Eastern Newt
Hemidactylium scutatum	Four-Toed Salamander

TABLE II cont.

Scientific Name	Common Name
Bufo americanus	American Toad
Bufo woodhousei	Fowler's Toad
Hyla chrysoscelis	Cope's Gray Treefrog
Pseudacris crucifer	Spring Peeper
Rana clamitans	Green Frog
Rana sylvatica	Wood Frog
Rana catesbeiana	Bullfrog
Rana palustris	Pickerel Frog
Chelydra serpentina	Snapping Turtle
Clemmys muhlenbergii	Bog Turtle
Terrapene carolina	Eastern Box Turtle
Sternotherus odoratus	Common Musk Turtle
Nerodia sipedon	Northern Water Snake
Thamnophis sauritus	Eastern Ribbon Snake

years (Herman 1992). This turtle is found in some foothill as well as mountain counties, and Tryon (1990) estimated the North Carolina population conservatively at 1500-2000.

3. Forested Wetlands

These forests grow on saturated, alluvial soils where they may be flooded from time to time. Dominant trees include red maple (Acer rubrum), birches (Betula lenta, Betula alleghaniensis) and eastern hemlock (Tsuga canadensis). Trees are either adapted to wet soil conditions or grow on hummocks. Small bogs on the wettest soils can cause canopy gaps, often dominated by brush. Deep mud and pools may occur. These wetlands can have many snags, hollow trees and fallen logs.

Wildlife species are adapted to forested habitats in the southern Appalachians and many use forested wetlands (Table IV). Species of particular interest in forested wetlands are Bog Turtle (Threatened), Mole Salamander (Special Concern), Longtail Salamander (Special Concern), Four-toed Salamander (Special Concern), Northern Saw-whet Owl (Special Concern), Southern Pygmy Shrew (Special Concern), Southern Water Shrew (Special Concern), and Southern Rock Vole (Special Concern).

4. Ponds

Both humans and Beaver create semi-permanent ponds in the region. Ponds often go through a succession of habitats that provide resources at one time or another for many of the wildlife species found around regional wetlands. Initial flooding following dam completion may kill areas of timber, creating snags and excellent habitat for snag using species. Tree death may result in areas of meadow and brush in flat terrain, as on floodplains. Old ponds may eventually succeed to dense brush and small trees, attracting wildlife typical of forested wetlands. Open water attracts some species not usually found in other types of wetlands in the region (Table V). Fish may or may not colonize the pond, depending on location and human activity.

The Beaver has such influence on small wetlands in western North Carolina that it deserves special consideration. The extirpation of the Beaver in western North Carolina facilitated the conversion of wetland habitats to agricultural and urban uses. Apparently Beaver was virtually, if not entirely, extirpated from the region by the early 1900's.

TABLE III

Some vertebrate species potentially found in wetlands with meadows and brush in western North Carolina

Scientific Name	Common Name
Circus cyaneus	Northern Harrier
Falco sparverius	American Kestrel
Gallinago gallinago	Common Snipe
Scolopax minor	American Woodcock
Tyto alba	Barn Owl
Asio otus	Long-Eared Owl
Contopus borealis	Olive-Sided Flycatcher
Empidonax alnorum	Alder Flycatcher
Tyrannus tyrannus	Eastern Kingbird
Hirundo rustica	Barn Swallow
Catharus guttatus	Hermit Thrush
Dumetella carolinensis	Gray Catbird
Toxostoma rufum	Brown Thrasher
Vireo griseus	White-eyed Vireo
Vermivora chrysoptera	Golden-Winged Warbler
Dendroica petechia	Yellow Warbler
Dendroica virens	Black-Throated Green Warbler
Wilsonia citrina	Hooded Warbler
Wilsonia canadensis	Canada Warbler
Pipilo erythrophthalmus	Rufous-sided Towhee
Passerculus sandwichensis	Savannah Sparrow
Melospiza melodia	Song Sparrow
Zonotrichia albicollis	White-Throated Sparrow
Zonotrichia leucophrys	White-Crowned Sparrow
Junco hyemalis	Dark-Eyed Junco
Agelaius phoeniceus	Red-Winged Blackbird
Euphagus cyanocephalus	Brewer's Blackbird
Snag and Hollow Tree Associated Species	
Strix varia	Barred Owl
Aegolius acadicus	Northern Saw-Whet Owl
Colaptes auratus	Northern Flicker
Progne subis	Purple Martin
Tachycineta bicolor	Tree Swallow
Sturnus vulgaris	European Starling
Sorex cinereus	Masked Shrew
Cryptotis parva	Least Shrew
Myotis lucifugus	Little Brown Bat
Lampropeltis triangulum	Milk Snake
Fallen Log Associated Species	
Eumeces anthracinus	Coal Skink
Elaphe obsoleta	Rat Snake
Lampropeltis getula	Common Kingsnake
Lampropeltis triangulum	Milk Snake
Virginia valeriae	Smooth Earth Snake
Sorex cinereus	Masked Shrew
Sorex longirostris	Southeastern Shrew
Cryptotis parva	Least Shrew
Reithrodontomys humulis	Eastern Harvest Mouse
Synaptomys cooperi	Southern Bog Lemming
Napaeozapus insignis	Woodland Jumping Mouse

TABLE III cont.

Scientific Name	Common Name
Mud, Rivulet, and Pool Associated Species	
Acris crepitans	Northern Cricket Frog
Pseudacris brachyphona	Mountain Chorus Frog
Pseudacris triseriata	Upland Chorus Frog
Scaphiopus holbrookii	Eastern Spadefoot
Charadrius vociferus	Killdeer
Tringa flavipes	Lesser Yellowlegs
Tringa solitaria	Solitary Sandpiper
Actitis macularia	Spotted Sandpiper

Visitors during the 1700's found Beaver and River Otter extremely common (Williams 1927). Hamnett (1953) discusses Beaver sightings in Yancey County around the time that restocking efforts got underway in 1938 in North Carolina's Sandhills region. The Yancey County Beavers were of unknown origin. Singer et al. (1981) state that Beaver colonized Great Smoky Mountains National Park in 1966.

Certainly the loss of Beaver from the region reduced by some unknown amount its wetland acreage. We do not know to what extent the decline of Beaver affected associated plant and animal species. The Beaver's recovery in the southern Appalachians offers the first possibility in several decades for substantial increases in wetland area, although residents may hold populations below carrying capacity. The North Carolina General Assembly in its last session modified depredation laws regarding Beaver. It is now legal for landowners with property damage to take Beaver on their property by any lawful method without obtaining a depredation permit from the state. Lawful methods include use of firearms, 330 connibear traps or snares. This law does not apply to four mountain counties.

Beaver activity will create new wetlands, but also may damage unique wetlands supporting rare species. Humans have so completely altered some watersheds that wetlands created through Beaver activity may bear little resemblance to the wetlands that once occurred. Nonetheless, Beaver probably will create habitats allowing the spread of much reduced species into new areas, for example bog species around pond margins. The possibility also exists, unfortunately, that the Beaver may flood the last habitats of vanishing species.

5. Summary

Clearly wetlands in western North Carolina provide valuable habitat to many wildlife species. Agricultural and urban development has particularly affected Swamp Forest Bog-Complex and Southern Appalachian Bog communities that occurred in valleys. Some wildlife species are completely dependent on mountain wetland habitats, while others use wetlands heavily because of the resources provided. At least ten state listed wildlife species probably occur in mountain wetland communities. Not only do western North Carolina wetlands support rare species, but valuable game species use area wetlands as well. Almost every wetland has a deer stand in it, for example, and area residents also realize that Ruffed Grouse, Wild Turkey, Raccoon, and American Woodcock inhabit wetlands too. Threats to wetlands include road construction, draining or other hydrologic alterations, filling, and flooding by humans or Beaver.

TABLE IV

Some vertebrate species potentially found in forested wetlands in western North Carolina

Scientific Name	Common Name
Ardea herodias	Great Blue Heron
Butorides striatus	Green-Backed Heron
Buteo lineatus	Red-Shouldered Hawk
Bonasa umbellus	Ruffed Grouse
Contopus borealis	Olive-Sided Flycatcher
Empidonax virescens	Acadian Flycatcher
Empidonax traillii	Willow Flycatcher
Cyanocitta cristata	Blue Jay
Catharus fuscescens	Veery
Catharus guttatus	Hermit Thrush
Hylocichla mustelina	Wood Thrush
Turdus migratorius	American Robin
Vireo olivaceus	Red-Eyed Vireo
Parula americana	Northern Parula
Dendroica virens	Black-Throated Green Warbler
Dendroica dominica	Yellow-Throated Warbler
Dendroica cerulea	Cerulean Warbler
Mniotilta varia	Black-And-White Warbler
Limnothlypis swainsonii	Swainson's Warbler
Seiurus motacilla	Louisiana Waterthrush
Oporornis formosus	Kentucky Warbler
Wilsonia citrina	Hooded Warbler
Wilsonia canadensis	Canada Warbler
Icteria virens	Yellow-Breasted Chat
Cardinalis cardinalis	Northern Cardinal
Pheucticus ludovicianus	Rose-Breasted Grosbeak
Molothrus ater	Brown-Headed Cowbird
Carpodacus purpureus	Purple Finch
Sylvilagus obscurus	Appalachian Cottontail
Ochrotomys nuttalli	Golden Mouse
Clethrionomys gapperi	Southern Red-backed Vole
Napaeozapus insignis	Woodland Jumping Mouse
Sus scrofa	Wild Pig
<u>Snag and Hollow Tree Associated Species</u>	
Eumeces inexpectatus	Southeastern Five-Lined Skink
Eumeces laticeps	Broadhead Skink
Aix sponsa	Wood Duck
Cathartes aura	Turkey Vulture
Otus asio	Eastern Screech-Owl
Bubo virginianus	Great Horned Owl
Strix varia	Barred Owl
Aegolius acadicus	Northern Saw-Whet Owl
Chaetura pelagica	Chimney Swift
Melanerpes carolinus	Red-Bellied Woodpecker
Picoides pubescens	Downy Woodpecker
Picoides villosus	Hairy Woodpecker
Dryocopus pileatus	Pileated Woodpecker
Myiarchus crinitus	Great Crested Flycatcher
Parus carolinensis	Carolina Chickadee
Parus bicolor	Tufted Titmouse
Certhia americana	Brown Creeper
Thryothorus ludovicianus	Carolina Wren
Troglodytes aedon	House Wren

TABLE IV cont.

Scientific Name	Common Name
Troglodytes troglodytes	Winter Wren
Sturnus vulgaris	European Starling
Sorex cinereus	Masked Shrew
Sorex fumeus	Smoky Shrew
Myotis lucifugus	Little Brown Bat
Tamiasciurus hudsonicus	Red Squirrel
Fallen Log Associated Species	
Ambystoma maculatum	Spotted Salamander
Ambystoma opacum	Marbled Salamander
Desmognathus fuscus	Dusky Salamander
Desmognathus ochrophaeus	Mountain Dusky Salamander
Eurycea l. longicauda	Longtail Salamander
Eurycea wilderae	Blue Ridge Two-lined Salamander
Eurycea guttolineata	Three-Lined Salamander
Gyrinophilus porphyriticus	Spring Salamander
Sternotherus minor	Loggerhead Musk Turtle
Eumeces anthracinus	Coal Skink
Eumeces inexpectatus	Southeastern Five-Lined Skink
Eumeces laticeps	Broadhead Skink
Elaphe obsoleta	Rat Snake
Lampropeltis getula	Common Kingsnake
Storeria occipitomaculata	Redbelly Snake
Crotalus horridus	Timber Rattlesnake
Cathartes aura	Turkey Vulture
Sorex cinereus	Masked Shrew
Sorex longirostris	Southeastern Shrew
Sorex fumeus	Smoky Shrew
Clethrionomys gapperi	Southern Red-Backed Vole
Microtus chrotorrhinus	Southern Rock Vole
Synaptomys cooperi	Southern Bog Lemming
Napaeozapus insignis	Woodland Jumping Mouse
Mud, Rivulet, and Pool Associated Species	
Ambystoma maculatum	Spotted Salamander
Ambystoma opacum	Marbled Salamander
Ambystoma talpoideum	Mole Salamander
Desmognathus fuscus	Dusky Salamander
Eurycea l. longicauda	Longtail Salamander
Eurycea guttolineata	Three-Lined Salamander
Gyrinophilus porphyriticus	Spring Salamander
Sternotherus minor	Loggerhead Musk Turtle
Eumeces anthracinus	Coal Skink
Eumeces inexpectatus	Southeastern Five-Lined Skink
Eumeces laticeps	Broadhead Skink
Elaphe obsoleta	Rat Snake
Lampropeltis getula	Common Kingsnake
Sorex palustris punctulatus	Southern Water Shrew

Human interest in conservation of these habitats is increasing. Wetland communities contribute to the biodiversity of the southern Appalachians. Biologists have not conducted many vertebrate and invertebrate surveys in the various wetland communities of western North Carolina. However, data from wildlife surveys would prove useful in wetland conservation efforts.

TABLE V

Some vertebrate species potentially associated with permanent pools in western North Carolina wetlands

Scientific Name	Common Name
Salvelinus fontinalis	Brook Trout
Cyprinus carpio	Common Carp
Notemigonus crysoleucas	Golden Shiner
Notropis volucellus	Mimic Shiner
Pimephales notatus	Bluntnose Minnow
Ambloplites rupestris	Rock Bass
Lepomis gibbosus	Pumpkinseed
Lepomis gulosus	Warmouth
Lepomis macrochirus	Bluegill
Micropterus dolomieu	Smallmouth Bass
Micropterus punctulatus	Spotted Bass
Micropterus salmoides	Largemouth Bass
Pomoxis annularis	White Crappie
Pomoxis nigromaculatus	Black Crappie
Podilymbus podiceps	Pied-Billed Grebe
Ardea herodias	Great Blue Heron
Butorides striatus	Green-Backed Heron
Branta canadensis	Canada Goose
Anas platyrhynchos	Mallard
Aythya collaris	Ring-Necked Duck
Fulica americana	American Coot
Pandion haliaetus	Osprey
Ceryle alcyon	Belted Kingfisher
Hirundo pyrrhonota	Cliff Swallow
Castor canadensis	Beaver
Ondatra zibethicus	Muskrat
Lutra canadensis	River Otter
Snag and Hollow Tree Associated Species	
Aix sponsa	Wood Duck
Tachycineta bicolor	Tree Swallow
Myotis lucifugus	Little Brown Bat
Lasionycteris noctivagans	Silver-Haired Bat
Fallen Log Associated Species	
Chrysemys picta	Painted Turtle
Mud and Rivulet Associated Species	
Catostomus commersoni	White Sucker
Moxostoma anisurum	Silver Redhorse
Moxostoma macrolepidotum	Shorthead Redhorse
Ictalurus punctatus	Channel Catfish
Noturus flavus	Stonecat
Ameiurus catus	White Catfish
Ameiurus nebulosus	Brown Bullhead
Necturus maculosus	Mudpuppy
Chrysemys picta	Painted Turtle

References

Barbour, R. W., and W. H. Davis. 1969. Bats of America. Univ. Press of Ky., Lexington, Ky. 286p.

Hamel, P. B. 1992. The land manager's guide to the birds of the South. The Nature Conservancy, Southeast Region, Chapel Hill, N.C. 437p.

Hamnett, William L. 1953. Tar heel wildlife. N.C. Wildl. Resour. Comm., Raleigh, N.C. 98p.

Martof, B. S., W. M. Palmer, J. R. Bailey, and J. R. Harrison III. 1980. Amphibians and reptiles of the Carolinas and Virginia. Univ. of N.C. Press, Chapel Hill, N.C. 264p.

Menhinick, E. F. 1991. The freshwater fishes of North Carolina. N.C. Wildl. Resour. Comm., Raleigh, N.C. 227p.

Potter, E. F., J. F. Parnell, and R. P. Teulings. 1980. Birds of the Carolinas. Univ. of N.C. Press, Chapel Hill, N.C. 408p.

Rohde, Fred C., Rudolf G. Arndt, David C. Lindquist and James F. Parnell. Freshwater fishes of the Carolinas, Virginia, Maryland and Delaware. 1994. The Univ. Of N.C. Press, Chapel Hill, N.C. 222p.

Schafale, Michael P., and Alan S. Weakley. 1990. Classification of the natural communities of North Carolina, third approximation. N.C. Div. Parks and Recreation, Raleigh, N.C. 325p.

Singer, Frances J., David Labrody, and Lorrie Sprague. 1981. Beaver reoccupation and an analysis of the otter niche in Great Smoky Mountains National Park. Resour. Manage. Rep. 40, USDI Nat. Park Serv., Gatlinburg, TN. 18p.

Terres, John K. 1991. Encyclopedia of North American birds. Wings Books, New York, NY. 1109p.

The Nature Conservancy. 1992. Biological Characterization Database. The Nature Conservancy, Inc., Arlington, VA.

Weakley, Alan S. and Michael P. Schafale. In press. Non-alluvial wetlands of the southern Blue Ridge-diversity in a threatened ecosystem. SAMAB Conf. on Wetland Ecol., Manage. and Conserv., Knoxville, TN.

Webster, W. D., J. F. Parnell, and W. C. Biggs, Jr. 1985. Mammals of the Carolinas, Virginia, and Maryland. Univ. of N.C. Press, Chapel Hill, N.C. 255p.

Williams, S. C. 1927. Lieutenant Harry Timberlake's memoirs, 1756-1765: with annotations, introduction and index. Johnson City Library, Johnson City, TN. 197p.

NON-ALLUVIAL WETLANDS OF THE SOUTHERN BLUE RIDGE -- DIVERSITY IN A THREATENED ECOSYSTEM

A.S. WEAKLEY and M.P. SCHAFALE

North Carolina Natural Heritage Program, Division of Parks and Recreation
P.O. Box 27687, Raleigh, North Carolina, U.S.A. 27611-7687

Abstract. The generally steep landscape of the Southern Blue Ridge is not conducive to the formation of extensive wetlands, but wetlands do occur. Wetlands in this region are mostly small in size (< 10 ha), and are found in locations where topography is unusually gentle or where seepage is unusually strong or constant. Despite their rarity and small size, such wetlands show great species and community diversity, and are one of the most important habitats for rare (endemic and disjunct) plants and animals in the region. Community species composition seems to vary primarily in relation to elevation, topographic position, hydrology, underlying bedrock composition, recent land use, and biogeographic history. Based on differences in vegetation structure and composition, landscape position, and hydrology, we recognize nine groups of non-alluvial wetlands in the Southern Blue Ridge. An inventory of non-alluvial wetlands in the mountains of North Carolina revealed that the majority of these naturally rare communities are now destroyed or severely altered. Bogs and fens of the North Carolina mountains have been reduced nearly six-fold from an original extent of about 2000 ha, so that only about 300 ha remain in reasonably intact condition, and most of the remnants are compromised by hydrologic alteration and nutrient inputs. Because wetlands tend to be concentrated in valley bottoms and at low elevations where most land is privately owned, efforts to assure their long-term viability will require innovative protection and restoration tools.

1. Introduction

Wetlands of the Southern Blue Ridge of North Carolina have been classified by Schafale & Weakley (1990) into 15 natural community types, based on differences in vegetation composition and physiognomy, topography, substrate, hydrology, soils, and other abiotic factors. Of these, eight types may be called "boggy" or non-alluvial wetlands. They can be differentiated from the other, more alluvial, wetlands by the substantial to dominant *Sphagnum* (versus little or no *Sphagnum*), as well as their non-alluvial hydrology dominated by seepage, spray, high water table, or paludification, with flooding by rivers or streams rare or nonexistent. Except for Spray Cliffs, they are also distinguished by occurring as more-or-less isolated wetlands, surrounded by terrestrial communities, and usually occurring on slopes or outer edges of stream valleys. Most of these communities have been commonly referred to as "bogs." This term will be retained here, though in many classifications of mire types, these wetlands would be considered fens (see discussion below). Although some occur along the floodplains of streams and small rivers, they occur away from the stream and above the normal reach of flooding and alluvial deposition. We here recognize an additional category, subdividing the Southern Appalachian Bog (Southern Subtype) into two types, French Broad Valley Bog and Low Mountain Seepage Bog, based on strong differences in vegetation structure and composition.

Despite the great scientific interest and conservation importance of these wetland communities, only a few isolated examples have been studied (Schafale & Weakley 1990, Richardson & Gibbons 1993, Stewart & Nilsen 1993). The rarity of these wetland types and their concentration of rare species found in them has led the North Carolina Natural Heritage Program (Division of Parks and Recreation) to make the identification prioritization, and protection of remaining sites a focus of its activity since 1985 (Weakley 1993, LeGrand 1993, Murdock, this volume). Field work conducted during the last decade has provided us with a more comprehensive listing of remaining sites, an improved understanding of the diversity of these wetland types, and the information needed to direct conservation efforts. Many questions remain, however. Quantitative vegetation data are essentially non-existent; many sites have received only a cursory survey, focused on

Water, Air and Soil Pollution 77: 359–383, 1994.

dominant species and rare species of special conservation concern. Information on soils and hydrology is poor or absent. Studies of the dynamics and successional patterns of these wetland types are non-existent. Information on potential impacts of grazing, nutrient input from atmospheric deposition and run-off from surrounding fields, and alteration of hydrology is anecdotal, yet effective management and stewardship of these ecosystems depends on an understanding of these forces.

This paper summarizes the available information on these interesting ecosystems. Its focus is on North Carolina, though much of its content applies to non-alluvial wetlands in the Southern Blue Ridge of northwestern South Carolina, northeastern Georgia, eastern Tennessee, and southwestern Virginia, as well. Bogs, fens, and other non-alluvial wetlands in the sedimentary rock provinces (Ridge and Valley, Alleghany Plateau) of the Central Appalachians of western Virginia and West Virginia show some similarities to the communities discussed here, but also differ in many ways (Walbridge, this volume). They have received more detailed study, much of it published in McDonald (1982). Non-alluvial wetlands north of the glacial boundary are much more extensive and have received much greater scientific attention, with numerous papers published on the successional dynamics of bogs and fens in New England. Despite the tendency to consider southern "bogs" and fens as essentially similar to (or poor cousins of) northern wetland types, non-alluvial wetlands of the southern mountains are not analogous to those of the northeastern United States in either floristics, climate, biogeochemistry, successional dynamics, or biogeographic history.

Most of the information on which this discussion is based is either in the form of unpublished reports and site survey forms in the files of the N.C. Natural Heritage Program, or unpublished observations by the authors. We acknowledge here the use of numerous reports and field forms prepared by a number of workers; only the more comprehensive of the unpublished reports will be cited (such as Gaddy 1981a and Smith 1993). Taxonomy and nomenclature of vascular plant species generally follow Kartesz (1994). Taxonomy and nomenclature of mosses generally follow Anderson, Crum, & Buck (1990) and Anderson (1990). Taxonomy and nomenclature of liverworts generally follow Hicks (1992). We present a discussion of broad-scale classification issues, followed by our classification of the community types of non-alluvial wetlands of the southern Blue Ridge. Type descriptions include physical characteristics, vegetation, range, distinguishing features, and examples. Following the classification we discuss distribution and abundance, biogeographic affinities, factors influencing compositional variation, natural dynamics and disturbance, and conservation status for these wetlands.

2. Broad Scale Classification

The word "bog" has been historically and traditionally applied to various non-alluvial wetlands in the Southern Appalachians, especially those with apparent *Sphagnum* and either herb- or shrub-dominated vegetation. Usage of this terminology has been nearly universal in popular, conservation, and scientific literature, and has been retained in recent scientific discussions of these communities (such as Schafale & Weakley 1990, Richardson & Gibbons 1993, Stewart & Nilsen 1993, and this paper). The term "fen" has also been used, but has usually been limited to the most minerotrophic examples, which contain some species characteristic of northern fens (Schafale & Weakley 1990, Richardson & Gibbons 1993).

Most international classifications of "mires" have emphasized a dichotomy into fens and bogs, a dichotomy which "reflects the historical lead of northwest continental Europe in peatland investigations" (Gore 1983). For instance, Gore (1983) follows an extensive and complex discussion of mire terminology with "a satisfactory definition of fens and bogs: fens are mires influenced by water derived predominantly from outside their own immediate limits; bogs are influenced solely by water that falls directly on to them as rain and snow." This definition is indicative of some conceptual problems in the simple dichotomy, since it leaves unaccounted for wetlands which receive some (but not predominant) flow from outside. Ingram (1983) states that "hydrologically, mires can be divided into two categories according to the nature of their water supply. Fens are mires developing in valleys or topographic basins. Part of their water recharge comes from atmospheric precipitation, but the remainder is telluric and it is this part which has the greater ecological effect.... Most bogs, by contrast, are recharged mainly by meteoric water and their surface vegetation is largely isolated from telluric influence."

Non-alluvial wetlands of the southern Blue Ridge vary considerably in the relative influence of seepage and precipitation. The relative influence can vary substantially even along a transect within a single small wetland. Even when telluric water (derived from surrounding rocks and soils) dominates the hydrology of a wetland in the southern Blue Ridge, there is considerable question about the chemical effect that it has. In the southern Blue Ridge, most wetlands are situated over nutrient-poor, felsic, metamorphic or igneous rocks. In general contrast to wetlands of northern Europe and northern North America, the Southern Appalachian landscape is an old one, with soils surrounding wetlands generally consisting of highly-leached, acidic Ultisols, often developed under vegetation dominated by soil-acidifying conifers and heaths. Seepage water is acidic and nutrient-poor.

Some workers have also questioned whether these wetlands have organic soils, and therefore whether they should indeed be considered mires or peatlands at all (Browning, this volume, T. Rawinski, pers. comm.). None of the named soil series for the southern Blue Ridge are classified as Histosols; soil series applied to these wetlands are classified as Inceptisols and Entisols (Browning, this volume). While it is true that the majority of non-alluvial wetlands of the southern Blue Ridge have soils which do not have enough organic carbon to meet the definition of Histosols, some sites, particularly those at medium to high elevations on flat valley bottoms in the northern portion of the southern Blue Ridge, do have soils which should apparently be classified as Histosols (Soil Survey Staff 1975, Soil Management Support Services 1985). For instance, Stewart & Nilsen (1993) studied one of only a few bogs in the mountains of east Tennessee and found 42 % soil organic matter in the top 20 centimeters; this would likely be a Histosol, depending on other factors such as depth of organic layer, whether the organic layer is over a lithic or paralithic contact or over fragmental material (Soil Management Support Services 1985). Based on our experience (but very little data), we believe that non-alluvial wetlands in the southern Blue Ridge span the distinction between organic and mineral soils, with most examples having shallow, organic-rich, mineral soils overlying bedrock.

Very few studies have touched on hydrology, soil chemistry, and water chemistry of non-alluvial wetlands in the southern Blue Ridge. In Mowbray and Schlesinger (1988), the only site studied was the Bluff Mountain Fen, the most minerotrophic mire in the southern Blue Ridge. In Stewart & Nilsen (1993), a single Blue Ridge site (impacted by cattle grazing and nearby agriculture) was compared to four sites on sedimentary rocks in West Virginia. Based on the very few measurements made (reported in published literature and unpublished information), some of these with questionable methods and accuracy, the documented pH's of bog soils in the southern Blue Ridge range from about 4.5 to about 6.5. Given that we lack substantial knowledge about the hydrology and chemistry of these wetlands, and that they do not appear to be closely analogous (in hydrology, chemistry, soils, history, or vegetation) to the better studied wetlands of northern Europe and northern North America, there seems little to be gained scientifically by forcing them into an artificial hierarchy.

All of these wetlands, however, are to one degree or another seepage-fed, and thus would be classified as fens in most mire classifications. The only strictly ombrotrophic wetlands (and therefore the only bogs, according to many mire classifications) in the southeastern United States are the remarkable peat-dome pocosins, primarily of eastern North Carolina. Lacking a better set of terminology (based on studies of the systems themselves), we here continue the use traditional for the region, using "bog" informally to refer to a wide range of moss-, herb-, and shrub-dominated wetlands with extensive *Sphagnum*, and restricting the term "fen" (in the phrase "Southern Appalachian Fen") to the most minerotrophic wetlands.

The most frequently used classification of wetlands in the United States is Cowardin et al. (1979). Non-alluvial wetlands of the southern Blue Ridge all belong to the Palustrine System. Within the Palustrine System, the following subclasses are represented in the southern Blue Ridge: Moss-Lichen Wetlands, Persistent Emergent Wetlands, Nonpersistent Emergent Wetlands, Broad-leaved Deciduous Scrub-Shrub Wetlands, Broad-leaved Evergreen Scrub-Shrub Wetlands, Needle-leaved Evergreen Scrub-Shrub Wetlands,

Broad-leaved Deciduous Forested Wetlands, and Needle-leaved Evergreen Forested Wetlands. This classification, like many hierarchical classifications employing vegetation physiognomy at a high level in the hierarchy, is not very satisfying or useful for Southern Appalachian bogs and other systems which consist of a complex of zones dominated by mosses, lichens, herbs, shrubs (deciduous and evergreen), and trees (deciduous and evergreen). Some wetlands, only a few hectares in size, would have all or nearly all of these subclasses represented. At the scale of mapping employed by the U.S. Fish and Wildlife Service in National Wetland Inventory maps, most non-alluvial wetlands in the southern Blue Ridge should probably be generalized as Palustrine Broad-leaved Deciduous Scrub-Shrub Wetlands, Palustrine Needle-leaved Evergreen Forested Wetlands, or Palustrine Broad-leaved Deciduous Forested Wetlands. Spray Cliffs, vertical wetlands hydrologically maintained by constant mist aerosol from waterfalls, are not explicitly accounted for in the Cowardin classification. Presumably, they would be considered Palustrine Moss-Lichen Wetlands.

3. Community Type Classification

In developing a classification of the diversity of non-alluvial wetlands in the southern Blue Ridge, we have chosen to emphasize a number of characteristics -- vegetational composition, vegetational structure, landscape position, elevation, and oligotrophy/minerotrophy. We believe that such a classification provides a framework for understanding the diversity of non-alluvial wetlands present in the southern Blue Ridge and is useful for conservation planning. A classification based more strictly on vegetation dominants or indicator species, in the tradition of European phytosociology (e.g. Rodwell 1991), would likely recognize many additional types, even in some individual bogs of a hectare or less. Similarly, classifications emphasizing physiognomy of vegetation would recognize numerous communities in many bogs. While such classifications are valid and valuable approaches, they are not as useful for our primary purposes. The classification presented here reflects fairly minor changes from that in Schafale & Weakley (1990), and is based primarily on vegetation data and considerations of landscape position. Statistical analyses of new data (in progress) suggest that further modifications will be needed to best reflect the range of variation of these communities. The Southern Appalachian Bog category (see below) is particularly broad and heterogeneous; further subdivisions would probably help clarify our understanding of the variation among these wetlands. Additional vegetation data, in conjunction with studies of the hydrology and chemistry of these wetlands, would greatly augment our understanding of these communities and would allow development of a more definitive classification.

3.1. SOUTHERN APPALACHIAN FEN

3.1.1. *Physical Characteristics*: The only known example occurs on a nearly flat plateau at an elevation of 1300 meters. Wetland hydrology is maintained by a series of seepages upslope, which provide relatively nutrient-rich waters of circumneutral pH (from percolation through soils derived from amphibolite, a mafic, metamorphic rock). The soils are shallow, organic-rich mineral soils overlying amphibolite bedrock. Although mapped as Toxaway series (a Cumulic Humaquept), they are almost certainly a different, undefined series.

3.1.2. *Vegetation*: Complex zonation of herbaceous vegetation appears to depend on small variations in hydrology and substrate. Species dominant in one or more of the zones include: *Rhynchospora alba, R. capitellata, Juncus subcaudatus, Cladium mariscoides, Carex stricta, Helenium autumnale, Schizachyrium scoparium, Sanguisorba canadensis, Solidago uliginosa*, and *Osmunda regalis* var. *spectabilis*. Other characteristic species include *Huperzia appalachiana, Eriophorum virginicum, Houstonia caerulea, Utricularia cornuta, Osmunda cinnamomea, Liatris aspera, Muhlenbergia glomerata, Tofieldia glutinosa, Carex conoidea, C. buxbaumii*, and *Parnassia grandifolia*. Occasional *Alnus serrulata* and other wetland shrubs occur. Characteristic bryophytes include *Sphagnum subsecundum, Rhytidium rugosum, Hypnum pratense, Campylium stellatum, Calliergon cordifolium*, and *Calliergonella cuspidata*.

3.1.3. *Range*: The only known well-developed example is on Bluff Mountain, Ashe County, North Carolina. A site recently discovered in Grayson County, Virginia has some similarities (T. Rawinski, pers. comm.)

3.1.4. *Distinguishing Features*: The Southern Appalachian Fen is distinguished from Southern Appalachian Bogs by the species composition including numerous northern calciphilic species such as *Muhlenbergia glomerata*, *Tofieldia glutinosa*, and *Sphagnum subsecundum*. A few bogs, occurring over hornblende-rich rocks, have fen-like zones in the center, but the bulk of the area is bog-like.

3.1.5. *Discussion*: Southern Appalachian Fens appear to result from the seepage of nutrient-rich, circumneutral waters, a situation rare in the southern Blue Ridge. Since this type is demonstrably seepage-fed and minerotrophic, it is the most fen-like of Southern Blue Ridge wetland communities. Bluff Mountain is the only well-developed example of the type here called Southern Appalachian Fen, but several other sites, particularly the bogs in Long Hope Valley, the Celo Bog, and fens in "The Glades" of Grayson County, Virginia, show some relationship to the Bluff Mountain Fen in nutrient status and species composition. Undiscovered small fens may exist, but none as large as Bluff Mountain are likely. Fens in other parts of the United States may be similar in some ways, but should not be considered the same natural community type because of their different climate, origins, and composition.

3.1.6. *Examples*: Bluff Mountain, Ashe County, North Carolina (Weakley et al. 1979, Tucker 1967, Mowbray & Schlesinger 1988).

3.2. SOUTHERN APPALACHIAN BOG

3.2.1. *Physical characteristics*: This type occurs at a variety of sites. Some are in flat areas, in portions of valley bottoms that are not subject to flooding. These sites receive little seepage and are presumably largely maintained hydrologically by rainwater and high water table. Others bogs of this type are in the upper portions of stream watersheds, on slight to moderate slopes, hydrologically maintained by very nutrient-poor to nutrient-rich seepage. Soils are organic or organic-rich mineral soils, presumably very acidic to slightly acidic or even circumneutral. In the northern part of the range, sites occur at elevations fróm 800-1400 m. In the southern part of the range, sites occur at elevations from 1000 to 1800 m.

3.2.2. *Vegetation*: The vegetation generally consists of a mosaic or zoned pattern of shrub thickets and herb-dominated areas, much of it underlain by *Sphagnum* mats. Trees such as *Acer rubrum, Pinus strobus, Tsuga canadensis, Pinus rigida*, and *Picea rubens* may be scattered throughout or may dominate in patches or on the edges. Shrubs may include *Alnus serrulata. Rosa palustris, Salix sericea, Aronia arbutifolia, A. melanocarpa, Rhododendron maximum, R. viscosum, R. catawbiense, Kalmia latifolia, K. carolina, Hypericum densiflorum, Lyonia ligustrina* var. *ligustrina, Ilex verticillata, I. collina, Spiraea tomentosa, S. alba*, and *Menziesia pilosa*. The herb layer may include *Carex leptalea, C. folliculata, C. gynandra, C. atlantica, C. echinata, Rhynchospora alba, R. capitellata, Scirpus expansus, S. cyperinus, S. polyphyllus, S. atrovirens, Osmunda cinnamomea, O. regalis* var. *spectabilis, Solidago patula, Senecio aureus, Thelypteris palustris* var. *pubescens, Juncus effusus, Juncus subcaudatus, Lilium grayi, Lysimachia terrestris, Vaccinium macrocarpon, Eriophorum virginianum, Oxypolis rigidior, Parnassia asarifolia, Saxifraga pensylvanica, Sagittaria latifolia* var. *pubescens*, and *Orontium aquaticum*. *Sphagnum* species include *S. palustre, S. affine, S. bartlettianum, S. recurvum*, and, rarely, northern disjuncts such as *S. warnstorfii, S. fallax, S. fuscum, S. subsecundum, S. angustifolium, S. capillifolium, S. subtile*, and *S. flexuosum*. Other important bryophytes include *Polytrichum commune, Rhizomnium appalachianum, Aulacomnium palustre*, and *Bazzania trilobata*.

3.2.3. *Range*: This type is scattered from southwestern Virginia and northeastern Tennessee south through the southern Blue Ridge to southwestern North Carolina. Most examples are north of the Asheville basin; some sites at moderate to high elevations south of the Asheville Basin are included in this type.

3.2.4. *Distinguishing Features*: Southern Appalachian Bogs are distinguished from Swamp Forest-Bog Complexes by their structure. Southern Appalachian Bogs are concentrically or patchily zoned, with herbs or shrubs dominating in the interior. Swamp Forest-Bog Complexes contain small areas of boggy vegetation in a matrix of forest. When both types occur together, *Sphagnum*-dominated areas greater than one acre in size should be considered Southern Appalachian Bogs. Southern Appalachian Bogs are distinguished from Southern Appalachian Fens by species composition, which is apparently correlated with Ph. Distinguishing species include several northern fen indicators such as *Muhlenbergia glomerata*, *Tofieldia glutinosa*, and *Sphagnum subsecundum*. As defined, the only known example of Southern Appalachian Fen is at Bluff Mountain. A few Southern Appalachian Bogs, occurring over hornblende-rich rocks, have fen-like zones along seepage rivulets, but the bulk of the area is more sphagnous and (apparently) more oligotrophic. The distinction between Southern Appalachian Bogs and High Elevation Seeps is not well defined. In general, High Elevation Seeps occur on upper slopes or ridgetops, while Southern Appalachian Bogs occur on non-flooded bottomlands or slope bases. Southern Appalachian Bogs tend to be larger, have well-developed *Sphagnum* mats, and have soils with more organic matter. Seeps tend to be dominated more by forbs, rather than by graminoids and shrubs. Southern Appalachian Bogs are distinguished from French Broad Valley Bogs and Low Mountain Seepage Bogs by floristic differences that correlate with geography and elevation. Southern Appalachian Bogs have a large component of species with northern affinities, including long-distance disjuncts, while French Broad Valley Bogs and Low Mountain Seepage Bogs contain many species more typical of the Coastal Plain (see Table 1). Species which occur in Southern Appalachian Bogs but are absent in French Broad Valley Bogs and Low Mountain Seepage Bogs include *Carex trisperma*, *C. buxbaumii*, *C. oligosperma*, *Rhynchospora alba*, *Ilex collina*, *Vaccinium macrocarpon*, *Saxifraga pensylvanica*. Species which occur in French Broad Valley Bogs and Low Mountain Seepage Bogs but are absent in Southern Appalachian Bogs include *Sarracenia jonesii*, *Sarracenia purpurea*, *Sarracenia oreophila*, *Smilax laurifolia*, *Viburnum nudum*, *Rhododendron arborescens*, *Dulichium arundinaceum*, *Carex collinsii*, *Helonias bullata*, *Gaylussacia dumosa* var. *bigeloviana*, *Chamaedaphne calyculata*, *Myrica gale*, *Juncus caesariensis*, *Rhynchospora rariflora*, *Woodwardia virginica*, and *Woodwardia areolata*.

3.2.5. *Variation*: Southern Appalachian Bogs vary with wetness and amount of organic matter accumulation, both within and among sites. Open areas tend to be more diverse than shrub-dominated areas. Some high elevation sites have northern bog species such as *Vaccinium macrocarpon* and the northern *Sphagnum* species listed above. Nutrient status also modifies species composition and aspect, with some Southern Appalachian Bogs (such as the Celo Bog and the Long Hope Valley bogs) approaching the Southern Appalachian Fen category in nutrients and species composition.

This remains a heterogeneous grouping, with examples occurring on a wide range of rock types, at various elevations, in most parts of the southern Blue Ridge, and in a variety of landscape positions. Some of the more distinctive variants are described below.

Long Hope Valley Variant. Long Hope Valley has at least twenty-three individual bogs, at elevations of 1250-1450 m, developed over amphibolite, a mafic metamorphic rock. They occur on slight to fairly steep slopes, are slightly to strongly seepage fed, generally have shallow soils over bedrock, and are surrounded by Swamp Forest-Bog Complex (Spruce Subtype), Northern Hardwoods Forests, or land cleared from one of these two communities. While they differ from one another considerably in size, slope, landscape position, and physiognomy, they are united by a very northern floristic assemblage. Some of them show a relationship to the Bluff Mountain Fen (also on amphibolite and less than 10 km distant) in their central zones, where rivulets carry relatively nutrient-rich seepage water. Species which are restricted to this variant are

Table 1. Biogeographic affinities of the flora of selected Southern Appalachian Bogs of the Southern Blue Ridge, by percentage of total flora at each site.

	Northern Species	Southern Appalachian Species	Widespread Species	Coastal Plain Species
Long Hope Valley Bogs (Southern Appalachian Bog)	56.6	12.1	31.3	0.0
Bluff Mountain Fen (Southern Appalachian Fen)	43.1	17.2	39.7	0.0
Sugar Mountain Bog (Southern Appalachian Bog)	46.8	19.1	34.1	0.0
Pineola Bog (Southern Appalachian Bog)	48.3	5.2	32.7	13.8
Big Pine Creek Bog (Southern Appalachian Bog)	46.4	10.7	37.5	5.4
Whiteoak Bottoms Bog (Southern Appalachian Bog)	39.1	19.6	35.9	5.4
Panthertown Valley Bog (Southern Appalachian Bog)	33.3	7.8	52.2	6.7
King Creek Bog (French Broad Valley Bog)	17.6	13.2	41.2	28.0
Eller Seepage Bog (Low Mountain Seepage Bog)	19.2	4.8	41.6	34.4

Lonicera canadensis, L. dioica, Menyanthes trifoliata, Taxus canadensis, Utricularia minor, Carex oligosperma, and *Ilex collina*. Other typical species of this variant which are found less frequently in other variants or types include *Cladium mariscoides, Betula allegheniensis, Carex trisperma, Rhynchospora alba, Utricularia cornuta, Hierochloe odorata, Picea rubens, Galium asprellum*, and *Parnassia asarifolia*.

Typic Variant. Bogs of this variant occur at relatively high elevations (from about 1000 to 1250 m) in the northern part of the southern Blue Ridge. Characteristic examples include Pineola Bog (Avery County, North Carolina), Sugar Mountain Bog (Avery

County, North Carolina), Invershiel Bog (Avery County, North Carolina), Beech Creek Bog (Watauga County, North Carolina). A few examples at very high elevations (ca. 1800 m) south of the Asheville Basin, such as the Flat Laurel Gap Bog, are tentatively placed here, though they may be better considered among the boggier examples of High Elevation Seep (see below). Rock types are usually felsic gneisses or schists, and seepage is acidic and nutrient-poor. Many examples are nearly flat and occur in the higher (rarely or never flooded) portions of the floodplains of creeks or small rivers, while others are on gentle slopes and are fed by oligotrophic seepage. Tree species occurring around the bog margin and as scattered, stunted individuals in the bog are characteristically *Picea rubens*, *Pinus strobus*, *Acer rubrum*, and *P. rigida*. At these sites *Picea rubens* occurs below its usual elevational range as an "elevational disjunct." While the flora is not as northern in its affinities as the flora of the Long Hope Variant, it is still a flora dominated by northern species (Table 1). Some examples, especially those with deeper peat (Pineola Bog), have a number of Coastal Plain species, while others (Sugar Mountain Bog) lack Coastal Plain species entirely. Possibly the deeper peat more closely resembles Coastal Plain habitats, or, alternatively, it may reflect a longer and more stable history more favorable for relictual distributions.

Low Elevation Variant. Bogs of this variant occur at lower elevations than the previous two variants (from about 750 to 950 m), in the northern part of the southern Blue Ridge, primarily in Alleghany and Ashe counties, North Carolina. They occur primarily in floodplain situations with little seepage. Their floristic composition is intermediate between the Typic Variant and the Southern Floodplain Variant.

Southern Floodplain Variant. Bogs of this variant occur south of the Asheville Basin at elevations of about 900 to 1200 m. Characteristic examples include Panthertown Bog, Greenland Creek Bog, Horsepasture River Bog, and the several Nantahala River Bogs. All or nearly all examples are nearly flat and occur in the higher (rarely or never flooded) portions of the floodplains of creeks or small rivers, and receive minimal seepage. Tree species occurring around the bog margin and as scattered, stunted individuals in the bog are characteristically *Pinus strobus*, *Acer rubrum*, *Liriodendron tulipifera*, and *P. rigida*; *Picea rubens* is never present.

3.2.6. *Examples*: Long Hope Valley bogs (23 bogs known), Watauga and Ashe counties, North Carolina (Weakley 1993b); Invershiel (Linville Gap) Bog, Avery County, North Carolina; Pineola Bog, Avery County, North Carolina; Celo Bog, Yancey County, North Carolina (McLeod 1983, McLeod & Croom 1983, McLeod 1988); Flat Laurel Gap Bog, Haywood County, North Carolina (Horton & Hotaling 1981, Shafer 1984, Shafer 1986); Quebec Branch Bog, Grayson County, Virginia; Big Wilson Creek Bog, Grayson County, Virginia; Panthertown Bog, Jackson County, North Carolina; Horsepasture River Bog, Jackson County, North Carolina (Smith 1993); Whiteoak Bottoms, Macon County, North Carolina (Gaddy 1981a); Sparta Bog, Alleghany County, North Carolina (Smith 1993).

3.3. FRENCH BROAD VALLEY BOG

3.3.1. *Physical characteristics*: This type occurs in flat or gently sloping areas, generally in valley bottoms that are not subject to flooding. Soils have not been carefully studied, but are likely to be acidic, organic-rich mineral soils developed over gravelly floodplain deposits. Most bog soils are mapped as Toxaway (Cumulic Humaquept), Wehadkee (Typic Fluvaquent), or Hatboro (Typic Fluvaquent) series. The hydrology is palustrine, permanently saturated to intermittently dry. Sites are generally in flat to slightly sloping areas near streams, and receive some seepage from adjacent slopes. The hydrology of this type appears to be less dominated by seepage than are Southern Appalachian Fens, Low Mountain Seepage Bogs, High Elevation Seeps, and most variants of Southern Appalachian Bogs. This type ranges from 500 meters to 700 meters in elevation.

3.3.2. *Vegetation*: This type supports a mosaic or zoned pattern of shrub thickets and herb-dominated areas, much of it underlain by *Sphagnum* mats. Trees such as *Acer rubrum, Pinus strobus, P. rigida, Nyssa sylvatica, Liriodendron tulipifera,* and *Tsuga canadensis* may be scattered throughout or may dominate on the edges. Shrubs may include *Alnus serrulata, Rosa palustris, Salix sericea*, *Aronia arbutifolia, Myrica gale,*

Gaylussacia dumosa var. *bigeloviana, Rhododendron maximum, R. viscosum, R. arborescens, Viburnum nudum, V. cassinoides, Kalmia latifolia, K. carolina, Hypericum densiflorum, Lyonia ligustrina* var. *ligustrina, Ilex verticillata*, and *Menziesia pilosa*. The woody vine *Smilax laurifolia* is often present, climbing high into the shrubs and trees. The herb layer may include *Carex leptalea, C. echinata, C. folliculata, C. gynandra, C. collinsii, Scirpus cyperinus, Osmunda cinnamomea, O. regalis* var. *spectabilis, Solidago patula, Senecio aureus, Thelypteris palustris* var. *pubescens, Isoetes caroliniana, Juncus effusus, J. caesariensis, Drosera rotundifolia, Woodwardia virginica, Woodwardia areolata, Dulichium arundinaceum, Sarracenia purpurea, S. jonesii, Sagittaria fasciculata, Eriophorum virginianum*, and *Parnassia asarifolia*. *Sphagnum* species include *S. palustre, S. affine, S. bartlettianum*, and *S. recurvum*.

3.3.3. *Range*: French Broad Valley Bogs are found at low elevations in the broad basin of the French Broad River in southern Buncombe, Henderson, and eastern Transylvania Counties, North Carolina. A concentration formerly occurred in Henderson County, North Carolina, but nearly all of these bogs have been destroyed by drainage.

3.3.4. *Distinguishing Features*: French Broad Valley Bogs are distinguished from Swamp Forest-Bog Complexes by their structure. French Broad Valley Bogs, like Southern Appalachian Bogs, are concentrically or patchily zoned, with herbs or shrubs dominating in the interior, though zonation is usually less well developed and overall this type is shrubbier. Swamp Forest-Bog Complexes contain small areas of boggy vegetation in a matrix of forest.

French Broad Valley Bogs are distinguished on the one hand from Southern Appalachian Bogs and on the other hand from Low Mountain Seepage Bogs by floristic differences that correlate with geography and elevation. Southern Appalachian Bogs have a number of northern disjunct species, while this French Broad Valley Bogs often contain species typical of the Coastal Plain (see Table 1). Species which occur in Southern Appalachian Fens and Southern Appalachian Bogs but are absent in French Broad Valley Bogs include *Carex trisperma, C. buxbaumii, Rhynchospora alba, Filipendula rubra, Dryopteris cristata, Thelypteris simulata, Spiraea alba, Schizachyrium scoparium, Lilium grayi, Pogonia ophioglossoides, Juncus subcaudatus, Ilex collina, Picea rubens, Vaccinium macrocarpon*, and *Saxifraga pensylvanica*. Species which occur in the French Broad Valley Bogs but are absent in Southern Appalachian Bogs and Southern Appalachian Fens include *Sarracenia jonesii, S. purpurea, Smilax laurifolia, Leucothoe racemosa, Viburnum nudum, Rhododendron arborescens, Dulichium arundinaceum, Carex collinsii, Helonias bullata, Woodwardia virginica*, and *W. areolata*. Species which occur in French Broad Valley Bogs but not in Low Mountain Seepage Bogs include *Carex collinsii, Chamaedaphne calyculata, Eleocharis tortilis, Gaylussacia dumosa var. bigeloviana, Myrica gale, Sarracenia jonesii*, and *Viburnum nudum*. See Low Mountain Seepage Bog for a list of species which occur in it alone.

3.3.5. *Discussion*: One of the most interesting features of French Broad Valley Bogs is the presence of a suite of species that are disjunct from a more northern Coastal Plain distribution. Most of these species do not occur at all in other bog types, though a few have limited occurrence in other bog types. These species include *Myrica gale, Chamaedaphne calyculata, Gaylussacia dumosa* var. *bigeloviana, Helonias bullata, Juncus caesariensis, Carex collinsii, C. barrattii, Narthecium americanum, Xerophyllum asphodeloides, Sarracenia purpurea*, and *Triadenum virginicum*.

3.3.6. *Examples*: King Creek Bog, Henderson County, North Carolina; Etowah Bog, Henderson County, North Carolina; East Flat Rock Bog, Henderson County, North Carolina (formerly the best example, now essentially destroyed).

3.4. LOW MOUNTAIN SEEPAGE BOG

3.4.1. *Physical characteristics*: Known examples occur on shallow slopes, at about 500 meters elevation, and have a palustrine hydrology, fed by acidic seepage.

3.4.2. *Vegetation*: Owing to disturbance of the only known examples, the original vegetation structure of this community type is unknown, but was probably primarily shrubby. Shrub species include *Alnus serrulata, Lyonia ligustrina, Aronia arbutifolia, A. melanocarpa, Rhododendron arborescens, Rosa palustris*, and *Sambucus canadensis*. Typical herb species include *Osmunda cinnamomea, Rhynchospora rariflora, Sarracenia oreophila, Eriocaulon decangulare, Thelypteris palustris* var. *pubescens, Sagittaria latifolia* var. *pubescens, Rhexia virginica, R. mariana, Eryngium integrifolium, Helianthus angustifolius, Eupatorium perfoliatum, E. pilosum, E. pubescens, E. fistulosum, Eriophorum virginicum, Sanguisorba canadensis*, and *Juncus caesariensis*.

3.4.3. *Range*: This type is known only from Clay County, North Carolina and Towns County, Georgia. It should be sought in nearby counties.

3.4.4. *Distinguishing features:* This type has numerous species which do not occur in other bog types in the southern Blue Ridge. Most of these species are more typical of Coastal Plain wetlands: *Sarracenia oreophila, Andropogon glomeratus, Aster dumosus, Cinna arundinacea, Eryngium integrifolium, Fuirena squarrosa, Helianthus angustifolius, Rhynchospora gracilenta, R. rariflora, Scleria ciliata, S. muhlenbergii, Gratiola pilosa, Xyris jupicai, Polygala cruciata, Drosera capillaris, Erianthus giganteus, Eupatorium pilosum, Juncus canadensis*, and *Panicum virgatum*. Many additional species occur in other bog types only very rarely or atypically, such as *Eriocaulon decangulare, Eleocharis tuberculosa*, and *Betula nigra*. While it shares some species with French Broad Valley Bogs, this type is not closely related.

3.4.5. *Discussion*: Fire may have been a natural disturbance in this community type. This is suggested by the large suite of species more typical of fire-maintained communities of the Coastal Plain. It is also possible that this suite of species is present in this community because of its low elevation and phytogeographic history.

3.4.6. *Examples*: Eller Seepage Bog, Clay County, North Carolina (Govus 1985, Govus 1987); Reed Creek Seepage Bog, Towns County, Georgia (Dennis 1980).

3.5. HIGH ELEVATION SEEP

3.5.1. *Physical characteristics:* High elevation seeps occur at a range of high elevation sites, in generally sloping areas, and receive nearly constant (or at least, regular) seepage. Soils are generally rocky, gravelly, or (less commonly) mucky soils. Their hydrology is palustrine, permanently saturated to intermittently dry, and varies in chemistry depending on the rock type. These communities occur at elevations from about 1200 m to about 2000 m.

3.5.2. *Vegetation*: The vegetation consists generally of an open to dense bed of wetland herbs. Seeps are often small enough that they may be substantially shaded by trees rooted in adjacent communities, but some are extensive and open, with trees scattered or absent. Herbs include *Chelone lyonii, Chelone cuthbertii, Veronica americana, Diphylleia cymosa, Saxifraga micranthidifolia, S. careyana, Cardamine clematitis, Parnassia asarifolia, Helenium autumnale, Chrysosplenium americanum, Boykinia aconitifolia, Polygonum sagittatum, Drosera rotundifolia, Cicuta maculata, Rhynchospora capitellata, Juncus subcaudatus, Carex leptalea, C. gynandra, Houstonia serpyllifolia, Aster puniceus, Impatiens pallida, Impatiens capensis, Hypericum prolificum, H. buckleyi, H. mitchellianum, H. graveolens, Viola cucullata, V. macloskeyi* ssp. *pallens, Hydrocotyle americana, Aconitum reclinatum, Hydrophyllum canadense, Monarda didyma, Solidago patula, Lycopus uniflorus, Rudbeckia laciniata, Houstonia serpyllifolia, Veratrum viride, Lilium grayi, Thalictrum clavatum, T. dioicum*, and *Trautvetteria carolinensis*. *Sphagnum* is often present and may occasionally have significant cover. Woody species include *Rhododendron maximum, Kalmia latifolia, Viburnum cassinoides, Lyonia ligustrina* var. *ligustrina*, and *Vaccinium corymbosum*. *Acer rubrum, Betula alleghaniensis, Picea rubens*, and *Amelanchier arborea* are common trees.

3.5.3. *Range*: High Elevation Seeps are scattered in small occurrences throughout the mountains at high elevations; they are fairly common, but never extensive. Presence is related to strike and dip of metamorphic foliation or fractures in the rocks.

3.5.4. *Distinguishing Features*: The distinction between High Elevation Seeps and Southern Appalachian Bogs is not well defined. In general, High Elevation Seeps occur on upper slopes or ridgetops, while Southern Appalachian Bogs occur on non-flooded bottomlands or slope bases (though sometimes at high elevations in hanging valleys). Southern Appalachian Bogs tend to be larger, to have well-developed *Sphagnum* mats, and to have a definite organic layer. Seeps tend to be more dominated by forbs, rather than by graminoids and shrubs, and to have bedrock or broken rock substrates which make microsite conditions heterogeneous. Seeps are less often strongly zoned.

3.5.5. *Variation*: This is currently a very diverse grouping, including a wide variety of small upland wetlands occurring within high elevation mountain communities. Division into subtypes is probably desirable, but further study will be required. Examples vary with underlying rock, topographic position, regularity and amount of seepage, elevation (climate), and soil development. Some examples show gradation to Southern Appalachian Bog.

3.5.6. *Discussion*: The classification of mountain wetlands is still somewhat tentative, because of their variable vegetation and because little is known about their hydrology and nutrient dynamics. High Elevation Seeps and Southern Appalachian Bogs grade conceptually into each other, although they are seldom associated at the same site and are very distinct at their extremes.

3.5.7. *Examples*: Roan Mountain, Mitchell County, North Carolina; Big Yellow Mountain, Avery County, North Carolina; Grandfather Mountain, Avery and Watauga counties, North Carolina; Bluff Mountain, Ashe County, North Carolina; Andrews Bald, Swain County, North Carolina; Long Hope Valley, Ashe and Watauga counties, North Carolina (Weakley 1993b); Ivestor Gap Cranberry Seep, Haywood County, North Carolina; Fork Ridge Balds and Seeps, Haywood County, North Carolina (Gaddy 1981b).

3.6. SWAMP FOREST-BOG COMPLEX (TYPIC SUBTYPE)

3.6.1. *Physical Characteristics*: This community type occurs in poorly drained bottomlands, generally with visible microtopography of ridges and sloughs or depressions. The soils show an alluvial origin, but are now removed from regular flooding. They are, however, seasonally to semipermanently saturated, apparently primarily from a high water table. Some examples also receive seepage from adjacent slopes.

3.6.2. *Vegetation*: Most sites are primarily forested, with a closed or open canopy and an open or dense shrub layer, interspersed with small boggy openings in depressions. *Tsuga canadensis* or *Acer rubrum* are usually the dominant trees. Other trees include *Salix nigra*, *Betula lenta*, *B. alleghaniensis*, *Quercus alba*, *Pinus strobus*, and various other alluvial species. The dominant shrubs are usually *Rhododendron maximum*, *Kalmia latifolia*, and *Leucothoe fontanesiana*. Common shrubs include *Salix sericea*, *Alnus serrulata*, *Ilex montana*, *Cornus amomum*, *Viburnum cassinoides*, and *Toxicodendron vernix*. Herbs in boggy openings include *Solidago patula*, *Aster novae-angliae*, *Dalibarda repens*, *Osmunda cinnamomea*, *Carex folliculata*, *C. gynandra*, *C. scabrata*, *C. leptalea*, *C. stricta*, *Sarracenia purpurea*, *Sagittaria latifolia* var. *pubescens*, and *Leersia virginica*. Herbs in the forest include *Glyceria melicaria*, *Lycopodium obscurum*, *Onoclea sensibilis*, *Maianthemum canadense*, *Thelypteris noveboracensis*, and *Osmunda regalis* var. *spectabilis*.

3.6.3. *Range*: This type is scattered throughout the southern Blue Ridge, sometimes associated with more open bogs, but often not.

3.6.4. *Distinguishing Features*: Swamp Forest-Bog Complexes are distinguished from

Southern Appalachian Bogs by their structure, which consists primarily of forested thickets with only small boggy openings. Boggy areas in Swamp Forest-Bog Complexes are less than one acre in size. Swamp Forest-Bog Complexes are distinguished from Montane Alluvial Forests and Acidic Cove Forests by being wetter, having open boggy vegetation in small depressions, and having scattered *Sphagnum* mats. The Floodplain Pool type occurs in deeper bottomland depressions, containing standing water for much or all of the year, and lacking dense boggy vegetation. The Typic Subtype may be distinguished from the Spruce Subtype by the composition of the forest canopy, which consists of *Tsuga canadensis*, *Acer rubrum*, and other lower elevation trees but not of *Picea rubens*.

3.6.5. *Variation*: Examples of this type vary with elevation and hydrology. Sites vary especially in the relative amounts of closed forest, shrubby openings, and boggy openings. Examples at lower elevations (especially south of the Asheville Basin) tend to be dominated by *Acer rubrum*, *Liriodendron tulipifera*, or *Nyssa sylvatica*, while examples at higher elevations (especially north of the Asheville Basin) are usually dominated by *Tsuga canadensis*. Such differences suggest that additional types, subtypes or variants should be recognized to better describe variation within this group. These communities have received almost no study -- even less than the more herbaceous bog and fen communities. More information is needed for a better understanding of these communities.

3.6.6. *Discussion*: The factors responsible for creating and maintaining these communities are not well known. Gaddy (1981a) suggested they were caused by paludification following tree blowdown or logging in wet alluvial forests. Some examples appear to be very old, however, and most logged bottomlands do not contain boggy vegetation. The boggy openings are generally associated with small depressions. They may be successional remnants of once more extensive bog areas. As in some Southern Appalachian Bogs, beaver activities may be a significant factor in these communities. The frequency and role of flooding in these communities is not known. They often occur near streams and undoubtedly are occasionally flooded. Some occur near the outer edge of floodplains and also receive seepage water from adjacent slopes. Others may receive seepage water flowing through Southern Appalachian Bogs.

3.6.7. *Examples*: Nantahala and Big Indian Creek Bogs, Macon County, North Carolina (Gaddy 1981a); Tulula Bog, Graham County, North Carolina (Gaddy 1981a); Etowah Swamp, Henderson County, North Carolina (Wickland & Horton 1977); Ochlawaha Bog, Henderson County, North Carolina; Panthertown Valley, Jackson County, North Carolina.

3.7. SWAMP FOREST-BOG COMPLEX (SPRUCE SUBTYPE)

3.7.1. *Physical Characteristics*: This community type occurs on poorly drained bottomlands of small streams at high elevations. The soils have been mapped as the Toxaway series (Cumulic Humaquept), but that classification should be considered preliminary. Soils are seasonally to semipermanently saturated. The frequency of flooding is not known, but appears to be very rare. Some areas also receive groundwater seepage from adjacent slopes. The best examples, in the hanging valley of Long Hope Creek, are at 1300 to 1400 m elevation.

3.7.2. *Vegetation*: This type supports forest with a closed or open canopy and open or dense shrub layer, interspersed with small, open, boggy patches in slight depressions. *Picea rubens* is the dominant tree, with *Tsuga canadensis*, *Betula alleghaniensis*, *Acer rubrum*, *Amelanchier arborea*, and other species sometimes present. A dense shrub layer of *Rhododendron maximum* and *Kalmia latifolia* is usually present. Other shrubs may include *Ilex verticillata*, *I. collina*, *Taxus canadensis*, *Viburnum cassinoides*, *Aronia melanocarpa*, and *Vaccinium* spp. Herbs are generally sparse under the canopy but may be dense in openings. Herb species include *Glyceria melicaria*, *Osmunda cinnamomea*, *O. regalis* var. *spectabilis*, *Maianthemum canadense*, *Carex trisperma*, *C. folliculata*, *Listera smallii*, and various species of the Southern Appalachian Bog type. *Sphagnum* patches may occur scattered beneath the canopy as well as in small depressions.

3.7.3. *Range*: Several examples of this type are scattered in the Southern Blue Ridge.

3.7.4. *Distinguishing Features*: Swamp Forest-Bog Complexes are distinguished from Southern Appalachian Bogs by their structure, which consists primarily of forested thickets with only small boggy openings. Boggy areas are less than one acre in size. They are distinguished from Red Spruce Forests (Schafale & Weakley 1990) by being wetter and having boggy openings and scattered *Sphagnum* mats. They also are generally at somewhat lower elevation than Red Spruce Forest. The Spruce subtype may be distinguished from the Typic Subtype by the composition of the forest, with *Picea rubens* as the dominant tree.

3.7.5. *Discussion*: The factors responsible for creating and maintaining these communities are not well known. Occurrence of spruce at unusually low elevations and the occurrence of northern disjunct species suggests that they are relicts from the Pleistocene glacial period, persisting in specialized environments. They may, however, represent a late stage of primary succession from more extensive open bogs.

3.7.6. *Examples*: Long Hope Valley, Ashe and Watauga counties, North Carolina (Weakley 1993b); Alarka Laurel, Swain County, North Carolina.

3.8. SPRAY CLIFF

3.8.1. *Physical Characteristics*: Spray cliffs occur on vertical to gently sloping rock faces that are constantly wet from the spray of waterfalls. Small pockets or mats of mineral or organic matter are interspersed with bare rock. These sites are palustrine, permanently saturated by spray, with or without seepage water.

3.8.2. *Vegetation*: Vegetation consists of a variable collection of mosses, liverworts, algae, vascular herbs, and occasional shrubs, most of them requiring constantly moist substrate and very high relative humidity. Many of the typical species of this community are bryophytes and ferns disjunct from tropical regions (Anderson & Zander 1973). There are also many endemic bryophytes. Vascular species include *Huperzia porophila, Asplenium montanum, A. trichomanes, A. rhizophyllum, A. monanthes, Cystopteris protrusa, Polypodium virginianum, Trichomanes boschianum, T. intricatum, Grammitis nimbata, Vittaria appalachiana, Hymenophyllum tayloriae, Phegopteris connectilis, Adiantum pedatum, Saxifraga careyana, S. caroliniana, Heuchera parviflora, Circaea alpina, Impatiens capensis, Houstonia serpyllifolia, Hydrocotyle americana, Thalictrum* spp., *Oxalis montana, Carex biltmoreana, Galax urceolata, Tsuga canadensis, Rhododendron maximum*, and *Kalmia latifolia*. Bryophyte species, many of them nearly or entirely limited to this community, include *Sphagnum quinquefarium, S. girgensohnii, Plagiomnium carolinianum, Mnium affine, M. marginatum, Isopterygium distichaceum, Bryocrumia vivicolor, Flakea papillosa, Hookeria acutifolia, Thamnobryum alleghaniense, Oncophorus raui, Hyophila involuta, Dichodontium pellucidum, Radula* spp., *Plagiochila sharpii, P. caduciloba, P. sullivantii, P. austinii, Fissidens osmundioides, Bazzania denudata, Conocephalum conicum, Pellia epiphylla, P. neesiana*, and *Riccardia multifida*.

3.8.3. *Range*: This type is scattered throughout the mountains, and is rare in the upper Piedmont. Spray Cliffs are most frequent and best-developed in the southern Blue Ridge Escarpment region of Transylvania and Jackson counties, North Carolina and adjacent portions of Oconee and Pickens counties, South Carolina and Rabun County, Georgia.

3.8.4. *Distinguishing Features*: Spray Cliffs are distinguished from other cliff communities by their association with waterfalls and their constant or near-constant wetness. Other cliff communities may have seepage areas, but the cliff as a whole is dry, is not subject to spray, and is frequently exposed to low humidity. Spray Cliffs are distinguished from forest communities by being steep, rocky, or wet enough to lack a closed tree canopy.

3.8.5. *Variation*: Examples of this type vary considerably, depending on amount and dependability of spray, elevation, rock type, orientation of rocks, degree of shading, and past and present climate. Some examples have well developed herb or bryophyte mats, while others are nearly barren. The most diverse Spray Cliffs are found in the Blue Ridge Escarpment gorges of Transylvania, Jackson, and Macon counties.

3.8.6. *Discussion*: This community occurs in unusually stable and equitable environments. The humidity is high and moisture supply is essentially constant. Temperatures are moderated by water, rock, and sheltering from sun and wind, resulting in only rare freezes or high temperatures (Billings & Anderson 1966, Bruce et al. n.d.). Potential disturbances include extreme droughts or freezes that may result in some die-off of sensitive species. Floods or rock falls may damage some parts, but in general Spray Cliffs are well sheltered from physical disturbance. This community type is considered distinct from other cliff communities (even those wetted by seepage), because of the very distinctive flora, featuring many endemic or tropically disjunct pteridophytes and bryophytes. Spray Cliffs differ from cliffs with seepage in having a more constant water supply, higher humidity in the air, and a more strongly moderated climate.

3.8.7. *Examples*: Upper Whitewater Falls, Whitewater River, Jackson and Transylvania counties, North Carolina; Rainbow Falls and Windy Falls, Horsepasture River, Transylvania County, North Carolina; Bearwallow Falls, Toxaway Falls, and Rock Creek Falls (Toxaway River tributaries), Transylvania County, North Carolina; Dry Falls, Crow Creek Falls, and Cullasaja Falls (Cullasaja River and its tributaries), Macon County, North Carolina; Scotsman Creek Falls, and other falls (Chattooga River tributaries), Macon and Jackson counties, North Carolina, Oconee County, South Carolina, Rabun County, Georgia; Schoolhouse Falls, Panthertown Valley, Jackson County, North Carolina; Linville Falls, McDowell County, North Carolina; Rocky Bottom Creek, Gorge of Eastatoe Creek (Eastatoe Creek tributaries), Pickens County, South Carolina.

3.9. UPLAND POOL

3.9.1. *Physical Characteristics*: This community type occurs in small upland depressions where water is ponded by an impermeable substrate, such as shallow bedrock. These areas are too small to be distinguished in standard soil mapping and most occur in areas that have not been mapped. The soils generally have a mucky surface layer and have a shallow clay hardpan or rock layer that prevents drainage. They are seasonally to semipermanently flooded, with rainfall apparently the main source of water, although some have small watersheds.

3.9.2. *Vegetation*: This community type is dominated by various wetland shrubs and herbs. Trees such as *Nyssa sylvatica*, *Quercus phellos*, *Acer rubrum*, and *Liquidambar styraciflua* may be present on the edge or scattered in the center. Shrub species include *Cephalanthus occidentalis*, *Vaccinium* spp., and *Leucothoe racemosa*. Herbs include *Osmunda regalis* var. *spectabilis*, various *Carex* species, *Juncus effusus*, and a variety of *Sphagnum* species, including *S. cuspidatum*, *S. palustre*, and *S. recurvum*.

3.9.3. *Dynamics*: Upland Pools are apparently stable over long periods, but presumably slowly fill with sediment or organic matter. They may tend to succeed to Upland Depression Swamp Forest (Schafale & Weakley 1990) over time. An ephemeral drawdown community may occur when water levels drop. Extended droughts may be necessary for establishment of some species.

3.9.4. *Range*: In the Blue Ridge, this community type is extremely rare, known only from very few scattered sites. This community has hydrologic affinities to depressional wetlands over sandstone on ridges in the Cumberland Plateau of Kentucky. Tennessee, and Alabama. The floristic affinities are unstudied.

3.9.5. *Distinguishing Features*: Upland Pools are easily distinguished from all other wetlands in the southern Blue Ridge. They have standing water for significant parts of the year and are kept wet by poor drainage of rainwater and local runoff rather than by seepage. They usually have a pronounced seasonal fluctuation in water level, filling in the winter and often drying completely in the summer. Although *Sphagnum* may be present, Upland Pools lack the peaty mats and associated bog species found in the bog types.

3.9.6. *Variation*: As presently defined this is a heterogeneous and somewhat discordant category which includes a variety of shrub- and herb-dominated "upland wetlands" with very different substrates. Factors affecting variation include depth of water, clay or rock substrate, and geographic location. Sites may have disjunct Coastal Plain species,

3.9.7. *Discussion*: This community type is bog-like only in the sense of being shrubby or open, non-alluvial wetlands. It does not have well-developed organic mats and lacks most bog species. This is an extremely rare type, with only a few known examples. Upland pools are often important breeding habitats for amphibians.

3.9.8. *Examples*: Linville Mountain Pond, Burke County, North Carolina; Shortoff Mountain Pond, Burke County, North Carolina; North Fork Watershed Pond, Buncombe County, North Carolina.

4. Distribution and Abundance

In North Carolina, non-alluvial mountain wetlands are found in all mountain counties. Their distribution is strongly skewed to particular regions, however, where topographic settings, elevation, and climate are more suitable. The more gentle topography and colder climate of Alleghany and Ashe counties have promoted the formation of bogs there, most of which have now been degraded by ditching and pasturing. Flat second terraces of the French Broad River and its tributaries in Henderson County, southern Buncombe County, and eastern Transylvania County once supported the largest concentration and acreage of non-alluvial wetlands in western North Carolina; flat land is at a premium in mountain counties, and nearly all of these wetlands have been ditched, drained, and converted to truck-farming fields, golf courses, industrial parks, and residential areas.

Secondary concentrations of non-alluvial wetlands are found in Avery and Watauga counties in northwestern North Carolina, and in Macon, Jackson, and Transylvania counties in southwestern North Carolina. Fewer non-alluvial wetlands occur in the Blue Ridge Province in the adjoining states of South Carolina, Georgia, and Tennessee, but the concentration (at least formerly) of bogs in Ashe and Alleghany counties, North Carolina extends northwards into Grayson County, Virginia.

Both the original (pre-settlement) and remaining extent of non-alluvial wetlands are difficult to determine. Because non-alluvial wetlands occur typically as small acreages surrounded by uplands they are often overlooked. Draft National Wetlands Inventory maps (based on interpretation of aerial photographs with limited ground-truthing) show less than 50 % of known sites, though field-oriented inventories such as those conducted in North Carolina have provided information to update and improve the accuracy of the mapping (see Hefner, this volume, for further discussion). Soil Conservation Service soil surveys, especially the older ones, also often fail to map these wetlands, treating them (because of their small acreage) as inclusions in other soil types. Older surveys also grossly generalized the classification of wetland soils in the Blue Ridge, treating all mappable wetlands (no matter how different) in one or two series (such as Toxaway), probably because once a soil was determined to be wet and "undesirable," there was little need to further consider its properties (see Browning, this volume, for further discussion).

Despite the difficulties, we have estimated the original and remaining extent of non-alluvial wetlands in the North Carolina Blue Ridge through use of a combination of soils maps, National Wetlands Inventory maps, information on known sites with relatively intact vegetation, information on drained or otherwise degraded sites, general observations of the landscape, and historical records of the presence of species indicative of a particular type of vegetation. For instance, *Sarracenia jonesii* is endemic to a small area along the North Carolina-South Carolina border. Herbarium records and interviews with botanists active in the 1930's document about twenty-five separate populations for this species in the general vicinity of Hendersonville, North Carolina; *S. jonesii* has been reduced in the last half-century to four populations. The only known habitat for it in the Hendersonville area (and the habitat in which it has always been described as occurring) is the French Broad Valley Bog type (as described below; partly equivalent to the Southern Appalachian Bog, Southern Subtype of Schafale & Weakley 1990). By examination of the mapped occurrences, accounts of earlier botanists, and examination of soils and wetland maps, one can deduce with reasonable accuracy the original extent of this wetland type in Henderson County and vicinity.

Another example is the frequent occurrence in Alleghany and Ashe counties of patches of *Juncus* and other wetland plants in what are now pastures. Such sites have been termed Meadow Bogs by Smith (1993) and Marsh-Bog Complexes by Schafale & Weakley (1985). Such areas almost certainly represent highly degraded remnants of either Southern Appalachian Bogs or Swamp Forest-Bog Complexes (Typic Subtype), and indeed some of them support populations of rare plant or animal species characteristic of one or both of those communities. Despite their open, sunny condition, the only species present at these sites which are characteristic of the more open community type (Southern Appalachian Bog) are those with good dispersal capabilities (such as the wind-dispersed plant *Epilobium leptophyllum* and the surprisingly mobile turtle *Clemmys muhlenbergi*). Based on this fact, the topographic situation and apparent hydrology of the sites, and the apparently greater rarity of Southern Appalachian Bogs in comparison to Swamp Forest-Bog Complexes, we believe the majority of "Meadow Bogs" were originally Swamp Forest-Bog Complexes, some of which have later been colonized by a few "open bog" species. Some "Meadow Bogs," however, were undoubtedly Southern Appalachian Bogs.

In the Blue Ridge of North Carolina, the acreage of non-alluvial wetlands with moderately intact vegetation is estimated to have been reduced from about 2000 ha to a bit more than 300 ha, an approximately six-fold loss (Table 2). Some of the types (described below) have experienced very little loss, either because they are very rare types with only one or a few occurrences which happen not to have been strongly altered (Southern Appalachian Fen, Spruce Subtype of Swamp Forest-Bog Complex, Upland Pool), or because they occur at elevations or in topographic settings that are less likely to have been extensively altered (High Elevation Seep, Spray Cliff). The main bog types (Northern and Southern Subtype of Southern Appalachian Bog, Typic Subtype of Swamp Forest-Bog Complex), however, tend to occur in relatively flat topographic situations, often along the toe of the slope in stream or small river valleys, and often at low to moderate elevations. In the mountains of North Carolina, such sites are among the most altered, by agriculture, pasturage, road construction, urbanization, and tourist development.

The particularly dramatic loss of the French Broad Valley Bog (from ca. 500 to ca. 5 hectares) is a result of the loss of its formerly extensive acreage concentrated along the French Broad River and its tributaries (Mud Creek, King Creek) in the vicinity of Hendersonville, East Flat Rock, Brickton, and Etowah, North Carolina. From the 1930's to the 1960's, this area was ditched and drained by the Soil Conservation Service, including channelization of many of the streams. These areas were then used for crop-farming and pasturage. More recently, additional areas have been lost to urbanization, suburbanization, and golf course development.

5. Factors Determining Variation

In our earlier classification (Schafale & Weakley 1990), we emphasized elevation, location within the state, source of wetland hydrology, and underlying rock type as the primary factors controlling the diversity of non-alluvial communities of the Southern Blue Ridge. Recent analyses (in progress and to be presented in a future paper) validate these factors as the most important determiners of variation. Elevation appears to be the most important factor, location within the state next most important, and underlying rock type third most important. Wiser (1993) found these same factors to be among the most important in explaining variation in high-elevation rock outcrops, another group of relictual Southern Appalachian communities occurring as herb- or shrub-dominated

Table 2. Loss of Natural Vegetation in Non-alluvial Wetlands in the North Carolina Blue Ridge.

NATURAL COMMUNITY TYPE	ESTIMATED ORIGINAL AREA IN NORTH CAROLINA (ha)	ESTIMATED CURRENT AREA IN NORTH CAROLINA (ha)	% OF ORIGINAL REMAINING
Southern Appalachian Fen	1	1	100 %
Southern Appalachian Bog	300	70	23 %
French Broad Valley Bog	450	5	1 %
Low Mountain Seepage Bog	20	2	10 %
Swamp Forest-Bog Complex, Typic Subtype	1000	100	10 %
Swamp Forest-Bog Complex, Spruce Subtype	50	45	90 %
High Elevation Seep	100	90	90 %
Spray Cliff	5	5	100 %
Upland Pool	5	5	100 %
Total	1931	323	17 %

"islands" in the matrix of a largely forested landscape.

Non-alluvial wetlands occur in the southern Blue Ridge at elevations ranging from roughly 500-2000 m. A few non-alluvial wetlands have been found in the neighboring Piedmont, and these show some relationship to the lowest elevation wetlands of the Blue Ridge. Climate is, of course, strongly correlated with elevation, and some of the variation in non-alluvial wetland communities is closely correlated to elevation. For instance, Coastal Plain species are much more prevalent at lower elevations. While some of this variation is probably related to physiological capabilities of the species, the phytogeographic history of the region may be more important. For instance, *Sarracenia purpurea* is a widespread species in bogs in northeastern United States and Canada, north to Labrador and Mackenzie, where it experiences colder conditions than occur even at the highest elevations in the south; its absence from bogs in the northern portion of the Southern Appalachians (north of the Asheville Basin) is more likely related to unknown phytogeographic factors. Geographic location within the southern Blue Ridge, though partially correlated with elevation and rock type, appears also to be an independent factor in determining the vegetation of wetland sites.

Landscape position and hydrology are clearly important factors. Upland Pools have seasonally ponded water on flat ridge tops. Spray Cliffs are steep, vertical, or overhanging, their water source being a constant bathing with a fine aerosol of water droplets from waterfalls, sometimes supplemented by seepage. High Elevation Seeps are always on relatively steep slopes and receive water from strong sources of seepage. Swamp Forest-Bog Complex types are on flat or very slightly sloping sites, usually in the infrequently flooded portions of floodplains of small to large streams or small rivers in mountain valleys. The only example of Southern Appalachian Fen occurs on a slightly sloping shelf on the upper portion of a mountain, fed by multiple seepages. Southern Appalachian Bogs occur in a wide variety of landscape positions, and with various sources of wetland hydrology. This variation (more or less correlated with variation in vegetation structure and composition) may be the basis for the future recognition of additional types or subtypes (see discussion of variation in Southern Appalachian Bogs, above).

Soils and geology are also responsible for variation within non-alluvial wetlands of the southern Blue Ridge, but this source of variation is completely unstudied. Among bog types, a few important factors of soil and geology can be deduced. The difference between bogs developed over mafic and felsic rocks is apparent, with the Southern Appalachian Fen type restricted to mafic rocks (amphibolite in this case), the Long Hope Valley Variant of Southern Appalachian Bog also occurring over amphibolite, and nearly all other bogs developed over felsic rocks. Seepage from mafic rocks is presumably higher in Ph, conductivity, and cations (such as calcium and magnesium) than seepage from felsic rocks, and this has effects on soil chemistry, nutrient availability, and species composition.

Soil depth and origin differ tremendously. Some non-alluvial wetlands have developed on alluvial deposits, but are now removed from the reach of flooding by stream down-cutting or levee deposition. These wetlands often have deep soils. For example, the Pineola Bog (Avery County, North Carolina), which is in the floodplain of the Linville River, may be on an old cutoff oxbow and has peaty soils apparently over two meters in depth. Other wetlands, fed by seepage and especially those occurring on substantial slopes, have very shallow soils with bedrock near the surface. Depth to bedrock in southern Blue Ridge bogs varies from less than 30 cm to well over 200 cm. Bogs with thin soils over bedrock generally lack trees and shrubs, presumably because of freeze-thaw cycles or alternating drought and flooding, and typically show a regular zonation of various herbs outwards from a rivulet to a narrow shrub-dominated zone. Bogs with deeper soils tend to have a complex mosaic of herb-, shrub-, and tree-dominated patches, rather than regular zonation.

6. Biogeographic Affinities

In order to understand non-alluvial wetlands of the Southern Blue Ridge it is prerequisite to consider their history. Present species composition often provides some information about the history and biogeographic affinities of sites. About 600 vascular plant taxa occur in non-alluvial wetlands in the Southern Blue Ridge of North Carolina. Each can be classified by its overall distribution into a number of biogeographically coherent classes. Five broad classes can be defined: northern species, widespread species, Southern-Central Appalachian species, Coastal Plain species, and tropical/extra-continental species. There are, in addition, interesting patterns within several of the classes.

The largest class is the widespread species, consisting of about 36 % of the total. These are primarily species with a wide distribution in eastern North America, occurring in wetlands (and sometimes uplands) in a wide variety of situations, generally not showing strong habitat or regional fidelity. The presence or absence of these species is usually not especially revealing about the history, edaphic features, or character of the wetland in which they occur. Examples are *Acer rubrum* var. *rubrum*, *Eupatorium perfoliatum*, *Rosa palustris*, *Juncus effusus*, *Sambucus canadensis*, and *Alnus serrulata*.

Northern species include about 31 % of the total. These are species distributed primarily in the northeastern United States and adjacent Canada (sometimes also in northern Eurasia). These are species at or near the southern limit of their distribution in their occurrences in wetlands of the Southern Blue Ridge. Some of these species are more or less limited to these wetlands (such as *Campanula aparinoides*, *Filipendula rubra*, *Juncus brevicaudatus*, *Caltha palustris*, *Eriophorum virginicum*, *Vaccinium macrocarpon*), while others are ecologically more widely distributed species which also occur in bogs as rare or typical components (such as *Lycopodium clavatum*, *Maianthemum canadense*, *Tsuga canadensis*, *Picea rubens*, *Viburnum cassinoides*, *Pinus strobus*, *Betula alleghaniensis*). While species in this class form a portion of the flora at all non-alluvial wetlands in the Southern Blue Ridge, they are particularly prevalent (often making up 40-60 % of the flora) in Southern Appalachian Fens and Southern Appalachian Bogs. Most of these species range south through the Appalachians in a more or less continuous distribution, and can thus be considered peripherals near their southern limit. Others, however, are disjunct to their sites in southern bogs (*Arethusa bulbosa*, *Carex oligosperma*, *Utricularia minor*, *Menyanthes trifoliata*). The intervening distributional gap is often on the order of 500 km. These disjunct species emphasize the relictual nature of many of these wetlands.

The third largest class contains the 20 % of the flora with Coastal Plain affinities These are species which have the bulk of their range in the Coastal Plain, most of them occurring in the mountains only as rare disjuncts in bog habitats, and are often not present (or very rare) in the Piedmont. Examples are *Fuirena squarrosa*, *Rhynchospora rariflora*, *Smilax laurifolia*, *Eleocharis tortilis*, *Woodwardia areolata*. This class is absent or nearly so in most Southern Appalachian Fens and Southern Appalachian Bogs, but forms 25-35 % of the flora in the French Broad Valley Bogs and Low Mountain Seepage Bogs. A very interesting set of species within this group are plants which are disjunct from a more northern Coastal Plain distribution to a montane distribution in the Southern Appalachians: *Thelypteris simulata*, *Triadenum virginicum*, *Gaylussacia dumosa* var. *bigeloviana*, *Helonias bullata*, *Juncus caesariensis*, *Carex collinsii*, *Narthecium americanum*. Two other species (*Myrica gale*, *Chamaedaphne calyculata*) could be classed in either the northern or Coastal Plain group; both are circumboreal bog shrubs, extending furthest south in eastern North America in the Coastal Plain (*M. gale* to Maryland, *C. calyculata* to eastern North Carolina) and disjunct in the southwestern Blue Ridge of North Carolina. All of these species except *Thelypteris simulata* are restricted to French Broad Valley Bogs (with *Juncus caesariensis* also found in Low Mountain Seepage Bogs).

The fourth largest class includes species primarily of the Southern and Central Appalachians, consisting of 13 % of the total. Examples are *Ilex collina*, *Diervilla sessilifolia*, *Rhododendron catawbiense*, *Lilium grayi*, *Houstonia serpyllifolia*, *Isoetes caroliniana*, *Sagittaria fasciculata*, and *Sarracenia jonesii*. Some of these, such as *Sarracenia jonesii* and *Sagittaria fasciculata*, are Southern Appalachian bog endemics, clearly evolutionary derivatives of more widespread, Coastal Plain species, and thus offer a relationship to the third category. Two other species, *Kalmia carolina* and *Chelone cuthbertii* show a similar relationship, as bimodal endemics of the Southern Blue Ridge and the Coastal Plain of the Carolinas and Virginia. Some of these are Southern Appalachian generalists, present in other habitats than non-alluvial wetlands, while others are bog specialists. Most of the endemics are found only in French Broad Valley Bogs (*Sarracenia jonesii*, *Sagittaria fasciculata*), or only in Southern Appalachian Bogs (*Ilex collina*, *Chelone lyonii*, *Lilium grayi*), while some occur in both (*Isoetes caroliniana*, *Houstonia serpyllifolia*, *Carex bromoides* ssp. *montana*).

The fifth category, tropical and extra-continental disjuncts, is a small one (1 %) among vascular plants, but considerably larger among nonvascular plants. These are species dependent on very humid habitats and are either disjunct from the tropics or from eastern Asia. The best examples among vascular plants are all pteridophytes: *Asplenium monanthes*, *Hymenophyllum tunbrigense*, *Hymenophyllum tayloriae*, *Trichomanes intricatum*, and *Grammitis nimbata* (Farrar 1967, Pittillo et al. 1975, Wagner & Sharp 1963). Dozens of bryophytes also fit this pattern (Anderson & Zander 1973). All of these species are limited to the Spray Cliff natural community, and are not found in more "conventional" non-alluvial wetlands. While some species may be recent colonists, most are certainly of ancient occurrence in these areas, emphasizing the strongly relictual nature of these peculiar, vertical wetlands.

At the height of Pleistocene glaciation, periglacial and tundra conditions prevailed at the highest elevations of the Southern Blue Ridge (Delcourt et al. 1993). Colder climates would have resulted in lower evapotranspiration rates than prevail today, and it is likely that non-alluvial wetlands occupied considerably greater areas in the southern mountains than they do today, at low to high elevations. Long Hope Valley, a north-facing, nearly flat-bottomed valley at 1340 m elevation, surrounded by peaks rising to 1700 m, would likely have supported tundra and krummholz vegetation.

7. Dynamics and Natural Disturbances

The factors responsible for creating and maintaining non-alluvial wetland communities of the Southern Blue Ridge are poorly known. Spray Cliffs are apparently very ancient relictual communities, often with dozens of bryophyte and pteridophyte species disjunct from other continents, many of them incapable of sexual reproduction (Anderson & Zander 1973, Billings & Anderson 1966). They are maintained by constant high humidity, steepness, general absence of soil, occasional soil slumping and tree fall, and periodic flood and freezing events which prevent succession to woody communities.

Upland Pools are very rare communities, with only three mountain examples known. The geologic factors which are responsible for their creation are unclear. The few known mountain examples occur in shallow natural depressions on upland ridges, where water is apparently perched over impermeable bedrock. Succession to forest is prevented by seasonal flooding of long duration.

The remainder of the communities, the more bog-like ones, show considerable variation in physiognomy, much of the it difficult to interpret. Anecdotal accounts indicate that most examples are experiencing (or have experienced in recent decades) invasion or increase of shrubs or trees at the expense of herbaceous zones. This vegetative succession threatens to completely close some bogs and eliminate many of the herbaceous species. While bogs may be undergoing primary succession that will eventually lead to a forest community, it seems unlikely that this process is proceeding so rapidly in recent decades after much longer periods of continued existence. This suggests either that the age of bogs is less than thought, or that some subtle change in conditions is promoting vegetative succession to woody species.

At least some bogs are known to be very old. Several bogs have been cored and dated, with dates of peat, sediment, or pollen accumulations at some sites dating to 10,000-12,000 years before present (Shafer 1984, Shafer 1986, M. Kneller, pers. comm.). Additional sites are unquestionably of similar ages, based on floristics, climate, elevation, and geographic position, such as the 23 bogs in Long Hope Valley (Ashe and Watauga counties) and the Bluff Mountain Fen (Ashe County). These sites have likely supported open, graminoid-dominated vegetation for the past 12,000 years (Pittillo, this volume). Many of these sites show little or no tendency towards woody encroachment.

Pineola Bog (Avery County), one of the largest of Southern Appalachian non-alluvial wetlands at 20 ha, is also very old, dating to 10,000 or more years before present, and has some of the deepest peat known in the Southern Appalachians, perhaps as deep as 2 m (M. Kneller, pers. comm.). This site, however, is showing considerable woody succession, and species requiring open, sunny conditions, such as *Vaccinium macrocarpon*, are much less common than in the 1930's and 1940's (L.E. Anderson, pers. comm; A.E. Radford, pers. comm.). Pineola Bog has been grazed fairly extensively in the past, has a U.S. highway bisecting it, has agricultural land nearly surrounding it, and has been encroached into and partially filled by construction activities. Any one of these disturbances (or a combination of them) may be responsible for changes in vegetation structure.

Other bogs may indeed be of recent origin, however. The Panthertown Valley bogs (Jackson County) may have formed as a result of logging and catastrophic fire, followed by beaver activity in the flat valley bottom of Panthertown and Frolictown creeks. The prevalence of widespread wetland species and the near absence of disjunct, relictual species in its flora suggests that it may be of relatively recent origin, its species having been recruited from the surrounding area (see Table 1).

Southern Appalachian bogs in some areas may have been maintained historically by beavers. Mitchell & Niering (1993) studied a wetland complex in Connecticut and concluded that "it is highly possible that prehistoric beaver flooding, as well as periodic droughts, in addition to more recent anthropogenic influences, have modified past vegetation. Disturbances associated with such periodic flooding and beaver activities may have also aided in the perpetuation of the bog flora. Thus, one might conclude that, historically, the system has been in a constant state of flux, including prolonged quiescent periods in which little change occurred, followed by more drastic oscillations... For most peatlands, this view of vegetation change appears to be a much more realistic model than any orderly successional dogma." Bogs of the Central Appalachians of West Virginia also show complex histories of beaver-caused modifications resulting in the maintenance (or expansion) of *Sphagnum*- or graminoid-dominated wetlands (Walbridge, pers. comm.). Although beaver-related successional changes have not been studied in the Southern Appalachians, we believe that conclusions similar to those of Mitchell & Niering (1993) likely apply to valley-bottom bogs of the southern Blue Ridge. The extirpation of beavers in the Southern Appalachians would have created a hiatus in this important wetland creating and altering force.

Along large stream valleys or small river valleys, beavers may have been responsible for a shifting mosaic of boggy habitats. Starting from open water, with *Sphagnum* mats around their margins, abandoned beaver ponds would have succeeded over a course of decades through herb-dominated communities to shrubby bogs, eventually ending in complexes of forested and shrub- or herb-dominated vegetation. This mosaic may have shifted slowly enough to allow rare and relictual species to disperse from successional bogs into newly created ones. Lending credibility to this theory is the recent recolonization by beavers of several areas with bogs, such as Panthertown Valley, Julian Price Park (Watauga County), and the Nantahala River bogs (S. Simon, pers. comm.; B. Teague, pers. comm.). Beaver activity appears to have increased the area of "boggy" habitats at these sites, but the short period of beaver activity (less than ten years) does not allow a confident prediction of the long-term effects.

Grazing has been nearly universal in bogs, and few examples exist in pristine condition (a few of the Long Hope Valley bogs may be the only examples in the Southern Blue Ridge which have never experienced pasturing). Grazing is presumed to have a number of effects on bogs. Browsing by cattle helps control woody species, but trampling and nutrient inputs modify bogs in numerous ways, and tend to destroy herbs and bryophytes, especially *Sphagnum*. *Sphagnum* appears to be the keystone of these systems, and nutrient inputs from cow dung are highly detrimental to *Sphagnum* cover, vigor, and diversity (L.E. Anderson, pers. comm.). Once *Sphagnum* cover is substantially reduced, hydrologic alteration of the bog is highly likely. While potentially reversible, the loss of *Sphagnum* and other bog species may result in a permanently altered successional trajectory. Thus, paradoxically, grazing by cattle may have been responsible for the more open condition described for Southern Appalachian bogs in the 1930's and 1940's, but may have set in motion modifications of hydrology and supply of nutrients promoting rapid succession by woody species.

Aerial deposition (changes in the chemistry of rain) is another potential nutrient input that may be altering woody succession. Additionally, most bogs were formerly surrounded by forested vegetation, but the great majority now have agricultural lands upslope. Many sites are closely surrounded by agricultural fields or pastures. Fertilizers certainly have impacts in some bogs. Alteration of the watershed has other effects. The formerly forested watersheds released water more slowly and constantly to the bog; a largely unforested watershed releases water rapidly, particularly in storm events (frequent in the Southern Appalachians). Flooding events tend to result in more rapid down-cutting of rivulets in the bog and of the characteristic bog-peripheral stream, leading to increased hydrologic head and drying of the bog. If associated with loss of water-retentive *Sphagnum* associated with grazing, down cutting could have particularly devastating effects on a delicately balanced water budget.

The presence in Southern Appalachian bogs of species commonly associated with fire-maintained ecosystems of the Coastal Plain has suggested that fire may have been a factor in the maintenance of open, graminoid-dominated systems. This suggestion seems somewhat plausible for a few sites, such as the Eller Seepage Bog (Clay County), which has a very large component of "Coastal Plain fire species," but it seems unlikely that most mountain bogs burned with any regularity. Most are in topographic situations among the least likely for fires to reach, often surrounded by mesic (or wetter), forested communities which would burn only under very rare circumstances. Most "Coastal Plain fire species" are not fire-dependent per se. Rather, they are species of open, sunny, moist to wet, acidic, peaty or sandy situations, largely maintained in the Coastal Plain by fire. Non-alluvial wetlands in the Southern Blue Ridge, generally occurring over felsic rocks which yield acidic, nutrient-poor soils, and with hydrology maintaining saturated and often peaty situations, may offer closely analogous habitats, even in the absence of fire. Northern bogs also often have species of Coastal Plain affinities, but are not fire-maintained habitats.

8. Conservation

The Blue Ridge region of western North Carolina, southwestern Virginia, eastern Tennessee, northwestern South Carolina, and northeastern Georgia has a large percentage of land in public ownership (Pisgah, Nantahala, Cherokee, Jefferson, Sumter, and Chattahoochee National Forests, Great Smoky Mountains National Park, various state parks, and other publicly-owned management areas). Non-alluvial wetlands of the southern Blue Ridge, however, are predominantly distributed at low elevations in floodplains of streams and rivers throughout the region, and on slopes in northwestern North Carolina, in areas generally in private ownership. Although the total current area of non-alluvial wetlands in the southern Blue Ridge is estimated to be less than 400 hectares, these communities have a disproportionate importance, by providing habitat for many rare and common species of plants and animals. They provide diversity in an otherwise overwhelmingly upland, terrestrial landscape (Pearson, this volume). Various conservation tools, including inventory, scientific study, education, landowner contact and voluntary conservation, regulation, acquisition, and management, can be applied to the problem of conserving mountain wetlands. A combination of all these tools will probably be needed if a significant portion of even the remnant of wetlands are to be preserved.

Inventory or survey of remaining wetlands is needed. All other conservation tools depend on the basic knowledge of the location and relative importance of remaining wetland sites. The U.S. Fish and Wildlife Service has conducted or is in the process of conducting National Wetlands Inventories throughout the southern Blue Ridge (Hefner, this volume). Unfortunately, the wetland types of the southern Blue Ridge are difficult to interpret from aerial photography. Moreover, many of the wetlands are too small to be recognized or mapped at the scale of 1: 24,000. While the National Wetlands Inventory provides valuable information on the status of wetlands, its utility in the southern Blue Ridge is limited by the nature of the wetlands present.

Inventories conducted by Natural Heritage Programs seek to identify the most ecologically significant sites in a given area. Natural Heritage Programs in all five states of the southern Blue Ridge have sought non-alluvial wetland areas and documented high quality sites. Even well-funded and detailed inventories are limited by the small size and relatively unpredictable distribution of wetlands in this region. Moreover, most bog wetlands in the southern Blue Ridge are surrounded by nearly impassable *Rhododendron maximum* thickets that render field survey time-consuming and difficult. Still, Natural Heritage Program inventories have located over a hundred bog sites in North Carolina alone (Gaddy 1981a, Smith 1993). Natural Heritage Program inventories generally result in basic site information, such as location, dominant plant species, species lists, rare species present, and ownership.

More detailed scientific study of wetlands in the region has been extremely limited. Detailed studies of soils, hydrology, nutrient cycling, dynamics, fauna (with the exception of bog turtles), and flora are almost non-existent. Detailed information of this sort is needed in order to provide better understanding of these systems, not only for pure scientific value but also to apply to their protection and management.

The southern Blue Ridge has historically been a rural region, dependent on subsistence agriculture and logging. While tourism and industry have become important parts of the region's economy, wetlands are still regarded by many landowners as impediments to cultivation and economic return and hence targets of drainage. Education about the significance and values of mountain wetlands is needed to prevent the ongoing destruction and degradation of wetland sites. Following inventory or identification of the location of wetland sites, one form of "targeted education" is landowner contact, which involves providing information to the owners of important natural areas about the resources on their property and options available for protecting them. Those options include voluntary conservation agreements, management agreements, leases, conservation easements, and bargain or market-value sale. Some of these options involve direct or indirect financial returns to the landowner, turning a perceived liability into an asset.

Placement of fill in wetlands is federally regulated under provisions of Section 404 of the Clean Water Act, but until recently, implementation of these regulations in the southern Blue Ridge was poorly enforced. The U.S. Army Corps of Engineers now has an active program regulating wetland fill in the southern Blue Ridge, but once again, the small size and isolated nature of many mountain wetlands make them difficult to effectively protect. In cases where fill is placed in wetlands without a permit, simple removal of the fill may not serve to restore the character and functions of the wetland. The restoration of severely damaged wetland sites, especially those with substantial *Sphagnum*, is difficult, if not impossible. Because of the sensitivity of *Sphagnum* to sediment and additional nutrients, the creation of boggy wetlands is also unlikely to be successful.

Acquisition of significant wetland sites by federal, state, or local conservation agencies or by private conservation organizations (such as The Nature Conservancy) offers the greatest potential for the protection of some sites. However, the resources available are limited, and only the most significant sites are likely to receive protection in this manner. Moreover, while wetland acreage is very limited, the effective and long-term protection and management of wetland sites requires some control over land management of much larger areas. A one-hectare wetland may be affected by pasturing, fertilizing, logging, or ground-disturbing activities anywhere in a watershed of tens or hundreds of hectares.

Nearly all remaining wetlands in the southern Blue Ridge have had some level of alteration in their watersheds. Restoration of the natural hydrologic regime (in terms of the amount, seasonality, and chemistry of water) is critical in maintaining relictual wetland systems. As discussed above, management issues are clouded by our poor understanding of the natural functioning and dynamics of these systems. There is a critical need for additional study to answer questions pertinent to the management and restoration of protected sites. An appreciation of the interest, functions, and values of the unique and endangered wetlands of the southern Blue Ridge has developed in time to allow the conservation of some remnants, but their protection will be achieved only with expense, difficulty, and perseverance.

Acknowledgments

The authors would like to express appreciation to Alan Smith, Chick Gaddy, Dennis Herman, Nora Murdock, Tom Govus, Dan Pittillo, Laura Mansberg, Merrill Lynch, Rob Sutter, Tom Rawinski, Doug Ogle, Craig Moretz, Julie Moore, Loyal Mehrhoff, Bob Currie, Lewis Anderson, Kevin Moorhead, and others who have gathered information on non-alluvial wetlands in southern Blue Ridge. Nearly all of this information is in the form of unpublished reports and field forms in the files of the N.C. Natural Heritage Program. We also thank Karen Patterson, Bob Peet, Lewis Anderson, and Mark Walbridge, for assistance with analysis, and for discussions about these interesting communities. The following individuals should be particularly commended for their efforts to protect these endangered communities: Dennis Herman, Nora Murdock, Janice Nicholls, Ann Prince, Inge Smith, Dick Everhart, Fred Annand, Merrill Lynch, Margit Bucher, Julie Moore, Rob Sutter, Bambi Teague, Sally Browning, Bob Johnson, Steve Chapin, Dave Baker, Ron Determann, and a number of individual landowners. Research was conducted primarily under the auspices of the N.C. Natural Heritage Program, Division of Parks and

Recreation. Funding for some of the field work was provided by the North Carolina Recreation and Natural Heritage Trust Fund, The Nature Conservancy (North Carolina Field Office), the H. Smith Richardson Trust, and the U.S. Fish and Wildlife Service.

References

Anderson, L.E. 1990. A checklist of *Sphagnum* in North America north of Mexico. Bryologist 93: 500-501.

Anderson, L.E., and R.H. Zander. 1973. The mosses of the southern Blue Ridge Province and their phytogeographic relationships. J. Elisha Mitchell Sci. Soc. 89: 15-60.

Anderson, L.E., H.A. Crum, and W.R. Buck. 1990. List of the mosses of North America north of Mexico. Bryologist 93: 448-499.

Billings, W.D. and L.E. Anderson. 1966. Some microclimatic characteristics of habitats of endemic and disjunct bryophytes in the southern Blue Ridge. Bryologist 69: 76-95.

Boufford, D.E., and E.W. Wood. 1975. Natural areas study of the Southern Blue Ridge of Georgia, North Carolina, and South Carolina. Report to Highlands Biological Station, Highlands, N.C..

Browning, S. (this volume).

Bruce, R.C., M.L. Hicks, D.A. Van Voorhees, and S.P. Yurkovich. (n.d.). The microenvironments of *Grammitis nimbata* and other ferns of tropical affinities in southwestern North Carolina. Unpublished report for the Highlands Biological Station, Highlands, N.C.

Cowardin, L.M., V. Carter, F.C. Golet, and E.T. LaRoe. 1979. Classification of wetlands and deepwater habitats of the United States. U.S. Fish and Wildlife Service Report FWS/OBS-79/31.

Delcourt, P.A., H.R. Delcourt, D.F. Morse, and P.A. Morse. 1993. History, evolution, and organization of vegetation and human culture. In W.H. Martin et al., eds. Biodiversity of the southeastern United States: lowland terrestrial communities. John Wiley & Sons, New York, NY.

Dennis, W.M. 1980. *Sarracenia oreophila* (Kearney) Wherry in the Blue Ridge Province of northeastern Georgia. Castanea 45: 101-103.

Farrar, D,R. 1967. Gametophytes of four tropical fern genera reproducing independently of their sporophytes in the southern Appalachians. Science 155 (3767): 1266-1267.

Gaddy, L.L. 1981a. The bogs of the southwestern mountains of North Carolina. Report to N.C. Natural Heritage Program, Raleigh, N.C.

Gaddy, L.L. 1981b. Fork Ridge balds and seep communities, North Carolina: an ecological evaluation of a potential national natural landmark. Report to National Park Service, Washington, D.C.

Gore, A.J.P. 1983. Introduction. In Gore, A.J.P., ed. Ecosystems of the world. 4A. Mires: swamp, bog, fen and moor, general studies. Elsevier Sci. Publ. Co., Amsterdam.

Govus, T. 1985. Natural area surveys in the southern mountains. Report to N.C. Natural Heritage Program, Raleigh, N.C.

Govus, T. 1987. The occurrence of *Sarracenia oreophila* (Kearney) Wherry in the Blue Ridge Province of southwestern North Carolina. Castanea 52 (4): 310-311.

Hefner, J. (this volume).

Hicks, M.L. 1992. Guide to the liverworts of North Carolina. Duke University Press, Durham, N.C.

Horton, J.H., and L.G. Hotaling. 1981. Floristics of selected heath communities along the southern section of the Blue Ridge Parkway. National Park Service Research/Resources Management Report No. 45.

Ingram, H.A.P. 1983. Hydrology. In Gore, A.J.P., ed. Ecosystems of the world. 4A. Mires: swamp, bog, fen and moor, general studies. Elsevier Sci. Publ. Co., Amsterdam.

Kartesz, J.T. 1994. A synonymized checklist of the vascular flora of the United States, Canada, and Greenland, second edition. Timber Press, Portland, OR.

LeGrand, H.E., Jr. 1993. Natural Heritage Program list of the rare animal species of North Carolina. N.C. Natural Heritage Program, Division of Parks and Recreation, Raleigh, NC.

McDonald, B.R., ed. 1982. Proceedings of the Symposium on Wetlands of the Unglaciated Appalachian Region. West Virginia University, Morgantown, WV.

McLeod, D.E. 1983. The vascular flora of a small southern Appalachian bog-fen [abstract]. Assoc. Southeastern. Biologists Bull. 30: 70-71.

McLeod, D.E. 1988. Vegetation patterns, floristics, and environmental relationships in the Black and Craggy Mountains of North Carolina. Ph.D. Dissertation, University of North Carolina at Chapel Hill.

McLeod, D.E., and J.A. Croom. 1983. Vegetation patterns and habitat types in a small Southern Appalachian bog-fen [abstract]. Assoc. Southeastern Biologists Bull. 30: 71.

Mitchell, C.C. and W.A. Niering. 1993. Vegetation change in a topogenic bog following beaver flooding. Bull. Torrey Bot. Club 120 (2). 136-147.

Mowbray, T., and W.H. Schlesinger. 1988. The buffer capacity of organic soils of the Bluff Mountain Fen, North Carolina. Soil Science 146: 73-79.

Murdock, N.A. (this volume).

Pearson, S. (this volume).

Pittillo, J.D. (this volume).

Pittillo, J.D., W.H. Wagner, Jr., D.R. Farrar, and S.W. Leonard. 1975. New pteridophyte records in the Highlands Biological Station area, Southern Appalachians. Castanea 40: 263-272.

Richardson, C.J., and J.W. Gibbons. 1993. Pocosins, Carolina bays, and mountain bogs. In W.H. Martin et al., eds. Biodiversity of the southeastern United States: Lowland terrestrial communities. John Wiley & Sons, New York, N.Y.

Rodwell, J.S., ed. 1991. British plant communities. Volume 2: Mires and heaths. Cambridge University Press, Cambridge.

Schafale, M.P., and A.S. Weakley. 1985. Classification of the natural communities of North Carolina, second approximation. N.C. Natural Heritage Program, Div. of Parks and Recreation, Raleigh, N.C.

Schafale, M.P., and A.S. Weakley. 1990. Classification of the natural communities of North Carolina, third approximation. N.C. Natural Heritage Program, Div. of Parks and Recreation, Raleigh, N.C.

Shafer, D.S. 1984. Late-Quaternary paleoecologic, geomorphic, and paleoclimatic history of Flat Laurel Gap, Blue Ridge Mountains, North Carolina. M.S. Thesis, Univ. of Tennessee, Knoxville, Tenn.

Shafer, D.S. 1986. Flat Laurel Gap Bog, Pisgah Ridge, North Carolina: Late Holocene development of a high-elevation heath bald. Castanea 51: 1-10.

Smith, A.B. 1993. Inventory of mountain wetlands. Report to N.C. Nat. Heritage Program, Raleigh, N.C.

Soil Management Support Services. 1985. Keys to soil taxonomy. Soil Management Support Services Technical Monograph No. 6.

Soil Survey Staff. 1975. Soil taxonomy; a basic system of soil classification for making and interpreting soil surveys. Soil Conservation Service Agriculture Handbook No. 436.

Stewart, C.N., Jr., and E.T. Nilsen. 1993. Association of edaphic factors and vegetation in several isolated Appalachian peat bogs. Bull. Torrey Bot Club 120 (2): 128-135.

Tucker, G.A. 1967. The Vascular Flora of Bluff Mountain. M.S. Thesis, University of North Carolina at Chapel Hill, Chapel Hill, N.C.

Wagner, W.H., Jr., and A.J. Sharp. 1963. A remarkably reduced vascular plant in the United States. Science 142 (3598): 1483-1484.

Walbridge, M. (this volume).

Weakley, A.S. 1993a. Natural Heritage Program list of the rare plant species of North Carolina. N.C. Natural Heritage Program, Division of Parks and Recreation, Raleigh, N.C.

Weakley, A.S. 1993b. The natural history of Long Hope Valley. Report to the H. Smith Richardson Trust and N.C. Natural Heritage Program, Raleigh, N.C.

Weakley, A.S., L.A. Mehrhoff, and L. Mansberg. 1979. Natural Area Inventory for Bluff Mountain, Ashe County, North Carolina. Report to N.C. Nature Conservancy, Carrboro, N.C.

Wickland, D.E., and J.H. Horton. 1977. A botanical evaluation of the French Broad River corridor in North Carolina. Report to Tennessee Valley Authority.

Wiser, S.K. 1993. Vegetation of high-elevation rock outcrops of the Southern Appalachians: composition, environmental relationships, and biogeography of communities and rare species. Ph. D. Dissertation, University of North Carolina at Chapel Hill.

RARE AND ENDANGERED PLANTS AND ANIMALS OF SOUTHERN APPALACHIAN WETLANDS

by Nora A. Murdock
U.S. Fish and Wildlife Service
Endangered Species Field Office
330 Ridgefield Court
Asheville, North Carolina, U.S.A. 28806

Abstract. At least one-third of the threatened and endangered species of the United States live in wetlands. Southern Appalachian bogs and fens, in particular, support a wealth of rare and unique life forms, many of which are found in no other habitat type. In North Carolina alone, nonalluvial mountain wetlands provide habitat for nearly 90 species of plants and animals that are considered rare, threatened, or endangered by the North Carolina Plant Conservation Program, the North Carolina Natural Heritage Program, the North Carolina Wildlife Resources Commission, or the U.S. Fish and Wildlife Service. These species include the bog turtle, mountain sweet pitcher plant, green pitcher plant, swamp pink, bunched arrowhead, and Gray's lily, all of which are either on the federal list of endangered and threatened species or under consideration for that list. Mountain wetland habitats for these species are being destroyed and degraded at an accelerating rate for highway construction and expansion and residential and recreational development, as well as for industrial and agricultural uses.

KEY WORDS: mountain bogs, wetlands, endangered and threatened species.

1. Introduction

In only two centuries, over one-half of the wetlands that originally existed in the United States, outside of Alaska and Hawaii, have been destroyed (Dahl, 1990). Despite protective legislation and recent efforts directed at slowing wetland destruction, wetlands losses in this country are continuing at nearly one-half million acres annually (Hefner, this volume; Dahl and Johnson, 1991).

Mountain wetlands, particularly nonalluvial wetlands, are one of the most important habitats for rare species in the Southeast. Until recently, however, they have received little attention from regulatory agencies, in part because those that remain are usually small (under 10 acres) and often difficult to map using standard wetland inventory techniques (Hefner, this volume; Weakley and Schafale, this volume). Much of the information presented here is from North Carolina, where the majority of these habitats remain, but the general trends can be applied to the remainder of the southern Appalachian area as well (southern Virginia, eastern Tennessee, western North Carolina, northern South Carolina, and northern Georgia). Tables I and II readily illustrate the importance of mountain wetlands, particularly bogs and fens, to rare plants and animals in this region. Almost one-fifth of the

Water, Air and Soil Pollution 77: 385–405, 1994.

TABLE I

RARE ANIMALS OF SOUTHERN APPALACHIAN WETLANDS

SPECIES	HABITATS					
	Bogs	Swamps	Wet meadows	Pools	Marshes	Bottomlands
star-nosed mole (Condylura cristata parva)	X	X	X			X
bog turtle (Clemmys muhlenbergii)	X		X		X	
mole salamander (Ambystoma talpoideum)				X		
four-toed salamander (Hemidactylium scutatum)	X	X		X		X
Baltimore butterfly (Euphydryas phaeton)	X		X		X	
Southern bog lemming (Synaptomys cooperi)	X		X		X	

TABLE II

RARE PLANTS OF SOUTHERN APPALACHIAN WETLANDS

SPECIES	HABITATS					
	Bogs	Fens	High elevation seeps	Seeps	Swamp forests	Wet meadows
bog rose (Arethusa bulbosa)	X					
bog jack-in-the-pulpit (Arisaema triphyllum ssp. stewardsonii)	X					
blunt-lobed grape fern (Botrychium oneidense)	X					
marsh marigold (Caltha palustris)	X					X
marsh bellflower (Campanula aparinoides)	X					
hay sedge (Carex argyrantha)						X
Barratt's sedge (Carex barrattii)	X					X
brown bog sedge (Carex buxbaumii)	X	X				
cone-shaped sedge (Carex conoidea)	X					
small crested sedge (Carex cristatella)	X					
few-seeded sedge (Carex oligosperma)	X			X		
necklace sedge (Carex projecta)	X				X	X
Schweinitz's sedge (Carex schweinitzii)					X	
sedge (Carex trichocarpa)	X					X

TABLE II (continued)						
SPECIES	HABITATS					
	Bogs	Fens	High elevation seeps	Seeps	Swamp forests	Wet meadows
three-seeded sedge (Carex trisperma)	X					
Cuthbert's turtlehead (Chelone cuthbertii)	X					
twig-rush (Cladium mariscoides)	X	X		X		X
robin runaway (Dalibarda repens)	X					
bog oatgrass (Danthonia epilis)	X			X		
purpleleaf willowherb (Epilobium ciliatum)	X			X		
narrow pipewort (Eriocaulon lineare)	X					
Texas hatpins (Eriocaulon texense)	X					
queen-of-the-prairie (Filipendula rubra)	X					X
fringed gentian (Gentianopsis crinita)		X				
yellow avens (Geum aleppicum)	X					
rough avens (Geum laciniatum var. tricho-carpum)	X					
littleleaf sneezeweed (Helenium brevifolium)	X			X		
swamp pink (Helonias bullata)	X					
holy grass (Hierochloe odorata)	X					
Appalachian fir-clubmoss (Huperzia appalachiana)		X		X		

TABLE II (continued)

SPECIES	HABITATS					
	Bogs	Fens	High elevation seeps	Seeps	Swamp forests	Wet meadows
long-stalked holly (Ilex collina)	X					
rough rush (Juncus caesariensis)	X					
rough blazing star (Liatris aspera)		X				
Heller's blazing star (Liatris helleri)		X				
yellow Canada lily (Lilium canadense ssp. canadense)	X					X
red Canada lily (Lilium canadense ssp. editorum)	X					X
Gray's lily (Lilium grayi)	X			X		X
American fly-honeysuckle (Lonicera canadensis)	X					
bog clubmoss (Lycopodiella inundata)	X			X		
large-flowered Barbara's buttons (Marshallia grandiflora)	X					
buckbean (Menyanthes trifoliata)	X					
bristly muhly (Muhlenbergia glomerata)		X				
sweet gale (Myrica gale)	X					
bog asphodel (Narthecium americanum)	X					
perennial sundrops (Oenothera perennis)	X					

TABLE II (continued)						
SPECIES	HABITATS					
	Bogs	Fens	High elevation seeps	Seeps	Swamp forests	Wet meadows
large-leaved grass-of-Parnassus (Parnassia grandifolia)		X		X		
swamp lousewort (Pedicularis lanceolata)					X	
northern rein orchid (Platanthera flava var. herbiola)	X					
large purple-fringed orchid (Platanthera grandiflora)	X		X	X		
white fringeless orchid (Platanthera integrilabia)	X					
purple fringeless orchid (Platanthera peramoena)	X					
bog bluegrass (Poa paludigena)	X					
northern white beaksedge (Rhynchospora alba)	X	X				
bunched arrowhead (Sagittaria fasciculata)	X				X	
Canada burnet (Sanguisorba canadensis)	X			X		
mountain sweet pitcher plant (Sarracenia rubra ssp. jonesii)	X					
green pitcher plant (Sarracenia oreophila)	X					
swamp saxifrage (Saxifraga pensylvanica)	X			X		
balsam ragwort (Senecio pauperculus)	X	X				

TABLE II (continued)						
SPECIES	HABITATS					
	Bogs	Fens	High elevation seeps	Seeps	Swamp forests	Wet meadows
bog goldenrod (Solidago uliginosa)	X			X		
Epling's hedge-nettle (Stachys eplingii)	X					
bog featherbells (Stenanthium robustum)	X					X
Canada yew (Taxus canadensis)	X				X	
bog fern (Thelypteris simulata)	X					
sticky bog asphodel (Tofieldia glutinosa)	X			X		
small bladderwort (Utricularia minor)	X					
cranberry (Vaccinium macrocarpon)	X			X		
American speedwell (Veronica americana)	X			X		
beargrass (Xerophyllum asphodeloides)	X					
yellow starry fen moss (Campylium stellatum var. stellatum)		X				
bog broom-moss (Dichranum undulatum)	X			X		
narrowleaf peatmoss (Sphagnum angustifolium)	X					
northern peatmoss (Sphagnum capillifolium)	X					
pretty peatmoss (Sphagnum fallax)	X					

TABLE II (continued)						
SPECIES	HABITATS					
	Bogs	Fens	High elevation seeps	Seeps	Swamp forests	Wet meadows
flexuous peatmoss (Sphagnum flexuosum var. flexuosum)	X					
brown peatmoss (Sphagnum fuscum)	X					
orange peatmoss (Sphagnum subsecundum var. subsecundum)	X	X				
fen peatmoss (Sphagnum warnstorfii)	X	X				
southern dung moss (Splachnum pennsylvanicum)	X					
liverwort (Cephaloziella hampeana)		X				
fringed brome (Bromus ciliatus)			X	X		
(Some of these species occur in other habitats in addition to mountain wetlands.)						

722 rare plant species monitored by the North Carolina Natural Heritage Program occur in nonalluvial mountain wetlands, and most of them are limited to these habitat types. When compared to the plant list, there are relatively few rare animal species known to be associated with southern mountain wetlands. However, systematic studies of the faunal assemblages in these habitats, particularly the invertebrates, have not been conducted. This paper summarizes existing information on the rare species of national significance that are native to the bogs, fens, and other nonalluvial wetlands of the southern Appalachian mountains. Particular emphasis is given to North Carolina, where most of these wetlands remain.

2. Rare animals

Systematic faunal surveys for rare species have not been conducted in mountain wetland habitats. This paper presents the limited existing information about the relatively few rare animal species known to inhabit this type of wetland.

The bog turtle (Clemmys muhlenbergii), although its range extends from New York south to northeastern Georgia, has declined seriously in recent decades due to habitat destruction and degradation, as well as collecting for the pet trade. It is now listed as endangered, threatened, or of special concern in every state within its range and is under consideration for the federal list of endangered and threatened species (Herman, 1992; U.S. Fish and Wildlife Service, 1991b). This is one of the smallest of the North American turtles, with adults reaching only 3 to 3 1/2 inches in length. The species is easily identified by its size and by the bright orange or yellow patches on the sides of its head. Bog turtles reach sexual maturity at about 6 years, and mating takes place from late April to early June. Eggs, usually 3 to 4, are laid from May to July in shallow nests in moss, grass, or soft soil (Conant and Collins, 1991; Martof et al., 1980; Mitchell et al., 1991; Herman, 1992). Despite the name, the bog turtle is not limited strictly to bogs but also inhabits relatively open swamps, marshes, wet meadows, and wet pastures. As stated by Conant and Collins (1991):

> "Although the bog turtle still occurs in certain areas, it is rare or completely absent in many regions where it once was fairly abundant. Sphagnum bogs, swamps, and clear, slow-moving meadow streams with muddy bottoms are preferred. The human propensity for draining and reclaiming such habitats has contributed to its disappearance."

Remaining bog turtle habitats are becoming increasingly isolated as more wetlands are destroyed. Although this turtle is capable of moving along streams and other wetland corridors in search of suitable habitat, threats to it increase as the distance between wetlands increases. In North Carolina (the southern stronghold for the species), many, if not most, of the remaining bog turtle sites are separated from the next nearest suitable habitat by at least one highway, where migrating turtles are often crushed by vehicles before they can get across the pavement. Bog turtles prefer relatively open habitat, and otherwise suitable habitat can be rendered unsuitable by dense growth of woody species. Historically, this was probably not a problem for the mobile bog turtle, since there were substantially more contiguous wetlands available in various stages of succession than there are now.

The southern Appalachian population of this species, most of which is in North Carolina, is disjunct from the northeastern portion of the range (Conant and Collins, 1991). In North Carolina, approximately 75 populations are known to survive, but many of these are at sites that are degraded and/or imminently threatened (Herman, 1992). There are two extant sites in Tennessee (Tryon, 1992), three in Georgia (D. Herman, Zoo Atlanta, personal communication, 1993) and two in South Carolina (Fontenot and Platt, 1993), although turtles have not been seen recently at the South

Carolina sites. Virginia has approximately two dozen extant populations; searches for additional populations are continuing there (J. Mitchell, University of Richmond, personal communication, 1993).

The Baltimore (Euphydryas phaeton), one of the checkerspot butterflies, is a rare invertebrate inhabitant of mountain bogs, marshes, and wet meadows. This striking butterfly has black wings with cream-colored spots on the outer half of both fore- and hind wings and an outer margin of red-orange crescents. The main food plant for the caterpillars is turtlehead (Chelone spp.), although beardtongue (Penstemon) and false foxglove (Aureolaria) are also eaten. The flight period in the South is from mid-May to late June, and there is one brood per year (Opler and Krizek, 1984; Opler, 1992). The invertebrate fauna of southern Appalachian bogs has not been systematically studied and may include additional species that are extremely rare and/or endemic to these habitats.

Floodplain pools in the mountains are an extremely important wetland habitat, and are even rarer than bogs. Little natural community structure remains where fertile and relatively rock-free bottomland wetlands once occurred. A higher percentage of this habitat type probably has been lost than other mountain wetlands. Floodplain pools are the primary breeding habitat for a number of amphibians, including the four-toed salamander (Hemidactylium scutatum) and the mole salamander (Ambystoma talpoideum). Other amphibians that are rare or declining in the mountains and use floodplain pools include the mountain chorus frog (Pseudacris brachyphona), the seepage salamander (Desmognathus aeneus), the longtail salamander (Eurycea longicauda), and the mud salamander (Pseudotriton montanus). These species and several other animals considered rare in the southern Appalachians are more thoroughly covered by Boynton (this volume). Amphibians, rare or not, dependent on ephemeral floodplain pools, are in decline because of habitat loss; additional species will become rare if current habitat loss rates continue.

Animal populations probably require larger areas to remain viable than many plant populations; therefore, the great reduction in mountain wetlands may have had a greater detrimental effect, percentagewise, on animals than plants. The presence of many disjunct populations of plants and animals currently known from mountain wetlands and the loss of much wetland habitat in the mountains suggests that many other disjunct populations and possibly some species may have been lost. A correlation between reduced recolonization potential and habitat fragmentation has probably contributed greatly to the accelerating decline of rare animals native to mountain wetlands.

In addition, acid precipitation, which seems to be more pronounced at higher elevations, could be having a greater effect on mountain wetlands than wetlands at lower elevations. However, there is a lack of historic or baseline data from these habitats in the southern Appalachians.

3. Rare plants

As evidenced by the 81 species in Table II, mountain wetlands are especially important habitats for rare plants. Some of these species are strictly endemic to southern Appalachian wetlands; others are species with more northern affinities that are extremely rare in the South. Both groups include several species that are now federally listed as endangered or threatened (U.S. Fish and Wildlife Service, 1993a; U.S. Fish and Wildlife Service, 1993b). This paper will concentrate on those that are rare throughout their range, most of which are on the federal list of endangered and threatened species or are candidates for that list.

Mountain sweet pitcher plant (Sarracenia rubra ssp. jonesii), an endemic to the bogs in the Blue Ridge Mountains and foothills of North Carolina and South Carolina, is federally listed as endangered. This insectivorous plant, with only 10 small populations surviving, is one of the rarest species in the genus. Four of the remaining populations are in the French Broad River drainage in Henderson and Transylvania Counties, North Carolina; five are in the Saluda River drainage in Greenville County, South Carolina; and one is in the Enoree River drainage in Greenville County, South Carolina. The species has also been reported from Buncombe County, North Carolina, but it is not currently known to survive there. Sixteen populations are known to have been destroyed by drainage, impoundment, and other forms of habitat degradation associated with highway and railroad construction, residential and industrial development, resort development, and conversion of habitat for agricultural purposes (Murdock, 1990). Most of the other eight Sarracenia species in the Southeast occur primarily on the coastal plain, and S. r. jonesii is widely disjunct from the other members of the S. rubra complex. Sarracenia purpurea is the only sympatric Sarracenia that shares the montane habitat of mountain sweet pitcher plant. The taxonomy of this genus is extremely complex, and extensive natural hybridization is documented (Bell, 1952; McDaniel, 1971).

Mountain sweet pitcher plant is a rhizomatous perennial that grows from 8 to 29 inches tall. The numerous erect leaves grow in clusters, are hollow and trumpet-shaped, and form slender, almost tubular, pitchers (inspiration for the most frequently used common name--pitcher plant) that are covered by a cordate hood. The pitchers are a waxy dull green, usually reticulate-veined with maroon/purple. The showy maroon flowers are fragrant, unlike most other pitcher plants. This species blooms from April to June, with fruits ripening in August (Massey et al., 1983; Radford et al., 1964; Wood, 1960; Wofford, 1989). Like many other species in this genus, mountain sweet pitcher plant shows some variation within and between populations, including a rare yellow-flowered form (Case and Case, 1976). Variations are probably due both to genetic and environmental differences. Plants growing in shade tend to be less erect and have much less conspicuously colored veins. Reproduction is by seeds or by fragmentation of rhizomes (Massey et al., 1983; Wood, 1960). Individual plants have been reported to live for as long as 20 to

35 years (McFarlane, 1908; McDaniel, 1971; F. Case, Saginaw, Michigan, personal communication, 1990).

Sarracenia rubra ssp. jonesii can be distinguished from other subspecies of Sarracenia rubra by its greater pitcher height, scape length equal to pitcher height, long petiole, abruptly expanded pitcher orifice, cordate and slightly reflexed hood, and petals and capsules that are usually twice as large as other Sarracenia rubra (Massey et al., 1983; Sutter, 1987; Case and Case, 1976; Wherry, 1929.)

The habitat of mountain sweet pitcher plant consists of mountain bogs and stream sides, usually on soils of the Toxaway silt loam or Hatboro loam series. These soils are deep, poorly drained combinations of loam, sand, and silt, with a high organic matter content in the upper layers and medium to highly acidic pH (U.S. Department of Agriculture, 1980). Most sites occur in level depressions associated with floodplains; however, a few occur in "cataract bog" or "waterslide" situations, where sphagnum and other typical bog species line the sides of waterfalls on granite rock faces. Bogs occupied by this species are typically dominated by herbs and shrubs but may have scattered trees such as red maple (Acer rubrum), hemlock (Tsuga canadensis), pitch pine (Pinus rigida), white pine (Pinus strobus), and, at high elevations, red spruce (Picea rubens). A dense shrub understory often alternates with open patches of sedges, forbs, and sphagnum. (See Weakley and Schafale, this volume, and Schafale and Weakley, 1990, for a more detailed description of this habitat type.)

Members of the genus Sarracenia form the exclusive food for a number of moths, including Olethreutes (which feeds upon the flowers and seeds), Papaipeme (a rhizome-borer), and three species of Exyra (which eat the leaves). Other insects are known to live in the pitchers, including two harmless species of mosquito, a Sarcophagan fly, a gnat, and a Sciarid fly (Wood, 1960; Folkerts, 1982). There may be some insect species restricted to mountain sweet pitcher plant that are also in danger of extinction (T. Gibson, University of Wisconsin, personal communication, 1990).

The most serious threats to mountain sweet pitcher plant are the destruction and/or degradation of its small wetland habitats, which have already resulted in the elimination of the species from 62 percent of the historic sites (Murdock, 1990).

The open, shrub-dominated communities occupied by this species are early successional types and must be maintained at this disclimax in order for mountain sweet pitcher plants to thrive. Historically, the bogs were probably kept open by water fluctuation, beaver activity, ice damage, and other forms of natural disturbance. Although the importance of periodic moderate fires to other members of this genus is well documented for coastal plain species (McDaniel, 1971; Folkerts, 1977; Barker and Williamson, 1988), the role played by fire in the montane habitat of S. r. ssp. jonesii is not known. Although wildfires may have served historically to create openings for colonization, there is some evidence that direct burning of this species may be detrimental (Case, personal communication, 1990; Weakley and Schafale, this volume).

Another insectivorous plant native to southern Appalachian bogs is the green pitcher plant (Sarracenia oreophila), also federally listed as endangered. This species is not as restricted as mountain sweet pitcher plant. It occupies areas of the Cumberland Plateau and the Ridge and Valley Province in northeast Alabama to the Blue Ridge of Georgia and North Carolina. Historically, the species also occurred in the coastal plain and piedmont of Alabama, as well as the Cumberland Plateau of central Tennessee, but these sites have been destroyed. There are indications that this species was once relatively common in parts of its range, but it has now been reduced to 35 populations in Alabama (Cherokee, DeKalb, Etowah, Jackson, and Marshall Counties), Georgia (Towns County), and North Carolina (Clay County). Half the remaining populations contain 50 or fewer individuals, and only five populations contain 500 or more individuals (Norquist, 1993).

Green pitcher plant, like mountain sweet pitcher plant, is a rhizomatous perennial. The leaves are of two types; the pitchers (hollow leaves) appear first in spring and grow to 28 inches tall. Pitcher leaves are green to yellow-green in color, and sunlit leaves sometimes appear suffused with maroon or are maroon-veined externally. By late summer the pitcher leaves have withered and have been replaced by falcate phyllodia (flattened leaves) that persist until the following season. The presence of numerous falcate phyllodia at the base of the plant serves as the main feature, distinguishing this species from the similar Sarracenia flava and S. alata (Norquist, 1993; McDaniel, 1971; Troup and McDaniel, 1980). The yellow flowers are borne on single stems 18 to 28 inches tall from April through May, with an occasional plant flowering out of season, usually in October (Norquist, 1993; Folkerts, 1992). Asexual reproduction by rhizomes appears to be the principal mode of reproduction (Troup and McDaniel, 1980; McDaniel, 1991). Rhizomes live for decades, and natural mortality is low (Folkerts, 1992). The main pollinator for this species is the queen bumblebee (Bombus). Folkerts (1992) noted that pollinator success was low at many sites, particularly at wooded sites, where 20 or fewer flowers were present. Patch size is known to be a major limiting factor for pollination success by bees (Rymal and Folkerts, 1982). Since the flight radius of Bombus species is no more than 1 mile, most of the green pitcher plant populations are thought to be genetically isolated (Folkerts, 1992; Norquist, 1993).

Green pitcher plant, unlike mountain sweet pitcher plant, grows in a variety of habitat types, including sandy stream banks, mixed oak flatwoods, and seepage bogs (Norquist, 1993; McDaniel and Troup, 1980). The two southern Appalachian occurrences for this species--northeastern Georgia and adjacent North Carolina--are seepage bogs, where associated species are more typical of coastal plain wetlands (Weakley and Schafale, this volume). Very few examples of this habitat type remain, and none of them remain in an undisturbed state. The presence of many fire-adapted species in this habitat suggests that it may have been maintained historically by periodic wildfires. Periodic fire is believed necessary for the survival of Cumberland Plateau populations of this species, and plants have responded well to prescribed burning at one Blue Ridge site (Norquist, 1993).

The greatest threat to the continued survival of this species is the clearing and improvement of land for residential, agricultural, silvicultural, and industrial purposes (Troop and McDaniel, 1980). Other factors involved in the destruction of historic sites include reservoir impoundment, intensive pasturing (causing trampling of plants and excess eutrophication of the sites), hydrological alteration, and woody succession related to fire suppression. As with many other carnivorous plants, over-collecting by amateur and commercial collectors has resulted in the depletion or complete extirpation of many populations (Troup and McDaniel, 1980).

Bunched arrowhead (Sagittaria fasciculata) is another plant endemic to bogs of the southern Appalachian Mountains and upper foothills. Nineteen colonies remain, all within Henderson County, North Carolina, and Greenville County, South Carolina, in the French Broad, Enoree, Tyger, and Reedy River drainages. At least eight populations are known to have been destroyed. Undisturbed sites (what few that remain) are typically located just below the origin of slow seeps that supply a continuous flow of cool, clean water (Rayner, 1981; Sutter, 1983) on gently sloping terrain in deciduous woodlands. Soils are usually of the Toxaway, Wehadkee, or Hatboro series (U.S. Department of Agriculture, 1980). Associated species include many with affinities to northern coastal plain sites (Weakley and Schafale, this volume).

Emergent leaves of this species are spatulate in shape and can reach 11 inches in length and 2 inches in width. Winter rosettes and submersed leaves are much smaller. The white flowers appear in mid-May, with flowering continuing to July. The lower flowers of the inflorescence are pistillate while the upper flowers are staminate. The fruit, a broadly winged achene, matures within a few weeks after flowering. Bunched arrowhead is easily distinguished from other species within the genus because it is the only Sagittaria species in the southern Appalachians with nonsagittate leaves. The achene is also distinctive (Massey et al., 1983; Sutter, 1983). This species' habitat, much of which has already been destroyed, is threatened by the same factors listed for mountain sweet pitcher plant above.

Southern appalachian wetlands are home to a number of rare species with distributions that extend into the northern states. Swamp pink (Helonias bullata), federally listed as threatened, is a distinctive perennial lily occurring in wetlands from New Jersey and Delaware south to northern Georgia. The largest population known for this species occurs in North Carolina on the Pisgah National Forest. The species occupies a variety of wetland types throughout its range, including Atlantic white cedar swamps, meadows, bogs, spring seepage areas, and swampy forested wetlands bordering meandering streams. In the southern Appalachians the species occurs in bogs and swamp forests in Augusta and Nelson Counties, Virginia (18 populations); Jackson, Henderson, and Transylvania Counties, North Carolina (8 populations); Greenville County, South Carolina (1 population); and Rabun County, Georgia

(1 population). As described in the recovery plan for the species (U.S. Fish and Wildlife Service, 1991a):

> "The most evident factor determining the suitability of habitat for Helonias is a constant water supply. The groundwater-influenced wetlands supporting the species are perennially saturated and rarely, if ever, inundated by floodwaters (Rawinski and Cassin, 1986). The water table is at or very near the surface and fluctuates only slightly during spring and summer months (Sutter, 1982)."

Soils in the southern part of the species' range are composed of a thin layer of decomposed organic matter overlaying silt loam (often of the Toxoway and Hatboro series). The subsoil is usually a mixture of sand, loam, and gravel. Canopy cover varies in density, with more open sites having less vigorous plants and suffering more intensive deer browsing (Sutter, 1982).

The oblong-spatulate or oblanceolate leaves are evergreen, forming a basal rosette. A thick, hollow flowering stem rises from the rosette, growing to a height of 4.5 feet. The inflorescence consists of 30 to 50 pink to lavender flowers. In the South, flowering usually takes place from April to May. Typically, only a small percentage of the plants in a population flower, and reproduction is primarily through clonal rhizomal growth (U.S. Fish and Wildlife Service, 1991a).

Helonias is threatened by habitat loss, fragmentation, and degradation, collection, and trampling (U.S. Fish and Wildlife Service, 1991a). Many populations have been destroyed by drainage and impoundment of habitat, as well as by conversion of sites for recreational, resort, residential, and industrial development. In addition, many other populations have been seriously impacted by the indirect impacts of upstream or upslope development. Swamp pink is one of many bog species that is especially sensitive to siltation. The original buffer zones designed to protect the species were later found to be insufficient; the recovery plan now recommends a minimum of 500 feet of vegetated buffer to protect these wetlands from runoff associated with the clearing and disturbance of adjacent areas (U.S. Fish and Wildlife Service, 1991a). The plan states that, in some cases, protection of the entire watershed may be necessary in order to adequately protect the species. In developed watersheds, where storm water is discharged through outfall structures, the frequency and duration of normal storm-related flooding is changed, leading to unnatural and detrimental impacts to wetlands from increased floodwater elevations, increased flow rates, and increased deposition of sediment. Swamp pink is apparently very slow, and perhaps unable, to recolonize openings in suitable habitat where there are no adjacent parent plants. This limited capacity for colonizing new sites emphasizes the need to protect existing sites (Virginia Natural Heritage Program, 1987; U.S. Fish and Wildlife Service, 1991a). Although habitat destruction and degradation are the most serious threats to this species, it is also threatened in some areas by collectors.

Bog asphodel (Narthecium americanum), a candidate for imminent federal listing, once occurred in mountain bogs in North Carolina and South Carolina, as well as wetland sites in Delaware and New Jersey. It has been completely eliminated from the southern states by habitat destruction and degradation.

The white fringeless or monkey-face orchid (Platanthera integrilabia) historically occurred in bogs in seven states. It is now restricted to just 17 populations in Georgia, Kentucky, South Carolina, and Tennessee, having been completely eliminated from the states of North Carolina, Alabama, and Mississippi. This species was once reported to be relatively common on the Cumberland Plateau, but with accelerating wetland destruction, it has become increasingly rare (Zettler and Fairey, 1990) and is now a candidate for federal listing as endangered or threatened.

Until recently, two other candidates for federal listing, rough rush (Juncus caesariensis) and bog bluegrass (Poa paludigena), were not known from the southern mountains. The previously known distribution for bog bluegrass was primarily in the Midwest and Northeast--Iowa, Illinois, Indiana, Michigan, Minnesota, New York, Ohio, Pennsylvania, Wisconsin, and Virginia. Rough rush was known only from New Jersey, Maryland, and Virginia. Both were recently discovered in mountain bogs in North Carolina (A. Weakley, North Carolina Natural Heritage Program, personal communication, 1993).

Certain bogs and other mountain wetlands are habitat for rare species that ordinarily do not occur in wet areas in the remainder of their range, including Heller's blazing star (Liatris helleri), which is federally listed as threatened. This North Carolina endemic usually occurs only on high cliffs and rocky summits, but it also grows at one wetland site in North Carolina's Bluff Mountain fen in Ashe County. However, this species is not typically considered a wetland species. Other rare species that occupy bogs, in addition to drier sites, include turkey beard (Xerophyllum asphodeloides), a species that is more widespread in northern states than here, and Gray's lily (Lilium grayi), a southern Appalachian endemic and a candidate for federal listing. Gray's lily also grows on mountain balds, high meadows, and other forest openings that are not wet (Massey et al., 1983).

Two other species, robin-run-away (or false violet, Dallibarda repens) and Canada burnet (Sanguisorba canadensis), are more common in the northern United States and Canada but are rare in the southern Appalachians, where they occur only in mountain bogs. As stated by Weakley and Schafale (this volume), the presence of these species is indicative of the relictual nature and great age of southern Appalachian bogs and fens.

4. Threats

The greatest threats to the rare species of mountain wetlands are habitat destruction and degradation. Collection of commercially valuable species, such as the carnivorous plants and the bog turtle, exacerbates the problem. The overt forms of habitat destruction are mentioned above within the detailed species accounts (e.g., drainage, impoundment, clearing for resort and residential development, and conversion to intensive agricultural use). However there are other more insidious threats to the rare species of these habitats. Although the role of fire in mountain wetlands is not understood, there is some evidence, as mentioned in the green pitcher plant account, that fire suppression may be detrimental to some types of bog. Channelization of adjacent streams can result in destruction of the hydrological integrity, even if the bog itself is not directly targeted. The deepening and widening of the stream channel often causes a lowering of the local water table, which results in drying of the bog habitat and acceleration of shrub succession. Site conversion to "productive" uses, such as row crops or improved pasture, usually follows. Many historically known populations of rare species have been eliminated in this manner. Many of the remaining mountain wetlands are in close proximity to agricultural fields, pastures, orchards, Christmas tree farms, and nurseries. Accidental herbicide drift or runoff from these areas or from adjacent power line maintenance operations could result in damage or destruction of populations of rare plants, many of which are very small. In addition, fertilizer runoff can contribute unwanted nutrients to bog systems, detrimentally affecting the native plants. This is particularly harmful to the carnivorous plants and to sphagnum. Sphagnum is the keystone species in many mountain wetlands, maintaining the hydrology of the site by holding water and slowly releasing it. Many of the rare species associated with these habitats, including the bog turtle, four-toed salamander, and pitcher plants, live in the moss. Turtles and salamanders lay their eggs inside moss clumps, and the rhizomes of pitcher plants are covered by sphagnum. Pitcher plant seedlings often get started in the edges of sphagnum mats. During dry periods, the sphagnum holds and conserves water and prevents the upper soil layers from completely drying out. Grazing of wetland habitats, while it serves to keep woody succession in check, thereby benefitting many rare species (particularly the bog turtle), can be detrimental in some cases. Intensive grazing can result in the trampling of rare plants and the addition of too many nutrients to the system, resulting in the decline of the sphagnum, followed by the rest of the system.

In view of the fact that some of the bogs are thousands of years old, the question arises as to why many of them are now succumbing relatively quickly to encroachment by woody species. These are complex and poorly understood ecosystems. Without further detailed studies of their hydrology and the life history of their inhabitants, the answer to that question can only be speculative. However, we do know that there are very few unaltered mountain wetlands left, and that relatively minor alterations, such as clearing the surrounding uplands or channelizing an adjacent stream, can substantially dry these habitats. In very dry times, such as the three consecutive drought years of 1986, 1987, and 1988, woody species can become established in areas where unaltered water tables would not have allowed them to grow. Once established, shrubs and trees consume a tremendous amount of water, further drying the habitat and accelerating the process of succession. There is some evidence that global or continental climatic change could be contributing to the alteration of these wetlands. Stahle et al. (1988) and Stahle and Cleaveland (1992) report that the droughts of the mid-1980s were extremely rare events, equaled only five times in the past 1,600 years. These authors further state that, given the continued accumulation of greenhouse gases, summer droughts are expected to increase over central North America during the next century. This could have serious implications for our few remaining mountain wetlands, many of which have already been seriously altered hydrologically. (See Weakley and Schafale, this volume, for a detailed discussion of threats to mountain wetland habitats.)

5. Conservation efforts

With the realization that most of our mountain wetlands are already gone, conservation agencies in southern Appalachian states have intensified efforts in recent years to protect what remains. Nevertheless, mountain wetlands continue to be lost and degraded, primarily because landowners do not understand their value or know little about the unique plants and animals that inhabit them. Attempts at restoring degraded wetlands have met with very limited success. In South Carolina, where one of the best remaining populations of the endangered bunched arrowhead was buried under several thousand tons of excess dirt from a highway project, the U.S. Fish and Wildlife Service; the U.S. Army Corps of Engineers, Charleston District; and the South Carolina Heritage Trust Program worked with the highway department to remove the fill and to try to restore the original hydrology of the site. In spite of some heroic efforts, restoration did not work in that situation. Once such a wetland is drastically altered, it is almost impossible to repair. The U.S. Fish and Wildlife Service is currently working with The Nature Conservancy and two private landowners to restore the hydrology of four damaged bogs in the mountains of North Carolina. These sites do not appear to be as seriously altered as the South Carolina bunched arrowhead site; however, much remains to be learned about these complex systems. Restoration may prove impossible, even for those affected by seemingly minor perturbations.

The U.S. Fish and Wildlife Service has jointly funded a project with the North Carolina Natural Heritage Program to conduct landowner research for all the remaining bogs in that state and to begin contacting the owners about protecting their sites. Past experience emphasizes the importance of protecting the few intact mountain wetlands that remain.

References

Barker, N., and Williamson, G.: 1988, Effects of a winter fire on Sarracenia alata and S. psittacina. American Journal of Botany 75, 138-143.

Bell, C. R.: 1952, Natural hybrids in the genus Sarracenia. I. History, distribution, and taxonomy. Journal of the Elisha Mitchell Scientific Society 68, 55-80.

Case, F., and Case, R.: 1976, The Sarracenia rubra complex. Rhodora 78, 270-325.

Conant, R., and Collins, J. T.: 1991, A field guide to reptiles and amphibians, eastern and central North America. Houghton Mifflin Company, Boston.

Dahl, T. E.: 1990, Wetlands losses in the United States 1780s to 1980s. U.S. Department of the Interior, Fish and Wildlife Service, Washington, DC.

Dahl, T. E., and Johnson, C. E.: 1991, Status and trends of wetlands in the conterminous United States, mid-1970s to mid-1980s. U.S. Department of the Interior, Fish and Wildlife ervice, Washington, DC.

Folkerts, G.: 1977, Endangered and threatened carnivorous plants of North America. IN Extinction is forever: status of endangered and threatened plants of the Americas, eds. Prace, G. T., and Elias, T. S.; New York Botanical Garden, Bronx, NY, pp. 301-313.

Folkerts, G.: 1982, The gulf coast pitcher plant bogs. American Scientist 70, 260-267.

------: 1992, Identification and measurement of damage caused by flower and seed predators associated with Sarracenia oreophila and recommended management/control measures deemed appropriate. Unpublished report submitted to U.S. Fish and Wildlife Service, Jackson, MS.

Fontenot, L. W., and Platt, S. G.: 1993, A survey of the distribution and abundance of the bog turtle, Clemmys muhlenbergii, in South Carolina. South Carolina Wildlife and Marine Resources Department, Heritage Trust Program, Clemson, SC.

Givnish, T. J.: 1988, Ecology and evolution of carnivorous plants. IN Abrahamson, W. B. (ed.), Plant-Animal Interactions. McGraw-Hill, New York, pp. 243-290.

Herman, D. W.: 1992, Status and distribution of the bog turtle, Clemmys muhlenbergii, in North Carolina. Report submitted to the North Carolina Wildlife Resources Commission, Raleigh, NC.

LeGrand, H. E., Jr.: 1993, Natural Heritage Program list of the rare animal species of North Carolina. N.C. Natural Heritage Program, Raleigh, NC.

Martof, B. S., Palmer, W. M., Bailey, J. R., and Harrison, J. R., III: 1980, Amphibians and reptiles of the Carolinas and Virginia. University of North Carolina Press, Chapel Hill.

Massey, J., Otte, D., Atkinson, T., and Whetstone, R.: 1983, An atlas and illustrated guide to the threatened and endangered vascular plants of the mountains of North Carolina and Virginia. General Technical Report SE-20. Asheville, NC: USDA Forest Service, Southeastern Forest Experiment Station, pp. 156-159.

MacFarlane, J.: 1908, Sarraceniaceae. IN Engler, A., Das pflanzenreich 4, 110.

McDaniel, S.: 1971, The genus Sarracenia (Sarraceniaceae). Bulletin of Tall Timbers Research Station 9, 1-36.

Mitchell, J. C., Buhlmann, K. A., and Ernst, C. H.: 1991, IN Virginia's endangered species; proceedings of a symposium. Terwilliger, K., ed. McDonald and Woodward Publishing Company, Blacksburg, VA.

Murdock, N.: 1990, Recovery plan for mountain sweet pitcher plant (Sarracenia rubra ssp. jonesii [Wherry] Wherry). U.S. Fish and Wildlife Service, Atlanta, GA.

Norquist, C.: 1993, Green pitcher plant (Sarracenia oreophila) recovery plan; draft second revision. U.S. Fish and Wildlife Service, Jackson, MS.

Opler, P.: 1992, A field guide to butterflies. Houghton Mifflin Company, New York.

Opler, P., and Krizek, G.: 1984, Butterflies east of the Great Plains. Johns Hopkins University Press, Baltimore.

Potter, E. F., Parnell, J. F., and Teulings, R. P.: 1980, Birds of the Carolinas. University of North Carolina Press, Chapel Hill.

Radford, A., Ahles, H., and Bell, C.: 1964, Manual of the vascular flora of the Carolinas. University of North Carolina Press, Chapel Hill.

Rawinski, T., and Cassin, J. T.: 1986, Range-wide status summary of Helonias bullata. Unpublished report to the U.S. Fish and Wildlife Service.

Rayner, D.: 1981, Sagittaria fasciculata in North Carolina. Unpublished proposal to the North Carolina Plant Conservation Program, Raleigh, NC.

Schafale, M. P., and Weakley, A. S.: 1990, Classification of the natural communities of North Carolina, third approximation. N.C. Natural Heritage Program, Division of Parks and Recreation, Raleigh, NC.

Stahle, D. W., Cleaveland, M. K., and Hehr, J. G.: 1988, North Carolina climate changes reconstructed from tree rings: A.D. 372 to 1985. Science 240, 1517-1519.

Stahle, D. W., and Cleaveland, M. K.: 1992, Reconstruction and analysis of spring rainfall over the southeastern U.S. for the past 1000 years. Bulletin of the American Meteorological Society 73(12), 1947-1961.

Sutter, R.: 1982, The distribution and reproductive biology of Helonias bullata L. in North Carolina. North Carolina Department of Agriculture, Plant Industry Division, Raleigh, NC.

------: 1983, Recovery plan for the bunched arrowhead (Sagittaria fasciculata). U.S. Fish and Wildlife Service, Atlanta, GA.

------: 1987, Unpublished Sarracenia jonesii species account. North Carolina Plant Conservation Program.

Troup, R., and McDaniel, S.: 1980, Current status report on Sarracenia oreophila. U.S. Fish and Wildlife Service, Atlanta, GA.

Tryon, B. W.: 1990, Bog turtles (Clemmys muhlenbergii) in the south--a question of survival. Bulletin of the Chicago Herpetological Society 25(4), 57-66.

------: 1992, The bog turtle, Clemmys muhlenbergii, in Tennessee, 1992. Report submitted to the Tennessee Wildlife Resources Agency and the Tennessee Nature Conservancy, Nashville.

U.S. Department of Agriculture, Soil Conservation Service: 1980, Soil survey of Henderson County, North Carolina.

U.S. Fish and Wildlife Service: 1991a, Swamp pink (Helonias bullata) recovery plan. Newton Corner, MA.

------: 1991b, Endangered and threatened wildlife and plants; animal candidate review for listing as endangered or threatened species, proposed rule. Federal Register 56 (225), 58804-58836.

------: 1993a, Endangered and threatened wildlife and plants; 50 CFR 17.11 and 17.12. Division of Endangered Species, Washington, DC.

------: 1993b, Plant taxa for listing as endangered or threatened species; notice of review. Federal Register 58 (188), 51144-51190.

Virginia Natural Heritage Program: 1987, Status survey report for Helonias bullata (swamp pink) in Virginia.

Weakley, A. S.: 1993, Natural Heritage List of the rare plant species of North Carolina. N.C. Natural Heritage Program, Raleigh, NC.

Webster, W. D., Parnell, J. F., and Biggs, W. C., Jr.: 1985, Mammals of the Carolinas, Virginia, and Maryland. University of North Carolina Press, Chapel Hill.

Wherry, E.: 1929, Acidity relations of the Sarracenia. Journal of the Washington Academy of Science 19, 379-390.

Wofford, B. E.: 1989, Guide to the vascular plants of the Blue Ridge. University of Georgia Press, Athens.

Wood, C.: 1960, The genera of Sarraceniaceae and Droseraceae in the southeastern United States. Journal of the Arnold Arboretum (XLI), 152-163.

Zettler, L., and Fairey, J. E., III: 1990, The status of Platanthera integrilabia, an endangered terrestrial orchid. Lindleyana 5(4), 212-217.

PART V

MANAGED WETLANDS

RIPARIAN FOREST BUFFER SYSTEM RESEARCH AT THE COASTAL PLAIN EXPERIMENT STATION, TIFTON, GA

R. K. HUBBARD AND R. R. LOWRANCE

USDA-ARS, Southeast Watershed Research Laboratory

P. O. Box 946, Tifton, GA 31793

Abstract. Recent attention has focused on riparian forest buffer systems for filtering sediment, nutrients, and pesticides entering from upslope agricultural fields. Studies in a variety of physiographic areas have shown that concentrations of sediment and agrichemicals are reduced after passage through a riparian forest. The mechanisms involved are both physical and biological, including deposition, uptake by vegetation, and loss by microbiological processes such as denitrification. Current research by USDA-ARS and University of Georgia scientists at Tifton, GA is focusing on managing riparian forest buffer systems to alleviate agricultural impacts on the environment. The underlying concept for this research is that agricultural impact on streams is best protected by a riparian forest buffer system consisting of three zones. In consecutive upslope order from the stream these zones are (1) a narrow band of permanent trees (5-10 m wide) immediately adjacent to the stream channel which provides streambank stabilization, organic debris input to streams, and shading of streams, (2) a forest management zone where maximum biomass production is stressed and trees can be harvested, and (3) a grass buffer strip up to 10 m wide to provide control of coarse sediment and to spread overland flow. Several ongoing projects at Tifton, GA are focusing on using riparian forest buffer systems as filters. A forest management project is testing the effects of different management practices on surface and ground water quality. This project includes three different forest management practices: mature forest, selectively thinned forest, and clearcut. In a different study a natural wetland is being restored by planting trees. The effectiveness of this wetland on filtering nutrients from dairy wastes which are being applied upslope is being evaluated. At this same site, a pesticide study is being conducted on the side opposite to where dairy wastes are applied. An overland flow-riparian buffer system using swine lagoon waste is evaluating the effectiveness of different vegetative treatments and lengths of buffer zones on filtering of nutrients. In this study three vegetative treatments are compared: (1) 10 m grass buffer and 20 m riparian forest, (2) 20 m grass buffer and 10 m riparian forest, (3) 10 m grass buffer and 20 m of the recommended wetland species maidencane. Waste is applied at the upper end of each plot at either a high or low rate, and then allowed to flow downslope. The three zone riparian forest buffer system is being used for the Riparian Ecosystem Management Model (REMM). This model, which is currently under development at Tifton, GA, is a computer simulation model designed to reduce soil and water degradation by aiding farmers and land use managers in decision making regarding how best to utilize their riparian buffer system. Both information currently being collected in field studies and development of the REMM are innovative farm-level and forestry technologies to protect soil and water resources.

1. Introduction

Management of land resources in the United States historically has been focused primarily on the uplands. Riparian zones and wetlands largely were viewed as wasted land to be drained, while stream and river systems were channelized to control flooding. Ecosystem concepts including sustainable environment and management of riparian

Water, Air and Soil Pollution **77**: 409–432, 1994.

zones and wetlands are both relatively new. The concept that what happens in one part of the landscape affects other landscape segments, and that overall these events affect the environment globally is the foundation for much of our environmental research today. The concept of managing wetlands is still under some debate, with opinions ranging from those who believe that wetlands should exist in pristine condition with no interference from agricultural or other activities, to those who believe that wetlands are natural sinks that can be used to effectively filter materials entering from upslope, and that emphasis should be placed on restoration of natural wetlands so that they can resume their filtering function in the landscape.

Research conducted in the coastal plain of Georgia during the 1970's provided early evidence that riparian zones are effective nutrient filters. The research at this time primarily focused on nitrogen (N). Extensive studies (Asmussen *et al.*, 1979; Yates and Sheridan, 1983) of nutrient budgets in riparian areas of the Tifton Upland in the coastal plain of Georgia examined nitrate (NO_3-N) losses from cropped areas and riparian wetland zones, and NO_3-N loads in streamflow. It was estimated that 96 percent of the NO_3-N was retained, utilized, or transformed in the heavily vegetated riparian forests of the Coastal Plain (Yates and Sheridan, 1983). Stream outflow loads of NO_3-N on a mixed-use agricultural watershed were found to be lower than NO_3-N loads input by rainfall (Asmussen *et al.*, 1979). Filtering of N by riparian systems was attributed to both denitrification and vegetative uptake (Lowrance *et al.*, 1984a,b).

Denitrification occurs under anaerobic conditions. Denitrifiers are facultative anaerobes which carry out anaerobic respiration by substituting NO_3-N or a related nitrogenous compound for oxygen as the terminal electron acceptor (Rowe and Stinnett, 1975). The pathway generally proceeds by reduction from NO_3-N to NO_2-N to N_2O-N to N_2 gas although some intermediates have been postulated (Rowe and Stinnett, 1975).

A number of factors influence denitrification rate. Denitrification rates are slower in acid soils than in alkaline soils. The relative amounts of N_2O-N and N_2 produced are affected by temperature, with N_2O being predominantly formed at lower temperatures, and N_2 at higher ones (Rowe and Stinnett, 1975). Conditions conducive to denitrification are commonly found in fine-textured, water logged soils with high organic content. Water apparently has a direct effect on denitrification: the closer the soil is to saturation, the more denitrification occurs. Little denitrification occurs in soils less than about 60 percent saturated (Broadbent and Clark, 1965). High organic content is conducive to denitrification, because heterotrophic denitrifiers need oxidizable organic material as a source of carbon for synthesis of protoplasm and as a source of electrons for the reduction of nitrogenous compounds (Rowe and Stinnett, 1975). Groffman *et al.* (1992) found that hydric surface (0-15 cm) soils (poorly and very poorly drained) consistently had higher denitrification enzyme activity than upland-wetland transition zone (moderately well and somewhat poorly drained) surface soils. Peterjohn and Correll (1984) estimated N dissimulation by denitrification to be 45 kg N ha^{-1} yr^{-1} for riparian forested areas.

The second mechanism by which riparian zones can reduce NO_3-N concentrations in water arriving from uplands is through vegetative uptake, particularly by the trees in forested riparian zones. Several investigators (Vitousek and Reiners, 1975; Lowrance *et al.*, 1983, 1984b) have suggested that select harvest of "mature trees in riparian forests is a method of perpetuating vigorous vegetative uptake of soil nutrients. Odum (1969) hypothesized that constant, pulsed, and annually increasing impacts of nutrients may keep the riparian forest in a "bloom" state, and the forest may respond by high and vigorous growth and nutrient uptake rates for a considerable period of time; much longer, perhaps, than the age generally considered as forest maturity. Work of Peterjohn and Correll (1984) using a nutrient mass balance approach indicated a net retention by a forested wetland of 75 kg total N ha^{-1} yr^{-1}. Nutrient assimilation and long-term storage in wood biomass ranged from 12 to 20 kg N ha^{-1} yr^{-1}.

The combined effects of denitrification and NO_3-N uptake in the riparian zone on N concentrations entering from the uplands have been documented by a number of investigators. Concentrations of N in many watersheds have remained nearly constant as loading of N from agricultural nonpoint sources has increased (Tomlinson, 1970; Thomas and Crutchfield, 1974; Hill and Wylie, 1977). Reddy and Graetz (1981) found that shallow reservoirs and flooded organic soils could be used for ammonium (NH_4-N) and NO_3-N removal from wastewater. Van Kessel (1977) measured NO_3-N removal rates in ditches, and found them to be as high as commercial treatment of sewage. Robinson *et al.* (1978), Hoare (1979), and Hill (1983) measured significant stream loss of NO_3-N by denitrification which greatly affected stream and watershed budgets. Brinson et al. (1981) showed that 75% of NH_4-N and 94% of NO_3-N was removed as floodwater moved through two riverside swamps. Other studies which have shown the role of forested wetlands as partial nutrient sinks include those of Kitchens *et al.* (1975), Boyt *et al.* (1976), Ewel and Odum (1978), Nessel (1978), Mitsch *et al.* (1979), Tuschall *et al.* (1981), Day *et al.* (1981), and Qualls (1984).

Research in the North Carolina coastal plain on soils having significant shallow subsurface flow showed than from 10 to 56 kg ha^{-1} yr^{-1} of NO_3-N moved from cropped fields in subsurface drainage water (Jacobs and Gilliam, 1985). Natural riparian vegetation downslope from the cropped fields resulted in a substantial portion of the NO_3-N in the drainage water being removed. Nitrate losses were believed to be due primarily to denitrification. Buffer strips less than 15 m wide caused significant losses of NO_3-N before runoff water or subsurface flow reached the stream. It was also noted in this study that while soybean production had increased 760 percent since 1945, with commensurately more N fixation, and while fertilizer use had increased 400 percent since 1945, no proportional increase in the N content of most coastal plain streams has been observed. Jacobs and Gilliam (1985) concluded that from an environmental standpoint, the most effective system for removing N is a natural drainageway bordered by poorly drained soil and dense riparian vegetation. Gilliam and Terry (1973) also found no increase in NO_3-N in streams in North Carolina over the last 50 years despite increases in fertilizer application.

A series of studies (Lowrance *et al.*, 1983; Lowrance *et al.*, 1984a; Lowrance *et al.*, 1984b; Lowrance *et al.*, 1984c; Lowrance *et al.*, 1985) in the Georgia coastal plain in the 1980's measured streamflow and shallow groundwater quality and found reduction in NO_3-N levels in waters leaving the riparian zone as compared to the agricultural upland. Total annual N inputs to the riparian zone averaged 12.2 kg ha^{-1} yr^{-1} in precipitation, 29 kg ha^{-1} yr^{-1} in subsurface flow, 10 kg ha^{-1} yr^{-1} in surface runoff, and 10.6 kg ha^{-1} yr^{-1} as N fixation for a total of 61 kg ha^{-1} yr^{-1}. Losses of N in streamflow averaged 13 kg ha^{-1} yr^{-1}. It was projected that in this physiographic region, total replacement of riparian forest with a mixture of crops similar to those grown on the upland would increase present mean annual NO_3-N concentrations in streamflow from 0.20 mg L^{-1} to an estimated 4 mg L^{-1}.

Both direct and indirect approaches have been taken in studying the role of riparian forests in agricultural settings. One direct method is use of transects running from the edge of the agricultural fields through the riparian forest (Doyle *et al.*, 1975; Lowrance *et al.*, 1984a; Peterjohn and Correll, 1984; Lowrance, 1992; Hubbard and Lowrance, 1993; Simmons *et al.*, 1993). Lowrance (1992), using a transect of wells from a row-crop field to a stream in the Georgia coastal plain, determined that NO_3-N in groundwater decreased by a factor of 7 to 9 in the first 10 m of forest. Within the next 40 m of forest, the mean NO_3-N concentration decreased from 1.80 to 0.81 mg NO_3-N L^{-1}. Simmons *et al.* (1992) assessed the removal of groundwater NO_3-N on a soil drainage sequence ranging from moderately well to poorly drained. To assess NO_3-N removal, the change in groundwater concentrations of NO_3-N relative to the concentration of the conservative tracer Br was observed in monitoring wells located in each soil drainage class. Removal of groundwater NO_3-N was consistently high in the wetland locations, generally in excess of 80% in both growing and dormant seasons. In the transition zones, attenuation was less than 36% during the growing season, and ranged from 50 to 78% in the dormant season. Attenuation in the transition zones was positively correlated with water table elevations. A second direct method is to utilize chemical budgets (Lowrance *et al.*, 1983; Todd *et al.*, 1983; Lowrance *et al.*, 1984b; Peterjohn and Correll, 1984).

More indirect methods have compared the nutrient concentrations in stream water from agricultural watersheds with varying amounts of riparian forest or have analyzed the predictive capability of models which include or exclude the presence or proximity of riparian forest (McColl, 1978; Schlosser and Karr, 1981a: Schlosser and Karr, 1981b; Omernik *et al.*, 1981, Yates and Sheridan, 1983). With the exception of the Omernik *et al.* (1981) study, all studies have reached the general conclusion that riparian forests effectively reduce the loss of N from agricultural lands to receiving waters.

Lowrance *et al.* (1983), working in the Georgia coastal plain, used a nutrient budget approach to determine the effects of riparian zones on water quality. In preparing the schematic shown as Figure 1, they assumed water and nutrient movements from different land uses to be proportional to the length of interface between each

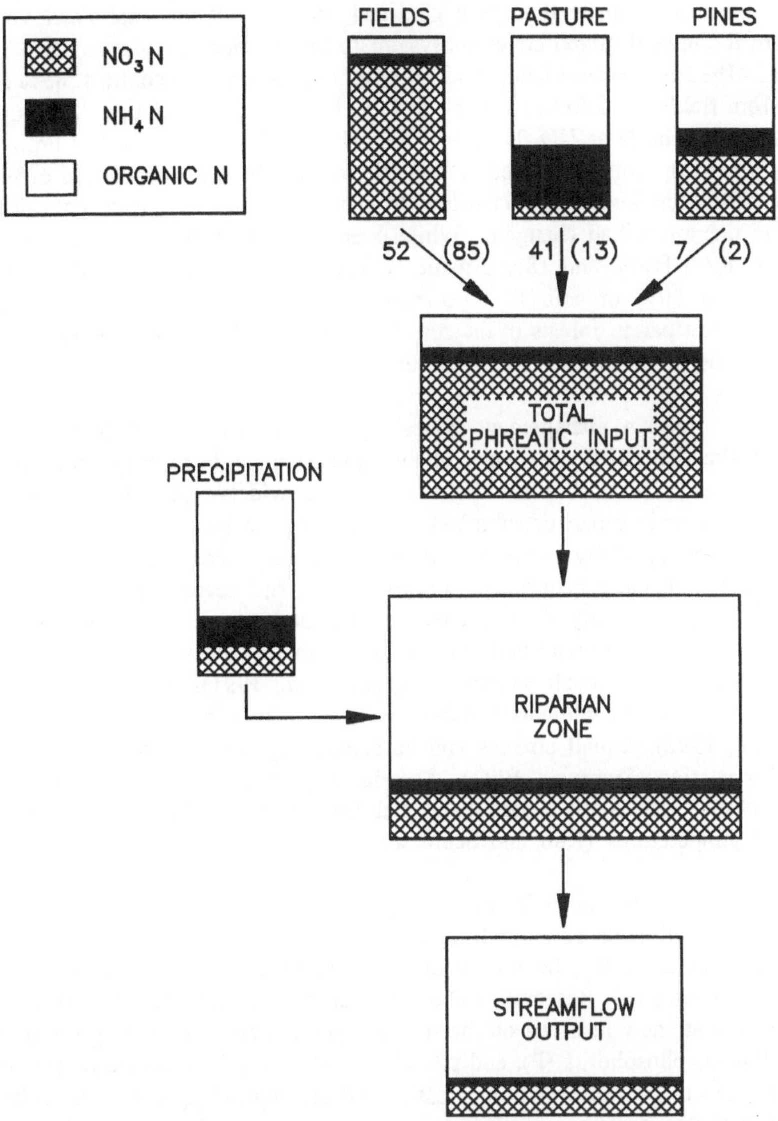

Figure 1. Change in N forms from subsurface and precipitation inputs to streamflow outputs. Bars represent relative amounts of nitrate, ammonium, and organic N making up inputs or outputs. Numbers on arrows represent percentages of water discharge and N (in parentheses) coming from fields, pastures, and pine forests. (From Lowrance et al., 1983)

land use and the riparian zone. Nitrogen movement from fields (row crops), pastures, and upland pine forest differed in both kind and amount. Of the total water volume moved from the upland mixed cover ecosystem to the riparian ecosystems, 52% came from fields, 41% from pasture lands, and 7% from pine forests. In contrast, 85% of the N moved from fields, 13% from pastures, and 2% from pines. Discharges from upland fields had 85% of the N as NO_3-N, 4% as NH_4-N and 11% as organic N. From pastures, only 9% of the total N moved as NO_3-N, while 31% was NH_4-N and 60% was organic N. Nitrogen losses in streamflow from the riparian zone were considerably smaller, and the forms had changed. While overall N inputs from the uplands were 74% NO_3-N, 8% NH_4-N, and 18% organic N, streamflow outputs were 18% NO_3-N, 2% NH_4-N, and 80% organic N. Lowrance *et al.* (1983) projected that partial conversion of the riparian forests of the coastal plain to cropland could increase NO_3-N and NH_4-N loads to streamflow by up to 800%.

The use of riparian management zones is relatively well established as a Best Management Practice (BMP) for water quality improvement in forestry practices (Comerford *et al.*, 1992), but has been much less widely applied as a BMP in agricultural areas or in urban or suburban settings. It is believed that riparian forest buffer systems are especially important on small streams where intense interaction of the terrestrial and aquatic ecosystems occurs. First and second order streams are estimated to comprise nearly three-quarters of the total stream length in the United States (Leopold, 1964). Fluvial activities influence the composition of riparian plant communities along these small streams (Gregory *et al.*, 1991). Likewise, terrestrial disturbances can have an immediate impact on aquatic populations (Sweeney, 1993; Webster *et al.*, 1992). Small streams can be completely covered by the canopies of streamside vegetation (Sweeney, 1992). Riparian vegetation is widely acknowledged to have beneficial effects on stream bank stability, stream biological diversity, and stream water temperatures (Karr and Schlosser, 1978).

In the late 1980's the need for more information on the filtering role of riparian zones relative to pollutants entering from agriculture was recognized. In response to this recognition USDA-ARS, beginning in 1989, committed monies and resources to the Southeast Watershed Research Laboratory at Tifton, GA, to expand existing research and initiate new research on the role of riparian zones in filtering not only N, but also sediment, phosphorus (P), and pesticides. The overall scientific objectives of the research were to gain information to assist in designing management strategies for riparian forest buffer systems so that permanent improvements in water quality can be achieved. Riparian forest buffer systems must be managed with an understanding of : 1) the processes which remove or sequester nonpoint source pollutants after they enter the system; 2) the effects of riparian management practices on the retention of nonpoint source pollutants; 3) the effects of riparian forest buffer systems on aquatic ecosystems; 4) the time till recovery after harvest of trees or re-establishment of riparian forest buffer systems; and 5) the effects of the underlying soil and geologic materials on chemical and biological processes.

A comprehensive program including both the United States Department of Agriculture - Agricultural Research Service (USDA-ARS) and cooperating scientists from the University of Georgia (UGA) was put together at Tifton, GA to intensively study riparian zone filtering processes and management. The program includes both research funded through USDA-ARS and UGA, and research funded through competitive grants. The projects include filtering of nutrients, sediment, and pesticides from row crop agriculture, restoration of a wetland to filter nutrients entering from an animal waste site, and comparison of different management techniques for filtering animal waste applied to riparian forest buffer systems by overland flow. The ultimate goal of the entire program of research is to determine management techniques for riparian forest buffer systems that contribute to a sustainable environment. This paper documents the projects in the riparian forest buffer systems research program.

2. Current Studies

Effect of Management of Riparian Forest Buffer Systems on Water Quality

In fall 1991 a project was started at a site near Tifton, GA to determine the effects of different riparian forest management practices on water quality (Figure 2). The work was started through funding by a CSRS National Research Initiative Competitive Grant (91-37102-6785) and with additional work at other sites will continue through 1996 with funding support by a new CSRS-NRI grant (93-37102-8955). The specific objectives of this research are to:

1) Determine rates of N, P, sediment, and bromide (Br) movement in surface and/or subsurface flow through managed and unmanaged areas of a riparian buffer system.

2) Determine the relative importance of denitrification, microbial immobilization, and root N uptake in NO_3-N removal in managed and unmanaged areas of a riparian forest buffer system.

3) Determine effects of the pesticide aldicarb (Temik) on soil microcosm processes that regulate N transformations in the riparian buffer system.

The investigations are being carried out at an existing riparian forest site located at the University of Georgia Coastal Plain Experiment Station near Tifton, GA (Figure 2). The soil of the riparian forest area is an Alapaha loamy sand (fine-loamy, siliceous, acid, thermic Typic Fluvaquents). The soil of the adjacent upland area is a Tifton loamy sand (fine-loamy, siliceous, thermic, Plinthic Kandiudults). The riparian forest trees are primarily slash pine (Pinus elliottii Engelm.) and long leaf pine (Pinus palustris Mill.). The 5 m nearest the stream channel supports different vegetation with more hardwoods including yellow poplar (Liriodendron tulipifera L.) and black gum (Nyssa sylvatica Marxh.). The forest provides a buffer which averages 55 m in width along an intermittent second-order stream channel.

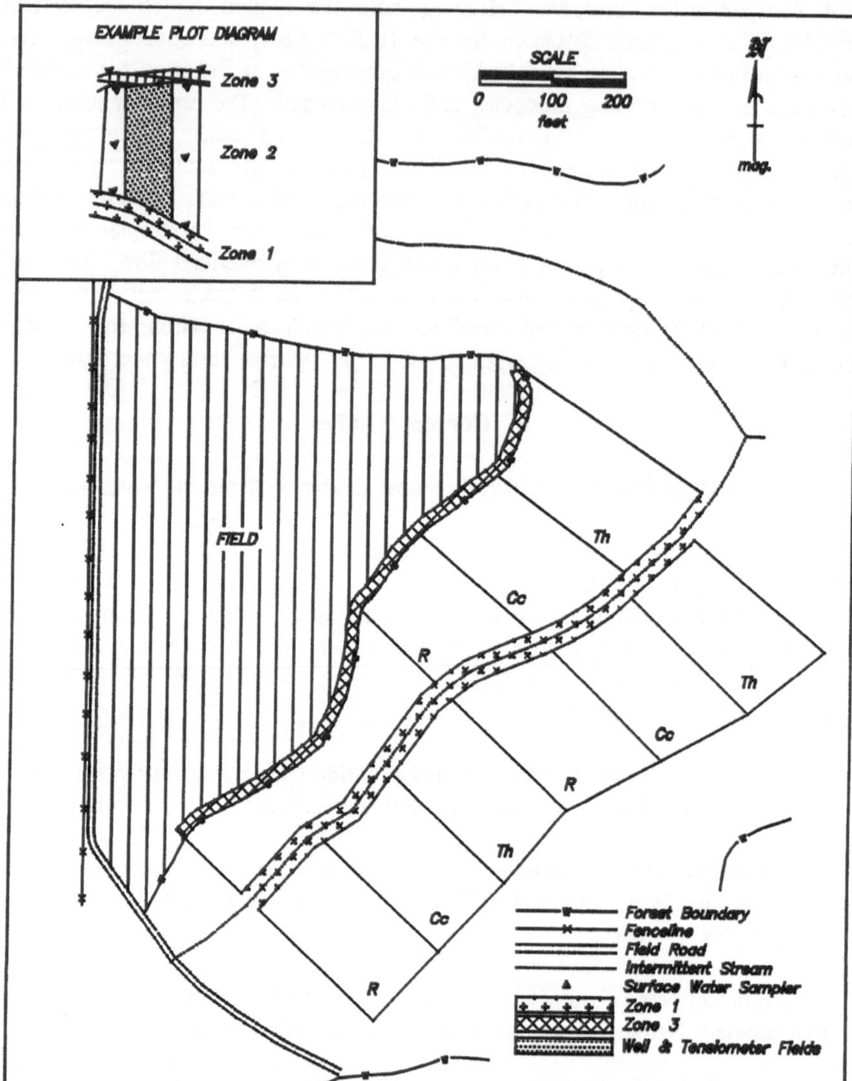

Figure 2. The three zone buffer system, the layout of Zone 2 treatment blocks, and the location of well and tensiometer fields and surface samplers within a representative block. Th = thinning cut; Cc = clear cut; R = reference-mature forest.

The riparian forest buffer system used for the study has three zones. The buffer system begins at the base of the field with Zone 3, a 5 m wide strip of Tifton 44 Coastal Bermudagrass (Cynodon dactylon L. Pers.). The Bermudagrass strip (Zone 3) is interplanted with abruzzi rye (Secale cereale L.) during winter to provide both biomass production and nutrient uptake. Zone 3 was established during fall 1991. The forest portion of the buffer consists of pines (Zone 2) and hardwoods (Zone 1). Zone 1 is a 10 m wide strip, adjacent to the stream channel, where trees will not be removed. Zone 2 is the remaining 50 m of forest between Zones 1 and 3. Zone 2 is being used for two forest management treatments and a reference. Treatment 1 is a clear-cut with all the merchantable tree biomass (either pulp wood or saw wood) removed. Removal of this material was accomplished during fall 1992. Limbs less than 6 cm dia. were left in the clear cut and thinned areas and were spread uniformly around the site. All other aboveground biomass was removed. Treatment 1 was reforested by planting of improved slash pine at a rate of 1,560 trees ha^{-1} during winter 1993. This rate was based on Georgia Forestry Commission (GFC) recommendations. Treatment 2, based on GFC recommendations, is a thinning cut with the standing biomass removed from all size classes to a target basal area of about 25 m^2 ha^{-1}. The thinning cut was done in fall 1992 at the same time as the clearcut. Woody and non-woody debris generated from the harvest treatments was allowed to remain in the respective treatments and was distributed within the sites at random by the log skidding operation. The control treatment for the study is the uncut forest. Each treatment is about 40 m wide and 50 m deep running downslope from the grass strip to the stream, on both sides of the stream channel.

The field above the buffer system on the west side of the stream was planted to corn (Zea mays L.) both in 1992 and 1993. The corn received 200 kg N ha^{-1} in fertilizer inputs. Atrazine and alachlor are being applied to the corn. The field above the buffer system on the west side of the stream will be in a three year rotation of corn-peanut-corn for 1994-1996. Conventional input corn (Zea mays L.) will be grown adjacent to the buffer area. The corn will receive 200 kg N ha^{-1} and 100 kg P ha^{-1} in fertilizer inputs in 1994 and 1996. No N will be applied to the peanuts, although fixation by high yielding peanuts can equal about 200 kg N ha^{-1} yr^{-1} (Hoyt, 1981). Fertilizer P additions will follow standard recommendations for peanuts of 50 kg P ha^{-1} yr^{-1}. Aldicarb will be applied to the peanuts to control thrips and nematodes.

Surface runoff measurements from the study to date show that sediment concentrations and loads were reduced by nearly 90% from the field output to the streamflow input. Mean sediment concentrations before timber harvest were 1281 mg L^{-1} entering the buffer system and 151 mg L^{-1} leaving the buffer systems.

Soil samplings from 1991 showed that in general, NO_3-N concentrations were high in the saturated layers at the edge of the upland field, but were significantly reduced within the first 5 m of the forest. Higher NH_4-N concentrations in the surface soils near the stream were consistent with the greater biomass N contents in this position.

2.1. WETLAND RESTORATION AND FILTERING OF ANIMAL WASTE

The project entitled "Development of an environmentally safe and economically sustainable year-round minimum tillage forage production system using farm animal manure as the only fertilizer" was started in 1991. This project, which was initially funded by the LISA (Low Input Sustainable Agriculture) program for a three year period, is being conducted by a multidisciplinary scientific team at a site where screened liquid dairy manure from a storage lagoon is applied on 5.6 ha by center pivot irrigation (Figure 3). The waste is applied to the east, west, south and north pivot quadrants at N application rates of 200, 400, 600, and 800 kg ha^{-1} yr^{-1}, respectively. These rates were selected so that the lowest rate is restrictive to plant growth and the highest rate is excessive for maximum plant growth based on N uptake rates calculated from previous experiments (Johnson *et al*., 1984). The cropping system consists of overseeding of abruzzi rye (Secale cereale L.) into Tifton 44 Bermudagrass (Cynondon dactylon L.) sod in the fall, followed by minimum tillage planting of silage corn (Zea mays L.) into the bermudagrass and rye stubble in the spring, followed by summer crops of hay or silage from the residual bermudagrass.

The north quadrant of the pivot, which receives an N rate of 600 kg ha^{-1} yr^{-1}, drains downslope into a wetland area (Figure 2). This area (0.92 ha), which was forested until 1985, was distinctly different from the upslope areas in vegetation at the start of the upland study. The vegetation before restoration was primarily wetland grasses (Paspalum sp.) and rushes (Juncus sp.).

The wetland is being restored to a forested condition which allows determination of the effects of the wetland on water quality during the restoration process. The specific objectives for the restoration were to: (a) measure nutrient (N, P) concentration changes in surface runoff and shallow groundwater as they move through the wetland; b) determine nutrient uptake and removal in the wetland by soil microbial processes and vegetation; and (c) evaluate the wetland as a potential bioremediation site.

The wetland was partially restored in February 1991 by reintroducing a combination of native trees over 0.47 ha (Figure 4). The trees will be grown for eventual harvest as pulpwood, timber wood, or both. Slash pine (Pinus elliottii Engelm.) and yellow poplar (Liriodendron tulipifera L.) were selected as a combination that would provide fast growth and year-round nutrient uptake. Slash pine is commonly used in the coastal plain in the landscape position analogous to Zone 2 of the riparian forest buffer system, while yellow poplar commonly grows in the wetter areas near the streams. The trees were planted with 1.5 m spacing within rows and 3 m spacing between rows to permit seasonal mowing of herbaceous vegetation.

Due to low survivorship of the poplars, swamp black gum (Nyssa sylvatica var. biflora Marsh.) and green ash (Fraxinus pennsylvanica (Borkh.) Sarg.) saplings were introduced to the hardwood area in April 1991. In April 1992, marsh cordgrass

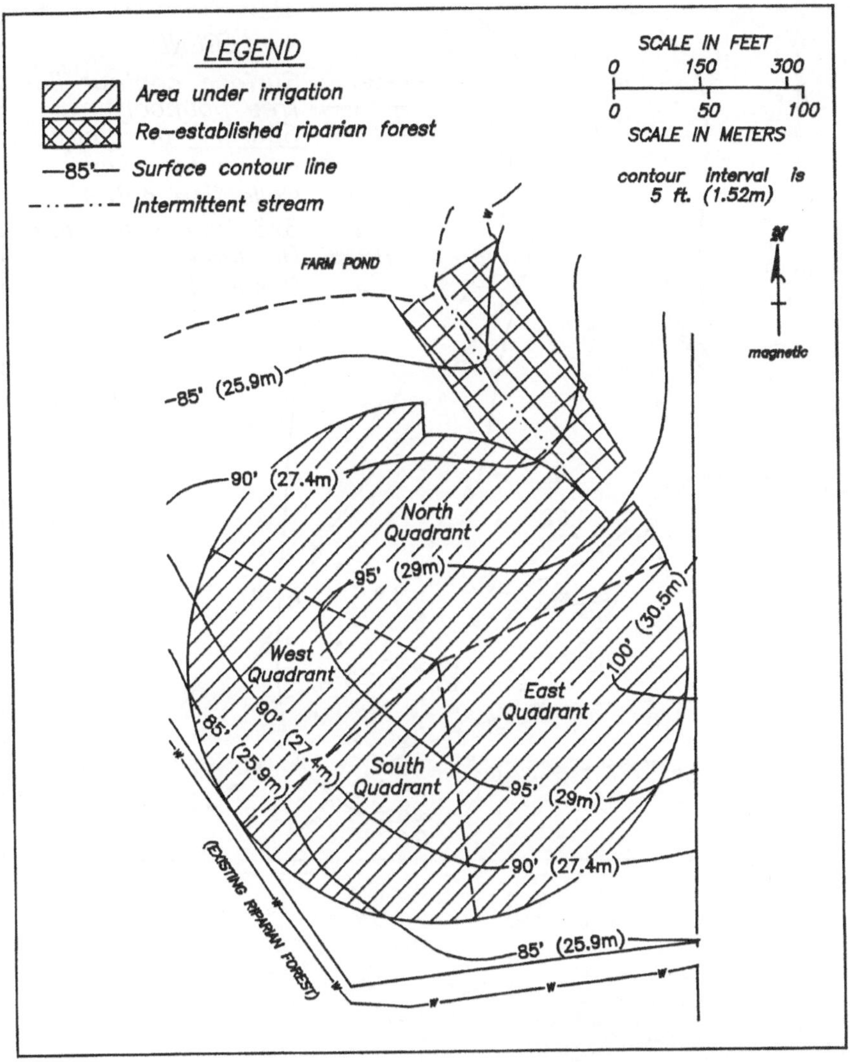

Figure 3. The dairy waste land application and dairy wetland restoration research facilities. The circle indicates the area irrigated by the center pivot waste application system. The wetland is indicated by the cross-hatched area.

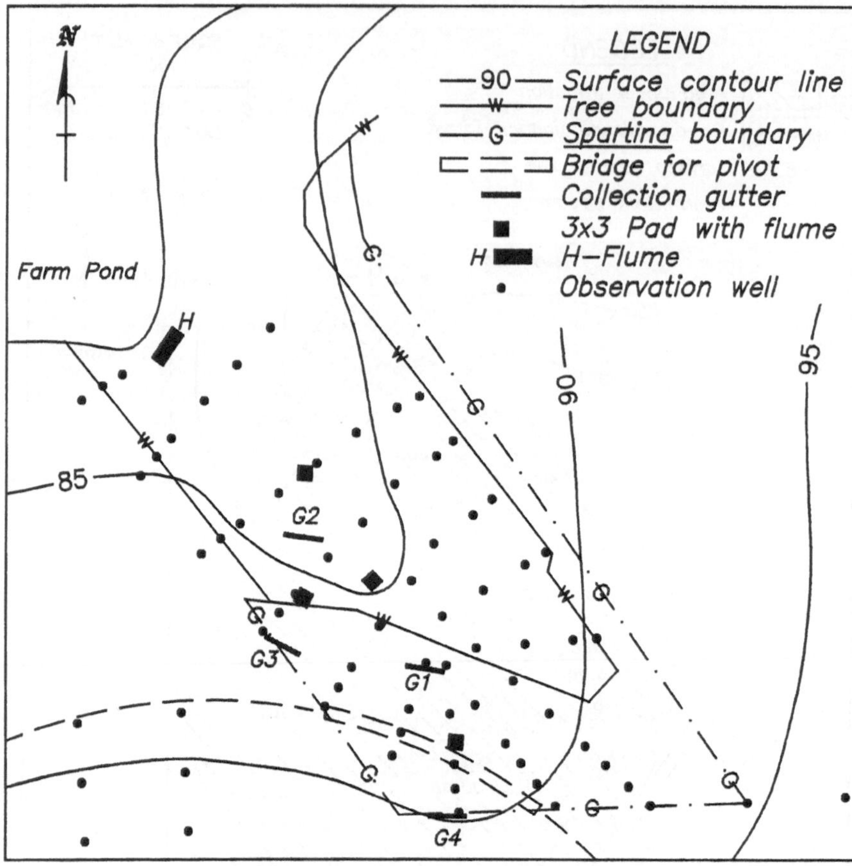

Figure 4. Map of the dairy wetland restoration site showing the network of monitoring wells and the location of the four collection gutters and flumes.

(Spartina patens (Aiton) Mulh.), a perennial grass, was established along the perimeter of the wetland (0.45 ha) to act as a transitional zone between the forage production system and the riparian forest.

Water quality measurements are made on water entering the wetland, moving through, and exiting. This is accomplished using a combination of monitoring wells, surface runoff collectors, and flumes. Nutrient movement and concentrations in the upland root zone and in shallow groundwater are being tracked in the north quadrant using both solution samplers and a network of 18 monitoring wells. The solution samplers are installed at 0.5 m, 1.0 m, 1.5 m, and 2.0 m depths. Nine of the 18 shallow groundwater wells in the north quadrant are at 3 m depth, while the other nine are at 6 m depth.

Groundwater in and around the perimeter of the wetland is monitored by 110 groundwater wells (Figure 4). On the side adjacent to the animal waste application site the wells are installed to a depth of 1 m and are fully slotted up to the soil surface. The opposite side from the animal waste site is being used for a pesticide study and the wells extend to a depth of 2 m. The upland wells were sampled and depth to the water table was measured biweekly from 1991 until June 1993 when a monthly sampling schedule was instituted. Water sampling of the suction lysimeters in the upland is on the same schedule as the sampling of the upland wells. A biweekly sampling schedule is used for measuring the depth to the water table and also collecting samples for nutrient and pesticide analyses in the wetland.

Surface runoff is sampled at two locations entering the wetland and at two locations near the stream flow (Vellidis *et al.*, 1992; Vellidis *et al.*, 1993). At each location, the runoff is collected in a gutter, passed through a 200 mm Modified Tucson Flume, and redistributed through a slotted gutter. Composite water samples are collected from the flumes with battery-powered peristaltic pumps.

Water samples from the wells in the upland are analyzed for NO_3-N, NH_4-N, total N, phosphate (PO_4), total P, calcium (Ca), potassium (K), magnesium (Mg), and sodium (Na) concentrations. Nitrate, NH_4-N, total N, PO_4, and total P analyses are performed by standard methods on the Lachat Flow Injection Analyzer (APHA, 1989). Analyses for Ca, K, Mg, and Na are done by atomic absorption. Nitrate concentrations in the wells on the north quadrant at 3 m have averaged about 15 mg L^{-1} since the start of the study (Figure 5). Water samples from the surface runoff collectors and shallow wells in the wetland are analyzed for NO_3-N, NH_4-N, total N, PO_4, and total P concentrations. The filtering effect of the wetland on NO_3-N is apparent from the contrast in NO_3-N concentrations in shallow wells at the upper edge of the wetland, and those found in wells in the stream area (Figures 6 and 7). Shallow lateral subsurface flow is the dominant loss pathway for water and solutes in the coastal plain of Georgia, as has been shown by numerous studies involving NO_3-N and conservative tracers such

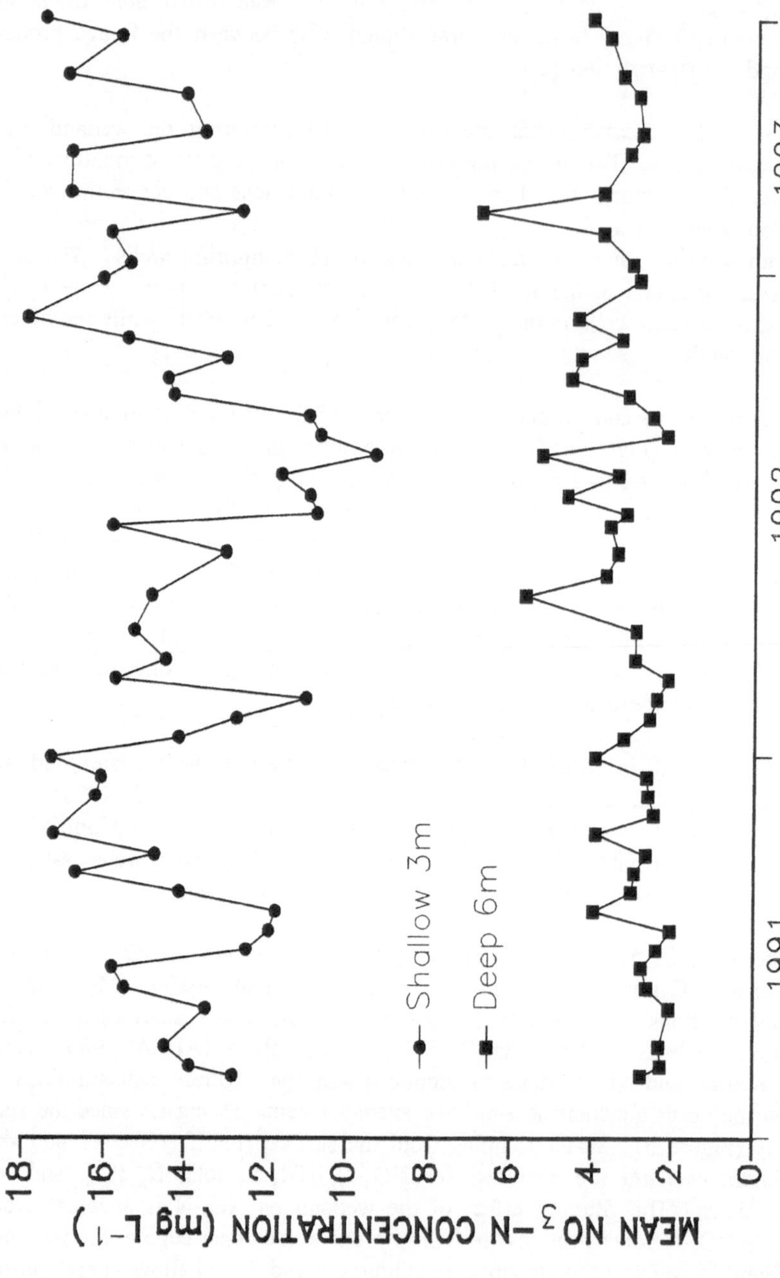

Figure 5. Mean NO_3-N concentrations in 3 and 6 m dairy upland wells in the quadrant draining into the wetland.

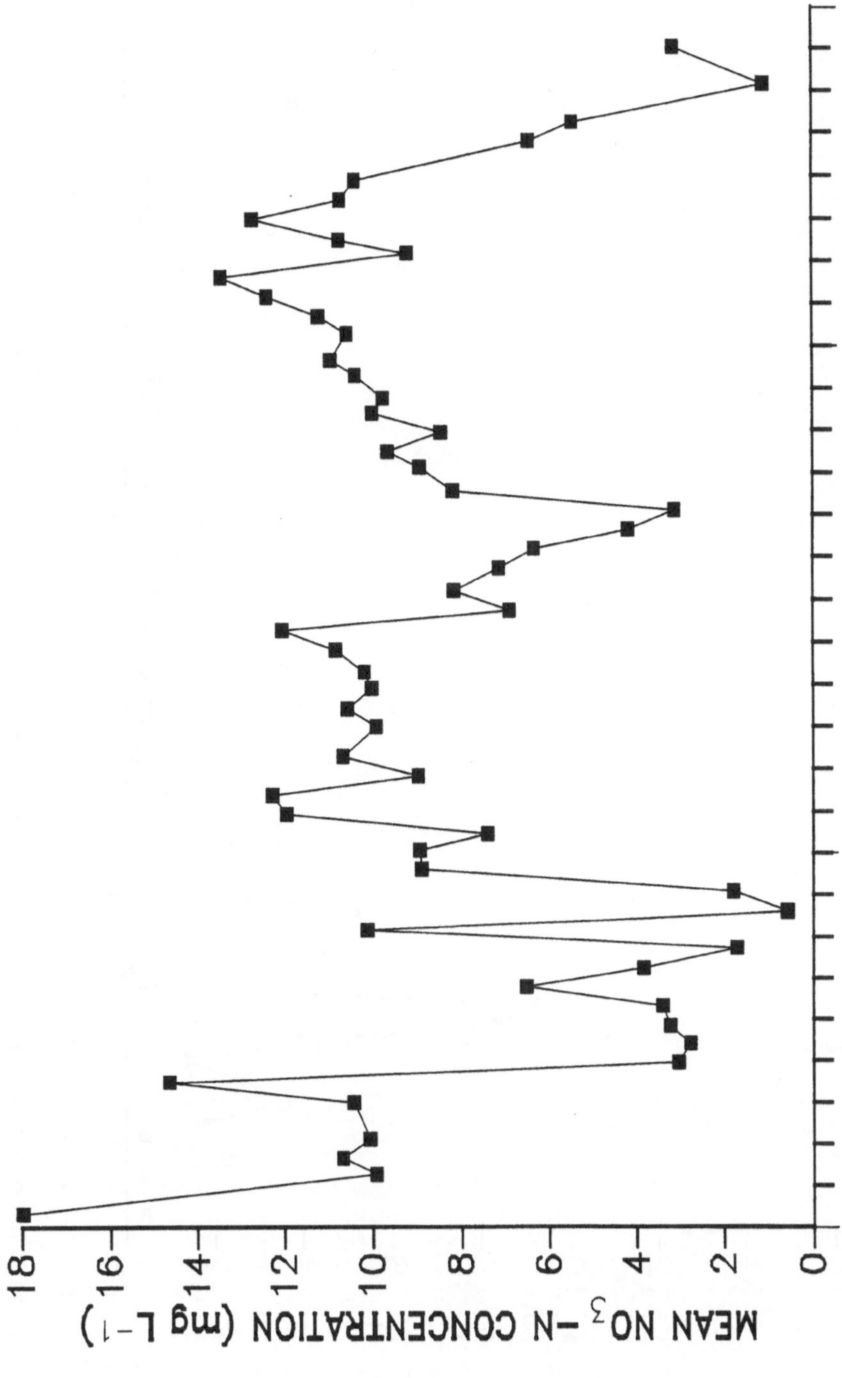

Figure 6. Mean NO_3-N concentrations in the wells in the uppermost landscape position at the dairy wetland.

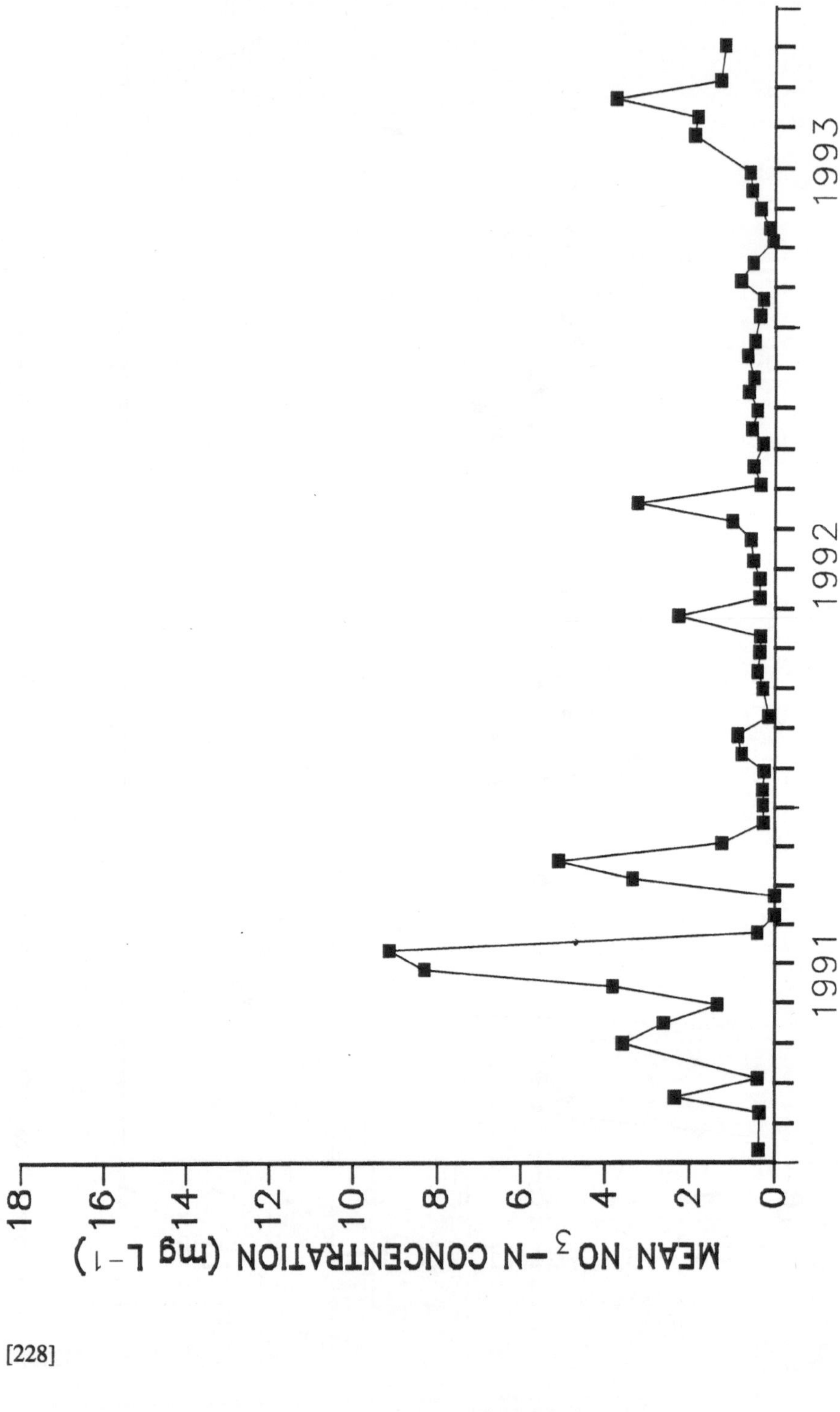

Figure 7. Mean NO_3-N concentrations in the wells in the lowest landscape position at the dairy wetland.

as Br (Hubbard and Sheridan, 1983; Hubbard and Sheridan, 1989; Hubbard and Lowrance, 1993).

Evaluation of the wetland as a bioremediation site will be accomplished by maintaining a nutrient budget for the riparian system over the life of the project (Hubbard *et al.*, 1992). This budget will include the observations of surface and shallow groundwater quality plus results from soil samples. Soil samples for denitrification and inorganic N measurements are being taken monthly at 5 depth increments to 0.3 m. Gaseous losses of N from the soil through denitrification are measured in intact core samples (Lowrance and Smittle, 1988). Nitrogen inputs by symbiotic N fixation will be estimated from the literature. It is anticipated that the effectiveness of the wetland ecosystem as a bioremediation system will increase as the trees mature.

2.2. OVERLAND FLOW - RIPARIAN ZONE TREATMENT OF SWINE WASTE

Work has been underway since January 1993 on the project entitled "Treatment of animal waste by overland flow through grass-riparian zone buffers". This work also is funded through a CSRS National Research Initiative Competitive grant (92-37102-7399). This project will (1) determine the ability of grass-riparian zone buffer strips to cleanse animal waste moving through the system via overland flow, and (2) compare the filtering effectiveness of naturally occurring riparian vegetation versus a recommended wetland species. Three different vegetative treatments are being used for the study: (1) 10 m grass buffer draining into 20 m of natural riparian vegetation forest vegetation, (2) 20 m grass buffer draining into 10 m natural riparian forest vegetation, and (3) 10 m grass buffer draining into 20 m of maidencane. Maidencane (Panicum hematomon) is a species recommended for constructed wetlands.

The animal waste used for the study is swine lagoon waste, which is applied to replicated plots of each vegetative treatment at either high or low wastewater application rates. Within the overall objectives the study will determine rates of sediment, N, P, copper (Cu), and zinc (Zn) removal and transformations. Movement of N (NO_3-N, NH_4-N, and total N), P (ortho, bioavailable, and total), sediment, Cu and Zn is measured in surface runoff-overland flow by collectors both within and at the bottom edge of the plots. Movement of N and ortho P is measured in the root zone and shallow groundwater within and below the plots using suction lysimeters and shallow wells. Soil samples were collected before the start of the study and will be collected again at the end of the study to determine nutrient and heavy metal accumulation over time.

Nitrogen retention in such systems is primarily dependent on denitrification and plant uptake. Rates of these processes will be measured simultaneously on a transect in each plot during the study. Denitrification is measured bimonthly using the acetylene inhibition technique on intact soil cores collected on transects from each plot. Nitrogen, P. Cu, and Zn uptake by the grass is measured periodically on cuttings.

Establishment and instrumentation of the 18 plots for this study were completed in September 1993 and wastewater application began in October 1993. It is anticipated that several years of data collection will be necessary to determine the vegetative treatment and swine lagoon waste application rate most effective in utilizing the waste without negatively impacting the environment.

2.3. MODELING

The information gained from these projects will help in understanding the processes occurring in riparian forest buffer systems. In addition, they will provide data for testing and validation of riparian forest buffer system models. In Tifton, a riparian forest buffer system model with the acronym REMM, Riparian Ecosystem Management Model, is currently under development (Altier, *et al.*, 1994).

The REMM uses a three zone concept. Zone 1 is a narrow band of permanent trees (5-10 m wide) immediately adjacent to the stream channel which provides streambank stabilization, organic debris input to streams, and shading of streams. Zone 2 is a forest management zone where maximum biomass production is stressed, within limits placed by economic goals. Zone 2 may be harvested on appropriate rotations (20-60 years). Zone 3 is a grass buffer strip up to 10 m wide to provide control of coarse sediment and spreading of overland flow.

The model comprises several interactive modules to keep track of water movement, nutrient cycling, and vegetative growth on a daily basis. Feedback between the different modules allows for considerable sensitivity to environmental changes. Algorithms are largely process-based so that the model can respond to diverse conditions. The soil is characterized in three layers through which vertical and horizontal movement of water and associated dissolved nutrients are simulated. The dynamics of carbon (C), N, and P are simulated by means of several organic matter pools characterized by different mineralization rates. Movement of N and P between pools is computed as a function of the mass of C and the C/N and C/P ratios in each pool. Inorganic P is in equilibrium between labile, active, and passive forms, each characterized by varying degrees of stability.

In the vegetation module, growth of upper and lower canopies are simulated concurrently. Carbon is allocated dynamically to the plant organs or held in reserve according to the phenological and nutritional status of the plants. Demand for N and P by the vegetation is determined as a function of biomass in leaf, stem, and root pools. The hydrology module has been run using upland surface and sub-surface input typical to the Southeastern Coastal Plain. At a riparian site in Tifton, GA, water depths in shallow wells have been recorded. Comparisons indicate a close correspondence between simulated and measured water levels.

The concepts of a three zone riparian forest buffer system have already been integrated into draft national specifications for riparian buffer systems by the USDA-Soil Conservation Service and Forest Service (USDA-SCS and USDA-FS). Completion of the REMM will result in a tool to assist agencies and land managers in developing best management designs for utilizing riparian forest buffer systems.

3. Summary

Ongoing studies by scientists of the Southeast Watershed Research Laboratory, USDA-ARS, and University of Georgia are examining the role of riparian forest buffer systems in filtering sediment, nutrients, and pesticides entering from upland agricultural fields. Previous studies have shown that N concentrations in shallow subsurface flow entering riparian forest buffer systems are substantially reduced by the forest. The current studies examine in detail the processes involved in the filtering of N, while also determining the fate of sediment, P, and pesticides. A study using swine lagoon waste also examines the fate of Cu and Zn.

A study with conventional upland agricultural practices (corn or peanuts) is examining the filtering of sediment and nutrients by three different forest management practices (mature trees, clearcut, or selective thinning). A study with liquid dairy manure is evaluating the filtering effectiveness of a reestablishing riparian wetland on nutrients entering from the waste application area via surface runoff or shallow subsurface flow. A pesticide study at this same site is in progress with a source applied to the wetland side not receiving animal waste. The feasibility of using grass-riparian forest buffer systems to filter nutrients from swine lagoon waste applied by overland flow will be investigated by a study examining both type of riparian vegetation and waste application rate.

The results of these studies will be evaluated by the Riparian Ecosystem Management Model (REMM). The model, currently under development at Tifton, uses a three zone concept (Zone 3, grass buffer; Zone 2, upland pine forest; Zone 1 streamside hardwood forest). The model has subcomponents for hydrology, sedimentation, and nutrients.

References

Altier, L. S., R. R. Lowrance, R. G. Williams, J. M. Sheridan, D. D. Bosch, R. K. Hubbard, W. C. Mills, and D. L. Thomas. 1994. An ecosystem model for the management of riparian areas. *Proceedings, Riparian Ecosystems in the Humid U. S. Conference.* (In press).

APHA. 1989. Standard methods for the examination of water and wastewater. 17th edition. *American Public Health Association*, Washington, D.C.

Asmussen, L. E., J. M. Sheridan, and C. V. Booram, Jr. 1979. Nutrient movement in streamflow from agricultural watersheds in the Georgia Coastal Plain. *Trans. ASAE* **22**:809-815, 821.

Boyt, F. L., S. E. Bayley, and J. Zoltec. 1976. Removal of nutrients from treated municipal wastewater by wetland vegetation. *J. Wat. Pollut. Control Fed.* **49**:789-799.

Brinson, M. M., H. D. Bradshaw, and E. S. Kane. 1981. Nitrogen cycling and assimilative capacity of nitrogen and phosphorus by riverain wetland forests. *Rep. No. 167, Wat. Resour. Res. Inst.* Univ. North Carolina. Raleigh, NC. 90 p.

Broadbent, F. E. and F. Clark. 1965. Denitrification. Bartholomew, W. V. (ed.). In: *Soil Nitrogen, Agronomy Monograph 10,* Madison, Amer. Soc. of Agron.: p. 347-362.

Comerford, N. B., D. G. Neary, and R. S. Mansell. 1992. The effectiveness of buffer strips for ameliorating offsite transport of sediment, nutrients, and pesticides from silvicultural operations. *Nat. Counc. Paper Ind. Air & Stream Improvement Tech. Bull. 631.* New York, NY.

Day, J. W., Jr., F. H. Schlar, C. S. Hopkinson, G. P. Kemp, and W. H. Conner. 1981. Modeling approaches to understanding and management of freshwater swamp forests in Louisiana (USA). *Coastal Ecology Laboratory. Center for Wetland Resources.* Louisiana State University, Baton Rouge.

Doyle, R. C., D. C. Wolf, and D. F. Bezdicek. 1975. Effectiveness of forest buffer strips in improving the water quality of manure polluted runoff. In: *Managing Livestock Wastes: 299-302. Proc. of the 1975 Int. Symp. on Livestock Wastes,* Am. Soc. of Agric. Eng., St. Joseph, MI.

Ewel, K. C. and H. T. Odum. 1978. Cypress swamps for nutrient removal and wastewater recycling. In: H. T. Odum, K. C. Ewel (prin. invest.), *Cypress Wetlands for Water Management, Recycling, and Conservation. Fourth Ann. Rept.*

to NSF and Rockefeller Found. Center for Wetlands, Univ. of Florida, Gainesville. pp. 16-34.

Gilliam, J. W. and D. L. Terry. 1973. Potential for water pollution from fertilizer use in North Carolina. *Ext. Circ. No. 550.* NC State Univ., Raleigh.

Gregory, S. V., F. J. Swanson, W. A. McKee, and K. W. Cummins. 1991. An ecosystem perspective of riparian zones. *BioScience* **41**:540-551.

Groffman, P. M., A. J. Gold, and R. C. Simmons. 1992. Nitrate dynamics in riparian forests: microbial studies. *J. Environ. Qual.* **21**:666-671.

Hill, A. R. and N. Wylie. 1977. The influence of nitrogen fertilizer on stream nitrate concentrations near Alliston, Ontario, Canada. *Prog. Water Technol.* **8**:91-100.

Hill, A. R. 1983. Denitrification: Its importance in a river draining an intensively cropped watershed. *Agric. Ecosyst. Environ.* **10**:47-62.

Hoare, R. A. 1979. Nitrate removal from streams draining experimental catchments. *Prog. Wat. Technol.* **11**,4-6:303-313.

Hoyt, G. D. 1981. Nitrogen cycling in a southeastern Coastal Plain agricultural ecosystem. *Ph.D. dissertation,* Univ. of Georgia, Athens, GA. 167.

Hubbard, R. K. and J. M. Sheridan. 1983. Water and nitrate-nitrogen losses from a small, upland coastal plain watershed. *J. Environ. Qual.* **12**:291-295.

Hubbard, R. K. and J. M. Sheridan. 1989. Nitrate movement to groundwater in the southeastern coastal plain. *J. Soil & Water Conserv.* **44**(1):20-27.

Hubbard, R. K., G. Vellidis, and R. Lowrance. 1992. Wetland restoration for filtering nutrients from an animal waste application site. *ASAE Symposium, Land Reclamation: Advances in Research and Technology.* 144-150.

Hubbard, R. K. and R. R. Lowrance. 1993. Spatial and temporal patterns of solute transport through a riparian forest. *Proceedings of the Riparian Ecosystems in the U. S. Conference.* (In press).

Jacobs, R. C. and J. W. Gilliam. 1985. Riparian losses of nitrate from agricultural drainage waters. *J. Environ. Qual.* **14**:472-478.

Johnson, J. C., R. E. Hellwig, G. L. Newton, J. L. Butler, and E. D. Threadgill. 1984. Use of liquid dairy cattle waste to produce Tifton 44 bermudagrass forage. *4th Annual Solar and Biomass Energy Workshop,* Atlanta, GA, April 17-19, 1984.

Karr, J. R. and I. J. Schlosser. 1978. Water resources and the land-water interface. *Science* **01**:229-234.

Kitchens, W. H., J. M. Dean, L. H. Stevenson, and J. H. Cooper. 1975. The Santee Swamp as a nutrient sink. In: F. G. Howell, J. B. Gentry, M. H. Smith (eds.), *Mineral Cycling in Southeastern Ecosystems. ERDA Symposium Series.* p. 349-366.

Leopold, L. B., M. G. Wolman, and J. P. Miller. 1964. Fluvial processes in geomorphology. W. H. Freeman and Company, San Francisco.

Lowrance, R. R., R. L. Todd, and L. E. Asmussen. 1983. Waterborne nutrient budgets for the riparian zone of an agricultural watershed. *Agr. Ecosystems Environ.* **10**: 371-384.

Lowrance, R. R., R. L. Todd, and L. E. Asmussen. 1984a. Nutrient cycling in an agricultural watershed: I. Phreatic movement. *J. Environ. Qual.* **13**:22-27.

Lowrance, R. R., R. L. Todd, and L. E. Asmussen. 1984b. Nutrient cycling in an agricultural watershed: II. Streamflow and artificial drainage. *J. Environ. Qual.* **13**:27-32.

Lowrance, R. R., R. Todd, J. Fail, Jr., O. Hendrickson, Jr., R. Leonard, and L. Asmussen. 1984c. Riparian forests as a nutrient filter in agricultural watersheds. *BioScience* **34**(6):374-377.

Lowrance, R., R. Leonard, and J. Sheridan. 1985. Managing riparian ecosystems to control nonpoint pollution. *J. Soil and Water Cons.* **40**:87-91.

Lowrance, R. and D. Smittle. 1988. Nitrogen cycling in a multiple-crop vegetable production system. *J. Environ. Qual.* **17**:158-162.

Lowrance, R. 1992. Groundwater nitrate and denitrification in a Coastal Plain riparian forest. *J. Environ. Qual.* **21**:401-405.

McColl, R. H. S. 1978. Chemical runoff from pasture: the influence of fertilizer and riparian zones. *N. Z. Journal of Marine and Freshwater Res.*, **12**:371-380.

Mitsch, W. J., C. L. Dorge, and J. R. Wiemhoff. 1979. Ecosystem dynamics and a phosphorus budget of an alluvial cypress swamp in southern Illinois. *Ecology.* **60**:1116-1124.

Nessel, J. K. 1978. Distribution and dynamics of organic matter and phosphorus in a sewage enriched cypress swamp. *Master's thesis.* Dept. Envir. Engrg. Sci., Univ. of FL, Gainesville.

Odum, E.P. 1969. The strategy of ecosystem development. *Science* 164:262-270.

Omernick, J. M., A. R. Abernathy, and L. M. Male. 1981. Stream nutrient levels and proximity of agricultural and forest land to streams: some relationships. *J. Soil Water Conserv.* **36**:227-231.

Peterjohn, W. T. and D. L. Correll. 1984. Nutrient dynamics in an agricultural watershed: observations on the role of a riparian forest. *Ecology* **65**:1466-1475.

Qualls, R. G. 1984. The role of leaf litter nitrogen immobilization in the nitrogen budget of a swamp stream. *J. Environ. Qual.* **13**:640-644.

Reddy, K. R. and D. A. Graetz. 1981. Use of shallow reservoir and flooded organic soil systems for waste water treatment: Nitrogen and phosphorus transformations. *J. Environ. Qual.* **10**:113-119.

Robinson, J. B., N. K. Kaushik, and L. Chatarpaul. 1978. Nitrogen transport and transformations in Canagigue Creek. *Internatl. Joint Commission,* Windsor, ON.

Rowe, M. L. and S. Stinnett. 1975. Nitrogen in the subsurface environment. EPA Report 660/3-75-030 Grant No. R801381. Corvallis, OR: Natl. Environmental Research Center, Office of Research and Development. *U. S. Govt. Printing Ofc.,* Washington, DC 20402.

Schlosser, I. J. and J. R. Karr. 1981a. Water quality in agricultural watersheds: Impact of riparian vegetation during base flow. Water Res. Bull. **17**:233-240.

Schlosser, I. J. and J. R. Karr. 1981b. Riparian vegetation and channel morphology impact on spatial patterns of water quality in agricultural watersheds. *Environ. Manag.* **5**:233-243.

Simmons, R. C., A. J. Gold, and P. M. Groffman. 1992. Nitrate dynamics in riparian forests: Groundwater studies. *J. Environ. Qual.* **21**:659-665.

Sweeney, B. W. 1993. Effects of streamside vegetation on macroinvertebrate communities of White Clay Creek in eastern North America. *Proc. Acad. Natur. Sci. Phila.* **144**:291-340.

Thomas, G. W. and J. D. Crutchfield. 1974. Nitrate-nitrogen and phosphorus contents of streams draining small agricultural watersheds in Kentucky. *J. Environ. Qual.* **3**:46-49.

Todd, R., R. Lowrance, O. Hendrickson, L. Asmussen, R. Leonard, J. Fail, and B. Herrick. 1983. Riparian vegetation as filters of nutrients exported from a coastal plain agricultural watershed. In: *Nutrient Cycling in Agricultural Ecosystems* (R. Lowrance, R. L. Todd, L. E. Asmussen and R. A. Leonard, eds.). p. 485-493. Special Publ. No. 23, Univ. of Georgia, College of Agriculture Experiment Stations, Athens, GA.

Tomlinson, T. E. 1970. Trend in nitrate concentrations in English rivers and fertilizer use. *Water Treat. Exam.* **19**:277-289.

Tuschall, J. R., P. L. Brezonik, and K. C. Ewel. 1981. Tertiary treatment of wastewater using flow-through wetland systems. In: *National Conference of Am. Soc. of Civil Engineers.* July 8-10, Atlanta, GA.

Van Kessel, J. F. 1977. Removal of nitrate from effluent following discharge on surface water. *Water Res.* **11**:533-537.

Vellidis, G., R. Lowrance, M. C. Smith, and R. K. Hubbard. 1993. Methods to assess the water quality impact of a restored riparian wetland. *J. Soil and Water Conserv.* **48**(3):223-230.

Vellidis, G., M. C. Smith, R. K. Hubbard, and R. Lowrance. 1992. Surface runoff samplers for nutrient assimilation measurement in a restored riparian wetland. *ASAE Symposium, Land Reclamation: Advances in Research and Technology.* 135-143.

Vitousek, P.M. and W.A. Reiners. 1975. Ecosystem succession and nutrient retention: A hypothesis. *BioScience* 25:262-270.

Webster, J. R., S. W. Golladay, E. F. Benfield, J. L. Meyer, W. T. Swank, and J. B. Wallace. 1992. Catchment disturbance and stream response: an overview of stream research at Coweeta Hydrologic Laboratory. p. 231-253. In P. J. Boon, P. Calow, and G. E. Petts (ed.) *River conservation and management.* John Wiley & Sons, New York.

Yates, P. and J. M. Sheridan. 1983. Estimating the effectiveness of vegetated flood plains/wetlands as nitrate-nitrite and orthophosphorus filters. *Agr. Ecosystems and Environ.* **9**:303-314.

RAPID WETLAND FUNCTIONAL ASSESSMENT: ITS ROLE AND UTILITY IN THE REGULATORY ARENA

William B. Ainslie*
U.S. Environmental Protection Agency, Region IV
345 Courtland Street, N.E.
Atlanta, Georgia 30365 (USA)

Abstract. Section 404 of the Clean Water Act (CWA) regulates the discharge of dredged or fill material, which is defined as a pollutant, into waters of the United States by requiring potential dischargers to obtain a permit for such activities. The Section 404(b)(1) Guidelines provide the substantive environmental criteria by which all dredge and fill permit applications are reviewed. The Guidelines consist of 4 basic steps: 1) evaluation of practicable alternatives; 2) evaluation of relation of discharge to other environmental standards; 3) assessment of significant degradation to waters of the U.S.; and 4) assessment of appropriate steps to minimize impacts. Wetland functional assessment is important in steps 1, 3, and 4. The use of wetland functional assessment techniques has typically been hindered by lack of time and resources, among other technical concerns, by the resource agencies implementing the Section 404 program. Functional assessment is critical to the Section 404 program since most decisions revolve around an assessment of wetland functions. The Hydrogeomorphic Classification for Wetlands (Brinson, 1993) and the developing functional assessment procedure shows potential for being rapid and inexpensive, scientifically-based and replicable. It is based upon functional indicators which can be recognized in the field and can form the basis for functional indices. The utility of the HGM procedure is illustrated using an example from West Kentucky.

1. Introduction

This paper is intended to place rapid wetland functional assessment in the context of the Section 404 regulatory program under the Clean Water Act (CWA). Wetland functional assessment is integral to the effective administration of the Section 404 program. The objective of the Clean Water Act (CWA) is to restore and maintain the physical, chemical and biological integrity of the nation's waters through the elimination of pollutant discharges (33 U. S. C. 466 et seq.). Wetlands are among the areas defined as waters of the United States [40 CFR 230.3(s)(7)) and dredged and fill material is defined as a pollutant [40 CFR 230.3(o)]. Section 404 of the CWA regulates the discharge of dredged or fill material into waters of the U.S. by requiring potential dischargers to obtain a permit for such activities. The objective of this paper is to review EPA regulations and policies as they pertain to the concept of wetland functional assessment, and to provide an example of how the hydrogeomorphic classification system for wetlands (Brinson, 1993) and the emerging wetland functional assessment, based on this classification scheme, can be utilized in the Section 404 process.

The U.S. Army Corps of Engineers (Corps) is the permitting authority under Section 404 of the Clean Water Act (CWA). In administering the permit program the Corps is required to assess the environmental impacts of a fill project utilizing the Section 404(b)(1) Guidelines (40 CFR 230) promulgated and overseen by the Environmental

*Mr. Ainslie wrote this paper in his private capacity. No official support or endorsement by the Environmental Protection Agency is intended or should be inferred.

Water, Air and Soil Pollution **77**: 433–444, 1994.

Protection Agency (EPA). The 404(b)(1) Guidelines (Guidelines) are the substantive environmental criteria/standards by which all permit applications are evaluated. A fundamental precept of the Guidelines is that dredged or fill material should not be discharged into the aquatic ecosystem, if it will have an unacceptable adverse impact either individually or in combination with known and/or probable impacts of other activities affecting the ecosystems of concern (Section 230.1(c)). In other words, fill material cannot be placed in a wetland if the fill will adversely impair or destroy ecological functions of the wetlands. Therefore, fundamental to implementing the Guidelines is the determination of wetland ecological function.

Wetland functional assessment enters the process primarily in the determination of significant degradation of waters of the United States but also plays a role in determining the level of alternatives analysis, appropriate compensatory mitigation for unavoidable losses, elevation of specific permit cases, and in some instances, may provide the basis for Section 404(c) veto actions. In addition, wetland functional assessment is the underpinning of advanced identification of wetlands (40 CFR 230.80).

2. Regulatory Basis

The evaluation of a permit application using the Section 404(b)(1) Guidelines involves a 4 step sequence:

1) evaluation of practicable alternatives;
2) evaluating the relation of the discharge to other environmental standards;
3) assessment of significant degradation to waters of the United States;
4) assessment of appropriate steps to minimize impacts of the discharge on the aquatic ecosystem.

Specifically, if a proposed activity is not water dependent the Guidelines require that "no discharge of dredged or fill material shall be permitted if there is a practicable alternative to the proposed discharge which would have less adverse impact on the aquatic ecosystem, so long as the alternative does not have other significant adverse environmental consequences" [40 CFR 230.10(a)]. This provision is commonly referred to as the "Alternatives Analysis" and is the first step in the Guidelines. Based on this provision, the applicant is required in every case to evaluate the use of non-aquatic areas and other aquatic sites which would result in a less environmentally damaging impact to the aquatic ecosystem. Yocom *et al.* (1989) and Kruczynski (1993) discuss the alternatives analysis of the 404 (b)(1) Guidelines in greater detail.

However, the amount of information required to assess the level of alternatives analysis is commensurate with the severity of the environmental impact (Memorandum to the Field on Appropriate Level of Analysis Required For Evaluating Compliance with the Section 404(b)(1) Guidelines Alternatives Requirements, August 1993). The alternatives analysis was not intended to be implemented with the same intensity of analysis for all proposed projects. Rather, the Guidelines suggest a correlation between the level of the alternatives analysis and the significance of the impact on the aquatic environment. Limited time and Agency resources dictate that an initial and abbreviated review of impacts be conducted to determine if the project will result in significant impacts and if practicable alternatives exist. If impacts are considered "minor" then the level of alternatives analysis is diminished. For instance, if a project involving a minor discharge into waters of the U.S., such as a boat launching ramp, will have little or no individual or cumulative impact on the aquatic ecosystem EPA will limit its review of alternatives to the project. Comments in this instance would be limited to minimizing (e.g. moving the boat ramp to lessen impacts on waters of the U.S.) and compensating for the unavoidable impact.

Although the alternatives analysis portion of the Guidelines encompasses more administrative and policy considerations, each project manager at EPA makes a determination of level of effort to be expended on a particular permit application based upon a desktop, best professional judgement assessment of the impacts to the aquatic resource. This assessment is often subjective and highly dependent on the permit reviewer's experience with the Section 404 program, the area to be impacted, knowledge of wetland ecology, and the effectiveness of communication between project managers in various agencies. Often commenting agencies will depend upon field information gathered by other agency personnel who conducted a field inspection. This information can be highly variable (depending upon the field inspector's experience, training, educational background, etc.) which can lead to inconsistent and sometimes erroneous conclusions about the wetland system to be impacted. In some instances, however, this approach can be highly effective. This in turn affects the recommendations made regarding the analysis of alternatives and the project in general. Whereas this approach is expeditious, it does little to ensure consistency between project managers within and between EPA Regions, much less with other agencies including the Corps. Once this preliminary assessment of impact is conducted, the level of alternatives analysis is determined and recommendations made to the Corps. As the level of significance of impacts increase, the level of alternatives analyzed by the applicant and scrutinized by EPA and the Corps will increase in an effort to avoid or reduce impacts.

The second step of the Guidelines is known as compliance with other environmental standards [40 CFR 230.10(b)] . The Guidelines require compliance with environmental standards to assure that proposed discharges will not violate water quality standards or endangered species. Permit applications must be evaluated in terms of direct and indirect impacts, to determine if they would "cause or contribute to" violations of state water quality standards which apply to wetlands and other surface water bodies. If the proposed discharge is potentially contaminated, it is subjected to various physical chemical and biological analyses to determine its suitability for disposal.

Under the Endangered Species Act, Federal Agencies have a mandatory obligation to consider the impacts of the projects they authorize or fund on threatened and endangered species. If the proposed project has the potential to affect an endangered or threatened species or its critical habitat the Corps must consult with the U.S. Fish and Wildlife Service (USFWS) and National Marine Fisheries Service (NMFS). Typically, EPA does not place much emphasis on reviewing compliance with environmental standards at this stage of the review process. During a Section 404(b)(1) review, state water quality standards and endangered species compliance determinations are conducted by the states and USFWS and NMFS, respectively.

The third step of the Guidelines involves the assessment of significant impacts, which includes "significant adverse effects" to wildlife, ecosystem integrity, recreation, aesthetics and economic values (40 CFR 230.10(c)]. The Guidelines prohibit discharges that will cause or contribute to significant degradation of the waters of the United States, which include wetlands. It is important to recognize that the Guidelines look at significance both directly (cause) as well as indirectly and cumulatively (contribute to) . Thus, a project could have significant impacts from the fill alone or it could cause significant indirect impacts (offsite). There are no regulatory threshholds for significance, and projects are evaluated on a case-by-case basis. This can lead to greater sensitivity in dealing with wetland resources in a particular region but can also lead to greater disparity in the magnitude and types of impacts allowed. It is this step in the Guidelines which requires an assessment of existing wetland functions and impacts/effects of the proposed project. The determination of whether a project will result in significant degradation is based upon best professional judgement of the project manager and other resource agencies and tends to place much emphasis on loss of wildlife habitat.

Finally, the fourth step in the Guidelines states that no discharge can be permitted unless all appropriate steps have been taken to minimize potential adverse impacts of the discharge on the aquatic ecosystem (40 CFR 230.10(d)). Once a project has satisfied the other provisions discussed above, appropriate steps must be taken to minimize adverse impacts. After impacts have been fully minimized, compensatory mitigation (e.g., restoration, enhancement or creation of wetlands) may be required to offset unavoidable losses. Compensatory mitigation can be used to "buydown" the impacts associated with a given project.

MEMORANDA OF AGREEMENT

In February 1990, EPA and the Corps signed a Memorandum of Agreement (MOA) that clarified the level and type of mitigation necessary to demonstrate compliance with the Guidelines. It affirmed the longstanding practice within EPA and many Corps Districts of approaching mitigation in a sequential manner involving avoidance, minimization, and compensation. It also stipulates that, in achieving the goals of the CWA, the Corps will strive to avoid adverse impacts and offset unavoidable impacts to existing wetlands. This MOA stipulates that functions and values should be assessed by applying wetland functional assessment techniques generally recognized by experts in the field and/or the best professional judgement of federal and state agency representatives, provided such assessments fully consider ecological functions included in the Guidelines. In addition, mitigation should be provided, at a minimum ratio of one to one functional replacement. However, without adequate functional assessment this stipulation is difficult to achieve. Scientifically valid and replicable assessment of function is essential to evaluate the ecological success of wetland mitigation projects.

In addition to the MOA on Mitigation, wetland functional assessment can play a significant role in effectively implementing another MOA on policies and procedures to implement Section 404(q) of the CWA which became effective in August 1992. Section 404 (q) provides for the Corps to enter into an MOA with appropriate federal agencies (i.e. EPA, USFWS, NMFS) to "minimize, to the maximum extent practicable, duplication, needless paperwork and delays in the issuance of permits." Under Section 404, the Corps has the primary authority for determining whether or not a permit for the discharge of dredged or fill material should be issued. After reviewing public comments (including those of resource agencies), if the Corps finds that the particular project in question is likely to receive a Section 404 permit, and EPA opposes issuance of the permit, administrative procedures, outlined in the "q" MOA, are available to EPA to address unresolved issues. The most recent MOA significantly modifies the 1985 MOA on permit elevation and restricts EPA's role in elevating Corps permit decisions to cases involving discharges that will have "a substantial and unacceptable impact on aquatic resources of national importance" (ARNI) . As Magney and Bogden (1993) pointed out, no criteria to determine if a wetland is of national importance have been developed, therefore guidance on what constitutes an ARNI remains to be determined through implementation of the elevation process. Further, what constitutes an ARNI will be highly dependent upon the wetland functional information gathered in the field and the interpretation and communication of that information in the permitting and 404(q) process.

3. Utility of Functional Assessment

The critical issue in wetland functional assessment and Section 404 permitting is how, and to what extent, to assess the function of wetlands for the determination of level of alternatives analysis; level of significant degradation; and wetland mitigation. It is clear that the Section 404 program is highly dependent upon wetland functional assessment methodologies, althouqh none of the standardized rapid assessment methods that have been developed are extensively used (Adamus, 1992). In addition, in-depth functional assessments are not regularly performed. It is estimated that only 2% of the permits reviewed by EPA nationally have a wetland functional assessment performed as part of the review process. Smith (1992) listed several reasons why specific wetland functional assessment techniques have not gained widespread acceptance. These are concerns over technical validity, usefulness of results in a permitting scenario, time requirements to develop assessment proficiency, limited number of functions and values assessed, restricted geographic applicability, proprietary restrictions, difficulty in implementing the methods within the time and resource constraints of the Section 404 Regulatory Program, and the inability to assess "value" of wetlands.

Kusler and Riexinger (1986) summarized some of the special wetland assessment needs of the regulatory community as: an assessment method must be rapid; inexpensive to run; replicable, scientifically based and defensible in court; applicable to portions of wetland systems as well as to entire systems; clear and simple enough to be performed by the regulated public and effectively reviewed by the agencies; and must provide functional indicators which can be used on mitigation sites to indicate success. The purpose of any functional assessment methodology should be to assist the project manager to efficiently and uniformly assess the specific functions when evaluating proposed activities in wetlands (Hausman, 1986).

The use of functional assessment in a routine manner in the Section 404 program is hindered by the short review period required and the immense load of permit applications. The Corps districts and EPA offices in the Southern Appalachian region process and review approximately 1500 individual permits a year. Comments on these permit applications must be returned to the Corps within 30 days. Within this 30 day time frame EPA must determine if a proposed project will adversely effect an ARNI and the extent of those effects. The Regulatory Guidance Letter (RGL) issued by the Corps May 13, 1992, on Federal Agencies Roles and Responsibilities stipulates that the Corps is solely responsible for making final permit decisions (i.e. the Corps is the project manager).

This role of the Corps as the "project manager" was reiterated in the President's Plan for Protecting America's Wetlands: A Fair, Flexible, and Effective Approach issued August 24, 1993. As such, they are responsible for collecting and evaluating information on all permit applications. EPA and other resource agencies are recognized as having the role in the process to provide substantive, project related information on the impacts of activities being evaluated by the Corps. Therefore, EPA must render substantive comments, within 30 days of issuance of the Public Notice, on the level of alternatives analysis necessary, the level of adverse effects to the wetland ecosystem, and the appropriate functional replacement for unavoidable wetland losses. The Corps must evaluate this information, in addition to information collected from other resource agencies on endangered species, water quality, historic site impacts, etc., and render a permit decision in a timely manner. The amount of time taken to make a permit decision is highly dependent on the specificity and factual gravity of comments received from the resource agencies. These constraints place a great deal of pressure upon a functional assessment technique to be rapid and accurate. Typically, if results of the functional assessment are compelling and other aspects of the Guidelines are at issue, EPA may choose to pursue an Section 404(q) elevation and/or a Section 404(c) veto action.

4. An HGM Example

A method which shows promise of becoming a useful tool in the permitting process is the Hydrogeomorphic Classification for Wetlands (HGM) (Brinson, 1993). This is a functionally-based classification of wetlands which emphasizes the hydrologic and geomorphic controls responsible for maintaining many of the functions of wetland ecosystems. The importance of abiotic features of wetlands for such functions as the chemical characteristics of water, habitat maintenance, and water storage and transport is also emphasized. The HGM approach can essentially be subdivided into 4 levels of information gathering each of which can be incorporated into the Section 404 permitting process at various points. Those 4 levels are: 1) Classification; 2) Profiling; 3) Assessment of function presence; and 4) Calculation of Functional Capacity Indices. (These levels are presented here as a means to organize this discussion and may not resemble other organizational schemes for HGM).

The first level is the wetland classification as put forth by Brinson (1993). The hydrogeomorphic classification system simplifies our concept of wetlands by placing each wetland into categories in which similar wetlands share similar functions. As Brinson pointed out, this simplification should improve communication among researchers and managers, and perhaps the public, by focusing on processes that are fundamental to the sustained existence of these ecosystems. This improved communication will aid the

Section 404 project managers by increasing their ability to effectively interact with the scientific community and understand the functioning of wetlands in a proposed project area. The approach is relatively simple to learn and in the process a great deal of insight can be gained into the ecological processes of wetland ecosystems.

The second level is the development of wetland functional profiles for each hydrogeomorphic class identified. A functional profile is developed which describes the geomorphic setting, source of water, hydrology and hydrodynamics, soils, plant species, and other characteristics of the wetland class in a particular geographic area. These functional profiles can be utilized directly in the Section 404 permitting process. For instance the functional profile can provide a description of the functional attributes of the site, and/or elucidate the importance of the wetland in the context of its position in the watershed and surrounding land uses. The level of detail involved in a functional profile will depend upon time, resources and the availability of information (Smith *et al.*, 1994). Time and resources may be short for a given Section 404 individual permit application. Therefore, the profile could take the form of a checklist or synopsis of important wetland charateristics. Over time and with additional information these profiles will lead to the establishment of reference sites or data sets which can be further utilized in developing functional assessments, mitigation standards, and training.

The third level involves the field assessment of a particular wetland area for indicators which document the presence of wetland functions. After construction of a profile, a list of functions and associated indicators could be derived as the basis for a functional assessment. This would entail more extensive literature review, field work and development of a logical rationale explaining which functions a wetland is likely to perform and why. The U.S. Army Corps of Engineers Waterways Experiment Station (WES) has developed a list of functions and associated indicators for riverine wetlands and for depressional/slope and fringe wetlands (Smith, pers. comm.). This could be useful in assessing what functions are present and determining which would be impacted by a particular proposed activity.

Level 3 may be bypassed in favor of level 4 which involves the development and measurement/estimation of "functional capacity indices" (Smith *et al.*, 1994). These functional capacity index models are currently under development by WES. [This step represents the most intense use of this approach]. These indices (level 4) will involve the collection of site specific information to produce an index which regional experts believe to be positively related to a quantitative measure of functional capacity (Smith *et al.*, 1994). The assumption is that the FCI is linearly related to a quantitative empirical measure of functional capacity. If the relationship between FCI and the quantitative measure of functional capacity is known to be nonlinear, it is converted to a linear function. This portion of the assessment process would give the project manager an index very similar to a Habitat Suitability Index (HSI) which can be multiplied by wetland area to give functional units. Functional capacity units can be used to estimate the loss of functional capacity resulting from a proposed project. This relates directly to a determination of significant degradation under Section 230.10(c).

An example of the potential application of these methods to Section 404 can be derived from Drakes Creek in Hopkins County, Kentucky. Drakes Creek is a small watershed (38,320 acres) which has been assessed as part of the ongoing West Kentucky advanced identification. The watershed is heavily impacted by coal mining and potential impacts to the wetland system are major. However, at this point the wetland system remains largely intact and continuous (4,777 acres or 12% of watershed). Utilizing the HGM this creek is a third order, low gradient, alluvial stream with a water source dominated by lateral surface or near surface transport from overbank flow. Flows are unidirectional and velocities correspond with low gradient landforms. The area functions in flood storage and conserves groundwater discharge, overbank flow contributes to both flashy hydroperiod and vertical accretion of sediments which creates rapid biogeochemical cycling and supplies nutrients, residence time of water allows long contact between water and sediment which promotes retention of dissolved elements and compounds, and low suspended load allows light penetration.

The ecological significance of these functions, derived from the classification, is that Drakes Creek may serve as major habitat for wildlife and biodiversity; exhibit strong biogeochemical activity and nutrient retention; maintain conditions for high primary productivity and complex habitat structure; and exhibit good conditions for trapping sediment and altering water quality. The area also has ecological significance as a nutrient trap, and provides strong food web support.

The above information represents rudimentary level 1 and 2 information (Brinson, 1993) combined with an onsite field inspection. In addition, the Drakes Creek watershed is heavily impacted by acid mine drainage which makes water quality enhancement a priority for the watershed. Although the information in the HGM manuscript is not intended for direct use, it is quite accurate in the Drakes Creek case and is illustrative of the HGM process. The HGM could greatly enhance the best professional judgement approach. Accurate best professional judgement (BPJ) decisions are typically based upon a great deal of field experience. Field experience is required to properly and accurately classify wetlands in a particular area. Once HGM and BPJ have been combined, a logical, technically-based classification is derived which forms the basis for technical decision-making for a given project. Based on this functional classification, the interpretation of ecological significance, and the position of the wetlands in a water quality impaired watershed a project manager could make a scientifically-based decision on the level of alternatives analysis required in a given Section 404 permit application. For instance, based on the wetland classification and ecological profile, a project on Drakes Creek would potentially impact several functions associated with wildlife, and water quality enhancement. These impacts could be severe, therefore the level of alternatives analysis would be high.

The functional classification also provides insight into and a departure point for the assessment of significant impacts involved in step 3 of the Guidelines. This assessment would involve level 3 and 4 information. In other words, detailed site specific information would be required to make an assessment of significant degradation. In the Drakes Creek example the site has been classified and an ecological profile developed. Based upon a list of functions and indicators developed for riverine wetlands by WES this area was field inspected. The results of the field inspection indicate that the following functions are present (level 3) in the Drakes

Creek wetland: short- and long-term water storage, energy dissipation, moderation of groundwater flow, biogeochemical transformations and processing, removal of dissolved elements and compounds, particulate retention, organic matter export, and maintenance of typical plant community dynamics, detrital biomass, vertical habitat structure, habitat interspersion and connectivity, and typical abundance of animals. The presence of each of these functions was determined by the presence of at least 3 indicators associated with any particular function.

The regulations at Section 230.10(c) stipulate that individual or cumulative effects contributing to significant degradation include: significant adverse effects on human health; life stages of aquatic life and other wildlife dependent upon aquatic ecosystems; aquatic ecosystem diversity, productivity, and stability; recreation, aesthetic and economic values. Based upon a comparison of the HGM classification and assessment information collected on Drakes Creek and the stipulations of Section 230.10(c) a project manager could maintain that a discharge of dredge or fill material would constitute significant degradation of waters of the United States. For example, Drakes Creek wetlands exhibit hydrologic function (water storage, flood velocity reduction, and moderation of ground water flow), water quality enhancement functions (biogeochemical transformations and processing, removal of dissolved elements and compounds, particulate retention and organic carbon export), and wildlife habitat functions (maintenance of plant community, characteristic detrital biomass, vertical habitat structure, habitat interspersion and connectivity, and typical abundance of animals). A discharge of fill material would effectively eliminate these functions on an individual site-specific basis and would impair these functions on a cumulative basis. In addition, significant adverse effects on aquatic ecosystem diversity, productivity and stability may include but are not limited to: loss of fish and wildlife habitat or loss of the capacity of a wetland to assimilate nutrients, enhance water quality or reduce water velocity. Clearly, the above mentioned functions relate directly to the language in the regulations.

If a project along Drakes Creek were deemed appropriate with mitigation, the MOA on mitigation stipulates that those functions unavoidably impacted must be replaced. The HGM, the derived Riverine functional assessment model, and other area specific information provide a list of the functions. The functional indicators used to determine the presence of a particular function may also be used to determine if a mitigation site is exhibiting the same function. For instance, the presence of debris jams, wrack or drift lines is one indicator for short term water storage, velocity reduction, and organic carbon export functions present in the impacted wetlands. Presence of this indicator in a mitigated site may also show the development of these functions. Since the HGM classification emphasizes the hydrologic and geomorphic factors which control wetland function it can be an effective tool for selecting appropriate mitigation sites. Drakes Creek was a low gradient, alluvial stream dominated by overbank flooding. Therefore, in assessing the appropriateness of mitigation siting (landscape position of the mitigation site) and the probability of achieving functional replacement the HGM approach provides a great deal of information. EPA could recommend to the Corps that an applicant attempting to mitigate for impacts to Drakes Creek should attempt to restore, enhance, or create a site adjacent to a low gradient stream that floods regularly. Once an appropriate site was chosen, it could be monitored for functional indicators that represent the functions lost (and those expected to be gained) and could be detected on a "young/immature" wetland site. The Louisville District of the Corps is currently taking this approach on mitigation siting and monitoring in their "Wetland Compensatory Mitigation and Monitoring Plan Guidelines for Kentucky" (J. Sparks, Louisville District, COE, pers.comm.).

A facet of the HGM functional assessment approach which will be difficult for the regulatory community to deal with is the establishment, monitoring, and protection of reference sites. Brinson (1993) states that by studying the functioning of various reference wetland types <u>in detail</u> one should be able to extrapolate to similar wetland types. This assumes that similar hydrogeomorphic types function similarly. Initially reference sites may have to be based on best professional judgement which is augmented by the calculation of an FCI. That is, resource agency personnel will utilize their field experience to visit sites which represent "good" wetlands and "degraded" wetlands. At each site they calculate the FCI. As more wetlands are evaluated the "reference universe" will expand adding a greater degree specificity to calculated FCIs. This kind of approach to establishing reference sites will have to be combined with more long term and detailed analysis of wetland functions. This task will most likely fall to the academic community and the private sector. Lack of technical, manpower, and financial resources will limit the ability of the resource agencies to establish long term reference sites. Shifting agency priorities makes the prospect of long range study of reference wetlands appear questionable. However, the resource agencies should establish a standard format for wetland functional assessment data and designate a repository for this data to be managed and maintained in the future.

5. Summary

The Section 404 (b) (1) Guidelines are the substantive environmental criteria by which all dredge and fill permit applications are evaluated for compliance with the objectives of the Clean Water Act. As such, the Guidelines require the generation of information on the effects of proposed projects on waters of the United States and the associated aquatic ecosystems, in particular, wetlands. Project managers in the Section 404 program must assess the level of proposed impacts to wetland ecosystems to determine the level of alternatives analysis required, the significance of degradation, and the appropriateness of mitigation. Wetland functional assessment is integral to the effective implementation of the Section 404 program. It's role in the program is implicit in the Guidelines, and reiterated in the MOA on Mitigation which stipulates that functions should be assessed using scientifically recognized techniques and/or best professional judgement. In addition the Section 404(q) MOA requires that permit elevations will be limited to those cases that involve aquatic resources of national importance.

The Hydrogeomorphic Classification for Wetlands has been field-tested in West Kentucky and shows considerable promise in being an effective tool in fulfilling the requirements of the Section 404 (b) (1) Guidelines and the permitting program. Information can be gathered at several levels of effort and utilized in the permitting process. The approach proves to be rapid, inexpensive, replicable, scientifically based, easily learned and is based on functional indicators which can be used in mitigation site assessment. It is based upon the establishment of reference wetlands which can be accomplished through the use of best professional judgement and the HGM functional assessment approach, the advanced identification process or by contracting with Universities or consultants for longer term studies of reference wetland functions. Ultimately there must be a repository for the information generated through the utilization of this technique and that repository should be established jointly by the resource/regulatory agencies involved in wetland regulation.

Acknowledgements

I thank the following colleagues for their time and patience in discussing the issue of functional assessment and the regulations: J. Sparks, D. Thompson, T. Welborn, and B. Kruczynski. I would also like to thank M. Brinson and D. Smith for lengthy discussions on the HGM process which aided my understanding.

References

Adamus, P.R.: 1992, Rapid Methods for Evaluating, Ranking or Categorizing Wetlands in *Statewide Wetland Strategies: A Guide to Protecting and Managing the Resource.* World Wildlife Fund. Island Press. Washington, D.C. 268 pp.

Brinson, M.M.: 1993, A Hydrogeomorphic Classification for Wetlands. Technical Report WRP-DE-4, U. S. Army Engineer Waterways Experiment Station, Vicksburg, MS.

Hausmann, S.: 1986, Special Assessment Needs and Issues: The Regulators Perspective in J.A. Kusler and P. Riexinger (eds), *Proceedings of the National Wetland Assessment Symposium* , Association of State Wetland Managers, pp. 124-125.

Kruczynski, W.L.: 1993, Is Comprehensive Alternatives Analysis Required in All Dredge and Fill Permit Decisions? Pages 134-138, in J. A. Kusler and Cindy Lassonde (eds), *Symposium on Mitigation Banking,* Association of State Wetland Managers, 220 pp.

Kusler, J.A. and P. Riexinger: 1986, The Regulators Perspective. *Proceedings of the National Wetland Assessment Symposium.* Association of State Wetland Managers, Association of State Wetland Managers. pp. 123.

Magney, D.L. and K.M. Bogdan: 1993, What are ARNIS? *National Wetlands Newsletter* 15(3).

Smith, R.D.: 1992, A Conceptual Framework for Assessing the Functions of Wetlands. Wetlands Research Program Technical Report TR-WRP-DE-3, U.S. Army Corps of Engineers Waterways Experiment Station, Vicksburg, MS.

Smith, R.D., A.P. Ammann, C. Bartoldus, and P. Garrett: 1994, A Procedure for Assessing the Functional Capacity of Wetlands. Wetlands Research Program Technical Report TR-WRP-DE-? U.S. Army Corps of Engineers Waterways Experiment Station, Vicksburg, MS. (in prep).

Yocum, T.G., R.A. Leidy and C.A. Morris: 1989, Wetlands Protection Through Impact Avoidance: A Discussion of the 404(b)(1) Alternatives Analysis. *Wetlands* 9(2),283-298.

U.S. Environmental Protection Agency and U.S. Department of the Army: 1990, Memorandum of Agreement between the Environmental Protection Agency and the Department of the Army concerning the determination of mitigation under the Clean Water Act Section 404(b)(1) Guidelines.

FOREST MANAGEMENT AND WILDLIFE IN FORESTED WETLANDS OF THE SOUTHERN APPALACHIANS

T. BENTLY WIGLEY
National Council of the Paper Industry for Air and Stream Improvement, Inc.
Dept. of Aquaculture, Fisheries, and Wildlife
Clemson University
Clemson, SC 29634-0362

THOMAS H. ROBERTS
Dept. of Biology
Tennessee Technological University
P.O. Box 5063
Cookeville, TN 38505

Abstract. The southern Appalachian region contains a variety of forested wetland types. Among the more prevalent types are riparian and bottomland hardwood forests. In this paper we discuss the temporal and spatial changes in wildlife diversity and abundance often associated with forest management practices within bottomland and riparian forests. Common silvicultural practices within the southern Appalachians are diameter-limit cutting, clearcutting, single-tree selection, and group selection. These practices alter forest composition, structure, and spatial hetereogeneity, thereby changing the composition, abundance, and diversity of wildlife communities. They also can impact special habitat features such as snags, den trees, and dead and down woody material. The value of wetland forests as habitat also is affected by characteristics of adjacent habitats. More research is needed to fully understand the impacts of forest management in wetlands of the southern Appalachians.

1. Introduction

Recent estimates indicate that there are about 7.4 million ha of forested wetlands in the southeastern states of Alabama, Florida, Georgia, Kentucky, North Carolina, South Carolina,Tennessee, Virginia, and West Virginia (Cubbage and Flather, 1993). The more mountainous states of Kentucky, North Carolina, Tennessee, Virginia, and West Virginia contain an estimated 1.6 million ha of forested wetlands (Cubbage and Flather, 1993). Precise estimates are not available for area in forested wetlands within the southern Appalachian ecoregion. However, most of the forested wetlands in the southern Appalachians probably are associated with streams.

The area of wet forest in a river reach varies with stream gradient and surrounding topography. Generally, however, the gradient of first- and second-order streams results in only a area that meets the jurisdictional definition of wetlands. In more mountainous portions of the southern Appalachians, watersheds are relatively intact and wetland forests are continuous with upland forests. In coves and valleys, and near the periphery of the southern Appalachian region, many of the forested

Water, Air and Soil Pollution **77**: 445–456, 1994.

wetlands and adjacent uplands have been converted to other non-forest land uses. Where wetland forests do occur, they frequently are only narrow fringes immediately adjacent to the stream or river. In managed forest landscapes, these narrow fringes usually are incorporated in streamside management zones (SMZs) (Wigley and Melchiors, 1993). Bottomland hardwoods (BLHs) occur in the region, but expansive floodplains such as those associated with lower gradient streams in the Gulf Coastal Plain and Mississippi Alluvial Valley are uncommon.

The relatively narrow configuration of forested wetlands in the southern Appalachians has important implications for their management. Actions that occur beyond the boundaries of the jurisdictional area can have significant effects on the quality of the riparian systems as habitat for terrestrial, semi-aquatic, and aquatic organisms. Therefore, when examining effects of management on wildlife, it is important to consider not only the management of wetland forests, but the management of adjacent upland forests as well.

Southern wetland forests provide habitats that generally support a rich wildlife community (Howard and Allen, 1989), including endangered or threatened species, and species that are candidates for listing (Ernst and Brown, 1989). For some species, the sometimes linear nature of forested wetlands in the southern Appalachians is important. Some species such as beavers (*Castor canadensis*), minks (*Mustella vison*), and prothonotary warblers (*Prothonotaria citrea*) are strongly associated with streamside habitats (Brinson *et al.*, 1981), and often use the streams and wetland forests as travel corridors. Sometimes, however, relatively few species from the total community show an distinct affinity for riparian areas. For example, Murray and Stauffer (1992) found that in Virginia, total bird density and species richness showed no riparian influence. Only two species, acadian flycatchers (*Empidonax virescens*) and Louisiana waterthrushes (*Seiurus motacilla*), were closely associated with habitats near streams.

Forested wetlands in the South have significant potential for timber production. Separate data are not available for the southern Appalachian region. However, between the years 1984 and 2030, hardwood timber removals from BLH forests across the South are predicted to increase by approximately 64% (USDA For. Serv., 1988). Concurrently, between 1990 and 2030, area of BLHs in the South is projected to decrease by about 13% (USDA For. Serv., 1988). Thus, management of forested wetlands across the South may intensify. Future trends for the southern Appalachians, however, are unclear.

For these reasons, we must understand the consequences of managing forested wetlands for multiple products, including timber and wildlife. It is increasingly difficult and impractical to exclusively rely upon public land acquisition and set-asides as a key conservation strategy. This is particularly true in the eastern states which encompass the southern Appalachians; there, about 91% of forests are privately held (Waddell *et al.*, 1989). Thus, it is evident that meaningful conservation strategies must consider and be compatible with the economic needs and goals of private landowners. Often those goals include the harvest of timber for economic or other benefits. In this paper we review and summarize key results from extant literature pertaining to timber harvesting in forested wetlands and its effects on wildlife communities. Unfortunately, data specific to the southern Appalachian ecoregion are limited. Therefore, the bulk of information in this paper is based on studies from related systems.

2. Forest Management Practices in Forested Wetlands

Management of forested wetlands in the southern Appalachians is highly varied with both even-aged and uneven-aged techniques employed. Toliver and Jackson (1989) provide an excellent overview of silvicultural practices appropriate for use in forested wetlands. One practice not discussed in their paper, however, is diameter-limit cutting. Diameter-limit cutting, while not usually recommended by foresters, is common within the southern Appalachians, particularly in forests held by non-industrial, private owners (D. H. Van Lear, personal commun.). Diameter-limit cutting removes all merchantable trees above a specified minimum diameter, commonly 30.5 cm. Early successional tree species often regenerate in the gaps created when large trees are harvested. This practice results in a highly variable stand structure, but usually in an uneven-aged stand with three or more canopy layers. Because diameter-limit cutting usually removes the largest and best quality trees, it can result in high-grading.

In parts of the southern Appalachians where forests are managed, many riparian forests are incorporated within SMZs. SMZs are strips of forest vegtetation maintained along waterways, and may include both wetland and upland habitats. SMZs commonly are recommended within state voluntary Best Management Practices (BMP) programs or required in states with forestry practices acts. BMP recommendations for SMZs vary among states. Over time, recommended widths have increased; thus, long-established SMZs usually are narrower than SMZs created more recently. Similarly, SMZ width generally increases with slope and width of the associated stream. This is particularly important in areas where slopes may be severe. For example, recommended SMZ width in Tennessee increases from about 14 to 20 m as slope increases from 10 to 20% (Tennessee Department of Conservation, 1985).

BMP programs commonly recommend SMZs with one or more zones. In the southern Appalachians, some timber harvesting activities usually are deemed most appropriate in the outer portions of the SMZ; this area sometimes is called the secondary SMZ. For example, in South Carolina, recommended SMZ width is 24.4 m on each side of a high-gradient stream capable of supporting trout. The inner half of the buffer is the primary SMZ and the outer half is the secondary SMZ. To minimize potential damage to the stream banks or channel, BMPs often recommend no site preparation or heavy equipment within the primary SMZ. Within the secondary SMZ, BMPs commonly specify minimum amounts of residual canopy cover or basal area. Thus, uneven-aged management such as single-tree or group selection is common within secondary SMZs.

3. General Effects on Wildlife Habitat

By altering existing habitat conditions, forest management affects wildlife at several temporal and spatial scales. At the stand level, species respond differently to the immediate consequences of management; populations of some species may be relatively unaffected, some may decrease, and some may increase. The response varies according to the forest community; the type, extent, and intensity of management; and the sensitivity of the wildlife species to habitat alteration. Other variables such as aspect, topographic position, and site quality can affect wildlife response to forest management.

Wildlife species associated with mature forests are likely to decline in abundance locally following intensive management such as clearcutting or seed-tree harvests which remove all or most of the overstory. Other types of forest management such as single-tree selection and group selection have less dramatic stand-level effects, and may not appreciably affect local species richness. However, more subtle effects can occur even with single-tree selection. For example, first-year data from an ongoing study in Arkansas suggest that parasitism of bird nests is higher in stands managed using single-tree selection than by other methods. Other treatments such as thinnings and improvement cuts also impact wildlife habitat. The degree to which the wildlife community is affected varies with the intensity of these operations. Generally, intermediate treatments have less impact on wildlife than regeneration operations.

Following is an overview of how the wildlife community is likely to be affected by forest management activities. We will discuss both even-aged (clearcutting, shelterwood, seed tree) and uneven-aged (single tree and group selection) harvest and regeneration methods. The consequences of diameter- limit cutting are not well understood, but likely are similar to those of uneven-aged management. Effects also vary based on the landscape-level context in which the management is occurring. These potential effects will be more fully addressed in a separate section.

3.1. CLEARCUT METHOD

Clearcutting is an intensive form of management in which all trees in an area are removed. Normally, clearcuts are a minimum of 1.2 ha to ensure that adequate light is available for establishment and growth of the new stand (Toliver and Jackson, 1989). In forested wetlands, clearcutting is a common management practice because it favors regeneration of commercially desirable tree species and minimizes the number of times a stand must be entered. The immediate stand-level effects of clearcutting is the reduction of habitat for some species, particularly those associated with mature forest. However, favorable conditions are created for species dependent on early successional stages (e.g., herbaceous or shrub stages). The degree to which clearcutting affects wildlife is related partially to size of the harvest unit. Clearly, smaller cuts have less immediate stand-level impact to species associated with mature forests. However, larger clearcuts are necessary if the management goal is to eventually develop extensive areas of mature, even-aged forest.

The loss of habitat features associated with clearcutting mature, deciduous stands may locally decrease habitat suitability for species such as the gray squirrel (*Sciurus carolinensis*), southern flying squirrel (*Glaucomys volans*), and raccoon (*Procyon lotor*) that rely to varying degrees on hard mast production and the availability of cavities in mature trees for dens. Similarly, abundance of species such as the ovenbird (*Seiurus aurocapillus*), red-eyed vireo (*Vireo olivaceous*), yellow-throated vireo (*Vireo flavifrons*), and white- breasted nuthatch (*Sitta carolinensis*) may decline locally when a mature overstory is removed (Thompson and Fritzell, 1990). Mitchell (1989) found that eleven species of birds were more abundant in a 127-year-old baldcypress-tupelo (*Taxodium distichum-Nyssa aquatica*) stand than in younger stands. In general, birds that glean insects from overstory foliage and limbs such as the red-eyed vireo, northern parula warbler (*Parula americana*), and American redstart (*Setophaga ruticilla*) will be negatively affected immediately following a clearcut. Richness and abundance of amphibians also can be reduced after clearcutting (Petranka *et al.*, 1993). However, even amphibian communities apparently return to pre-harvest structure and composition within 50-70 years (Petranka *et al.*, 1993). Thus, Enge and Marion (1986) concluded that amphibian and reptile communities would be most diverse in a landscape dominated by a variety of stand types and structures.

Populations of some species not usually associated with mature forests often benefit from clearcutting. Increased forage in the form of green foliage and soft mast, and cover from ground-level vegetation and logging slash will improve habitat conditions for species associated with brushy areas and early successional habitats. Capture rates for many small mammal species may be higher in harvested tracts than in uncut stands. As an example, Kirkland (1977) found that clearcutting deciduous forests in the northern Appalachians altered the density and relative abundance of small mammals. Some species that were substantially more abundant following harvest included the masked shrew (*Sorex cinereus*), smoky shrew (*S. fumeus*), woodland jumping mouse (*Napeozapus insignis*), red-backed vole (*Clethrionomys gapperi*), meadow vole (*Microtus pennsylvanicus*), and bog lemming (*Synaptomys cooperi*). Although species richness initially decreased following cutting, it increased during the sapling and young-pole stage. Several species attained their highest densities during the five-year period following cutting. Buckner and Shure (1985) reported that densities of both the deer mouse (*Peromyscus maniculatus*) and white-footed mouse (*P. leucopus*) were higher in openings than in control forests. The two species showed different responses to opening size; the former reached highest levels in 10-ha openings, while the latter was most abundant in smaller ones (<2.0 ha).

Richness and diversity of forest birds has been found to be higher where clearcutting creates a variety of stand conditions and edge habitat (Mitchell and

Lancia, 1990). Welsh and Healy (1993) found that bird species diversity was greater on areas under even-aged management than on unmanaged areas. Managed areas contained all species present on unmanaged areas plus 20 additional species. Because their study was conducted in New Hampshire, however, it is not known if its results apply to the southern Appalachians. Generally, species that prefer dense thickets of saplings, shrubs, or vines such as white-eyed vireo (*Vireo griseus*), hooded warbler (*Wilsonia citrina*), golden-winged warbler (*Vermivora chrysoptera*), and northern cardinal (*Cardinalis cardinalis*) will benefit from activities that open the canopy. Larger openings created through timber harvesting may support some species such as the common yellowthroat (*Geothlypis trichas*), yellow-breasted chats (*Icteria virens*), prairie warblers (*Dendroica discolor*), and indigo bunting (*Passerina cyanea*) not normally associated with mature forests. Some species associated with these openings, e.g., prairie warblers, golden-winged warblers, are relatively uncommon and may be declining.

3.2. SEED-TREE AND SHELTERWOOD METHOD

The seed-tree and shelterwood regeneration methods affect wildlife much as clearcutting does. The primary difference is the maintenance of some overstory trees for a period of years allowing some species that use the bole and canopy of mature trees to continue using the site. These methods currently are not employed extensively in southern Appalachian wetlands, but may be soon on national forests.

3.3. GROUP SELECTION METHOD

Group selection normally involves harvesting small groups of trees (typically from 0.4 to 2.0 ha), resulting in an uneven-aged stand. In the harvested area, the effects are similar to that of a clearcut operation; however, habitat within the entire stand is less dramatically affected. The effects of group selection on wildlife communities have not been well studied. Thus, the following discussion of group selection is based partially on studies of small clearcut operations.

Because the groups selected for harvest or intermediate treatment are small and scattered throughout the forest, populations of many species found in the original stand may be largely unaffected by group selection. For example, few if any, negative effects would be expected to occur to populations of gray squirrels, eastern wild turkeys (*Meleagris gallopavo*), or pileated woodpeckers (*Dryocopus pileatus*), all species associated with mature forest conditions. However, the effects of small scattered openings on other forest-dwelling species, particularly "interior" neotropical migrant songbirds, is unknown.

Species associated with early successional habitat such as eastern cottontail (*Sylvilagus floridanus*), white-tailed deer (*Odocoileus virginianus*), ruffed grouse (*Bonasa umbellus*), prairie warbler, common yellowthroat, golden-winged warbler, and yellow-breasted chat use regeneration areas due to the abundant food and cover resources created (U.S. Forest Service, 1980). Openings created by group selection in an otherwise forested landscape may be extremely important to some game species. For example, wild turkeys often nest in or near such areas (Speake *et al.*, 1975; Wesley *et al.*, 1981), and broods require insects associated with herbaceous vegetation (Hurst and Stringer, 1975; Healy and Nenno, 1983). Ruffed grouse also use openings as brood habitat (Cade and Sousa, 1985). The American woodcock (*Scolopax minor*), a game species of minor importance in the southern Appalachians, is known to make extensive use of regenerating bottomland hardwoods in other portions of the South (Roberts *et al.*, 1984).

3.4. SINGLE-TREE SELECTION

Single-tree selection results in the removal of individual trees. As with group selection, foresters usually determine the desired structure and composition of the residual stand using systems such as BDq regulation (Marquis, 1978). With BDq regulation, the desired residual basal area (B), maximum retained diameter class (D), and q-factor (q) are determined. The q-factor is the negative exponential constant between diameter classes. It is used as a multiplier to yield the target number of trees in each diameter class. Trees are removed from the stand within the constraints of those 3 variables. Guldin (1991) provides an excellent overview of the technical application of this regulation method.

This system of harvest and regeneration is sometimes compared to the natural, gap-phase processes by which small gaps are created in the canopy, and subdominant trees are released. Most wildlife species associated with the existing stand will continue using the site, and the small regeneration areas may temporarily provide habitat for early successional species. For example, hooded warblers which often are associated with canopy gaps in mature forests (James, 1971). Other species that likely would respond similarly include the northern cardinal, carolina wren (*Thryothorus ludovicianus*), and white-eyed vireo.

Species associated with mature forests such as the gray squirrel and eastern wild turkey likely will be unaffected by single-tree selection, and in fact, may benefit from increased soft mast production associated with gaps. For example, Nixon *et al.* (1980) found that while food and cover resources for gray squirrels were reduced immediately following harvest, population levels one year later were the same as pre-cut levels. Squirrels frequently foraged in logged areas, presumably feeding on abundant fungi and insect populations.

Of the regeneration methods, single-tree selection alters stand structure and existing habitat conditions least. Thus, it is viewed by some as the most desirable method for use. In spite of this, research has documented that shade-intolerant tree species that often are the target of management (e.g., oaks [*Quercus* spp.]) do not reproduce well under this system. Long-term management by single-tree selection can, in fact, result in significant and sometimes unwanted changes in the makeup of the forest community, i.e., toward more shade tolerant species. This undesirable side-effect, along with other factors such as the necessity for repeated entry into the stand, extensive road construction, and potential damage to residual trees caused by harvest negates many of its positive features.

4. Landscape Considerations

Harvesting and regeneration also affect wildlife at a larger, landscape scale. The habitat quality of any particular forest stand sometimes is more affected by its context than its within-stand characteristics. For example, the wildlife community in a 20-ha, sawtimber-size, floodplain stand will vary depending upon whether it is located within a larger, similar landscape or within a landscape dominated by early successional habitats. Wildlife response to forest management activities in such a stand also will vary depending upon the landscape context.

Often, wildlife species richness increases with area of a habitat; this phenomenon also has been observed in riparian forests (Stauffer and Best, 1980; Keller *et al.*, 1993). Species gained as forest area increases are sometimes referred to as "area-sensitive" or "interior" species. Large blocks of mature BLH forest often contain large numbers of habitat specialists, including forest-interior, neotropical migrants (Hamel, 1989; Mitchell *et al.*, 1990). Fragmentation, resulting in small, isolated forest patches has been found to be related to declines of some forest-interior species and local extinctions of others (Finch, 1991).

Most of the studies related to fragmentation, however, have been conducted in regions of the country lacking extensive forest cover. The authors of a recent study in heavily forested areas of Missouri (Thompson *et al.*, 1992) concluded that, at the landscape-scale, clearcutting was compatible with maintaining populations of forest-interior, neotropical migrants although some species would likely become less abundant. Likewise, research in northern hardwoods of New Hampshire (Welsh and Healy, 1993) suggests that even-aged management actually increases forest songbird diversity without the loss of any species. It is unknown whether these conclusions are applicable to BLH and riparian forests in the southern Appalacians.

The relationship between wildlife species richness and forest area has not been well studied, but there is some indication that it is not linear. In the study by Stauffer and Best (1980), most of the gains in bird species richness occurred in the narrower riparian width classes. In fact, 70-78% of the breeding bird species occurred in riparian strips that were 17% of the maximum width sampled of approximately 250 m. Some species occurring only in the widest riparian strips are considered by many to be interior species, e.g., American redstart, scarlet tanager (*Piranga olivacea*). Others, however, such as the rufous-sided towhee (*Pipilo erythrophthalmus*) are not. Moderate-width riparian forests may provide for the habitat needs of many species. For example, Keller *et al.* (1993) recommended that riparian forests be at least 100 m wide to provide nesting habitat for interior species.

It cannot be inferred that increased species abundance always is related to increased habitat area. Many other habitat variables such as forest species composition and richness, average dbh, snag availability, and understory density are all important in avian communities. One recent study showed that the character of adjacent habitats can have a major influence on wildlife communities (Tappe *et al.*, 1993). Unfortunately, few studies have critically evaluated how stand characteristics within riparian and other wetland forests interact with other forest characteristics and those of adjacent habitats.

SMZs generally are considered to contribute to the maintenance of diverse wildlife communities in managed forest landscapes. They offer an opportunity to provide habitat features such as den trees and snags that are less abundant in intensively managed stands (Wigley and Melchoirs, 1993). Triquet *et al.* (1990) concluded that SMZs provide habitat for some species of mature-forest and edge-dwelling songbirds that otherwise would be absent. Burk *et al.* (1990) found that eastern wild turkeys used SMZs more than expected in a managed forest landscape. Likewise, studies have demonstrated the value of SMZs to squirrels (Warren and Hurst, 1980; McElfresh *et al.*, 1980; Dickson and Huntly 1987; Fischer and Holler, 1991).

SMZs also are thought to function as travel corridors (Harris, 1989), connecting habitats dissected by other forest types, non-forest land uses, or other barriers. Simberloff *et al.* (1992), however, note that a "remarkable publicity campaign, much of it outside the bounds of mainstream science, has promoted corridors for conservation." In truth, there are few data regarding the efficacy of forested corridors or whether SMZs function as such. Simberloff *et al.* (1992), however, observed that the utility of riparian forests, such as those contained within SMZs, often is important independent of their value for movement.

5. Conclusions

Timber harvesting and other forest management activities can be compatible with the goal of maintaining a diversity of wildlife species and communities in BLH and riparian forests. At the stand level, forest management influences wildlife diversity and abundance in these forests by affecting a multitude of habitat features. Any perturbation, such as timber harvesting, likely will reduce habitat suitability for some species but improve suitability for others. However, it is important to consider the spatial and temporal aspects of stand-level changes. Negative impacts to populations of many species are frequently short-lived and a diversity of seral stages created by harvesting, including some of all age and structural classes, are needed to provide for the full spectrum of biota.

Harris and Gosselink (1990) suggested that an ideal approach to the maintenance of biotic diversity would consider all types of species such as interior, edge, and wide-ranging species. Thus, mature stands are needed for interior or area-sensitive species. Similarly, early successional stages are needed for specialists dependent upon those habitats. The necessity of having a diverse forest ecosystem is highlighted by trends in the Breeding Bird Survey which show that neotropical migrant songbird species asssociated with both early- and late-successional habitats may be experiencing long-term population declines. And, managed forests can accomodate a suprisingly large array of species. Even those species sometimes considered "old-growth" dependent, can sometimes be accomodated in managed forests through use of creative silvicultural practices (Lennartz and Lancia, 1987). For most wildlife species, reversing the decline in BLH and riparian area is probably a more important concern than the spatial or temporal arrangement of a forested mosaic.

References

Brinson, M. M., Swift, B. L., Plantico, R. C. (and others).: 1981, Riparian ecosystems: their ecology and status. *U. S. Fish and Wildlife Service, FWS/OBS-81/17.*

Buckner, C. A., and Shure, D. J.: 1985, The response of Peromyscus to forest opening size in the southern Appalachian Mountains. *J. Mammal.* **66**, 299-307.

Burk, J. D., Hurst, G. A., Smith, D. R., Leopold, B. D. and Dickson, J. G.: 1990, Wild turkey use of streamside management zones in loblolly pine plantations. *Proc. Nat. Wild Turkey Symp.* **6**, 84-89.

Cade, B. S., and Sousa, P. J.: 1985, Habitat suitability index models: ruffed grouse. *U.S. Fish and Wildlife Service Biol. Rep. 82(10.86).*

Cubbage, F. W., and Flather, C. H.: 1993, Forested wetland area and distribution: a detailed look at the South. *J. For.* **91**, 35-40.

Dickson, J. G., and Huntley, J. C.: 1987, Riparian zones and wildlife in southern forests: the problem and squirrel relationships. in: J. G. Dickson and O. E. Maughan (eds.), Managing southern forests for wildlife and fish--a symposium. *USDA For. Serv. GTR-SO-65*, pp. 37-39.

Enge, K. M., and Marion, W. R.: 1986, Effects of clearcutting and site preparation on herpetofauna of a north Florida flatwoods. *For. Ecol. and Manage.* **14**, 177-192.

Ernst, J. P., and Brown, V.: 1989, Conserving endangered species on southern forested wetlands. in: Hook, D. D., and Lea, R. (eds.), Proc. of the symp.: the forested wetlands of the southern United States. *USDA Forest Service GTR-SE-50*, pp. 135-145.

Finch, D. M.: 1991, Population ecology, habitat requirements, and conservation of neotropical migratory birds. *USDA For. Serv. GTR-RM-205.*

Fischer, R. A., and Holler, N. R.: 1991, Habitat use and relative abundance of gray squirrels in southern Alabama. *J. Wildl. Manage.* 55:52-59.

Guldin, J. M.: 1991, Uneven-aged BDq regulation of Sierra Nevada mixed conifers. *West. J. Appl. For.* **6**, 27-32.

Hamel, P. B. 1989, Breeding bird populations on the Congaree Swamp National Monument, South Carolina. in: Sharitz, R. R., and Gibbons, J. W. (eds.), Freshwater wetlands and wildlife. *DOE Symposium Series No. 61.*, pp. 617-628.

Harris, L. D.: 1989, The faunal significance of fragmentation of southeastern bottomland forests. in: Hook, D. D., and Lea, R. (eds.), Proc. of the symp.: the forested wetlands of the southern United States. *USDA Forest Service GTR-SE-50*, pp. 126-134.

Harris, L. D., and Gosselink, J. G.: 1990, Cumulative impacts of bottomland hardwood forest conversion on hydrology, water quality, and terrestrial wildlife. in: Gosselink, J. G., Lee, L. C., and Muir, T. A. (eds.), Ecological processes and cumulative impacts: illustrated by bottomland hardwood wetland ecosystems. Lewis Publishers, Inc., Chelsea, MI., pp. 259-322.

Healy, W. M., and Nenno, E. S..: 1983, Minimum maintenance versus intensive management of clearings for wild turkeys. *Wildl. Society Bull.* **11**, 113-120.

Howard, R. J., and Allen, J. A.: 1989, Streamside habitats in southern forested wetlands: their role and implications for management. in: Hook, D. D., and Lea, R. (eds.), Proc. of the symp.: the forested wetlands of the southern United States. *USDA Forest Service GTR-SE-50*, pp. 97-106.

Hurst, G. H., and Stringer, B. D., Jr.: 1975, Food habits of wild turkey poults in Mississippi. *Proc. Nat. Wild Turkey Symp.* **3**, 76-85.

James, F. C.: 1971, Ordinations of habitat relationships among breeding birds. *Wilson Bull.* **83**, 215-236.

Keller, C. M. E., Robbins, C. S., and Hatfield, J. S.: 1993, Avian communities in riparian forests of different widths in Maryland and Delaware. *Wetlands* **13**, 137-144.

Kirkland, G. L.: 1977, Responses of small mammals to the clearcutting of northern Appalachian forests. *J. Mammal.* **58**, 600-609.

Lennartz, M. R., and Lancia, R. A.: 1987, Old-growth wildlife in second-growth forests: opportunities for creative silviculture. in: Proc. national silviculture workshop: silviculture for all resources. *USDA For. Serv. Timber Management, Washington, DC, USA.*, pp. 74-103.

Marquis, D. A.: 1978, Application of uneven-aged silviculture and management on public and private lands. *USDA For. Serv. GTR-WO-24.*

McElfresh, R. W., Inglis, J. M., and Brown, B. A.: 1990, Gray squirrel usage of hardwood ravines within pine plantations. *Louisiana State Univ. Annu. For. Symp.* **29**, 79-89.

Mitchell, L. J.: 1989, Effects of clearcutting and reforestation on breeding bird communities of baldcypress-tupelo wetlands. *M.S. Thesis. North Carolina State Univ., Raleigh, NC.*

Mitchell, L. J., and Lancia, R. A.: 1990, Breeding bird community changes in a baldcypress-tupelo wetland following timber harvesting. *Proc. Annu. Conf. Southeast. Assoc. Fish and Wildlife Agencies* **44**, 189-201.

Murray, N. L., and Stauffer, D. F.: 1992, Habitat use by nongame birds in central Appalachian riparian forests. *Final Report to USDA For. Serv.*

Nixon, C. M., Havera, S. P., and Hansen, L. P.: 1980, Initial response of squirrels to forest changes associated with selection cutting. *Wildl. Soc. Bull.* **8**, 298-306.

Petranka, J. W., Eldridge, M. E., and Haley, K. E.: 1993, Effects of timber harvesting on southern Appalachian salamanders. *Conserv. Biol.* **7**, 363-370.

Roberts, T. H., Hill, E. P. and Gluesing, E. A.: 1984, Woodcock utilization of bottomland hardwoods in the Mississippi Delta. *Proc. Annu. Conf. Southeast. Assoc. Fish and Wildlife Agencies* **38**, 137-141.

Simberloff, D., Farr, F. A., Cox, J., and Mehlman, D. W.: 1992, Movement corridors: conservation bargains or poor investments? *Conserv. Biol.* **6**, 493-504.

Speake, D. W., Lynch, T. E., Fleming, W. J., Wright, G. A., Hamrick, W. J.: 1975, Habitat use and seasonal movements of wild turkeys in the Southeast. Proc. Nat. Wild Turkey Symp. **3**, 122-130.

Stauffer, D. F., and Best, L. B.: 1980, Habitat selection by birds of riparian communities: evaluating effects of habitat alterations. *J. Wildl. Manage.* **44**, 1-15.

Tappe, P. A., Thill, R. E., Melchiors, M. A., and Wigley, T. B.: 1993, Wildlife values of streamside management zones in the Quachita Mountains, Arkansas. Proc. conf. on riparian ecosystems in the humid U. S.: functions, values, and management. (in press).

Tennessee Department of Conservation, Division of Forestry: 1985, A plan for management of water quality affected by sivicultural and other forest activities in Tennessee.

Thompson, F. R., III, and Fritzell, E. K.: 1990, Bird densities and diversity in clearcut and mature oak-hickory forest. *USDA For. Serv. Research Paper NC-293.*

Thompson, F. R., Dijak, W. D., Kulowiec, T. G., and Hamilton, D. A.: 1992, Breeding bird populations in Missouri Ozark forests with and without clearcutting. *J. Wildl. Manage.* **56**, 23-30.

Toliver, J. R., and Jackson, B. D.: 1989, Recommended silvicultural practices in southern wetland forests. in: Hook, D. D., and Lea, R. (eds.), Proc. of the symp.: the forested wetlands of the southern United States. *USDA Forest Service GTR-SE-50*, pp. 72-77.

Triquet, A. M., McPeek, G. A., and McComb, W. C.: 1990, Songbird diversity in clearcuts with and without a riparian buffer strip. *J. Soil and Water Conserv.* **45**, 500-503.

USDA For. Serv.: 1980, Wildlife management handbook: southern region. *USDA For. Serv. FSH 2609.23R.*

USDA For. Serv.: 1988, The South's fourth forest: alternatives for the future. *USDA For. Serv. For. Resour. Rep. No. 24.*

Waddell, K. L., Oswald, D. D., and Powell, D. S.: 1989, Forest statistics of the United States, 1987. *USDA For. Serv. Resour. Bull. PNW-RB-168.*

Warren, R. C., and Hurst, G. A. 1980, Squirrel densities in pine-hardwood forests and streamside management zones. *Proc. Annu. Conf. Southeast. Assoc. Fish and Wildl. Agencies* **34**, 492-498.

Wesley, D. E., Perkins, C. J., and A. D. Sullivan, A. D.: 1981, Wildlife in cottonwood plantations. *South. J. Applied For.* **5**, 37-42.

Welsh, C. J., and Healy, W. M.: 1993, Effect of even-aged timber management on bird species diversity and composition in northern hardwoods of New Hampshire. *Wildl. Soc. Bull.* **21**, 143-154.

Wigley, T. B., and Melchoirs, M. A.: 1993. Wildlife habitat and communities in streamside management zones: a literature review for the eastern United States. *Proc. Conf. Riparian Ecosystems in the Humid U. S., Functions, Values, and Management.* (in press).

BEST MANAGEMENT PRACTICES FOR FORESTED WETLANDS IN THE SOUTHERN APPALACHIAN REGION

W. Michael Aust
Department of Forestry, 228 Cheatham Hall, Virginia Polytechnic Institute & State University, Blacksburg, VA 24061-0324

Abstract. Forestry best management practices (BMPs) have been developed for all of the states included in the Southern Appalachian Region (Alabama, Georgia, Kentucky, North Carolina, South Carolina, Tennessee, Virginia, West Virginia). All of the state forestry BMPs were developed to reduce nonpoint source pollution from forestry operations. However, the states have developed BMPs that differ substantially with regard to methodology, particularly for forested wetlands. The state BMP guidelines vary in several major areas, including wetland types, BMP manual detail, streamside management zones, harvesting operations, site preparation operations, regeneration systems, road construction, and timber removal activities. An understanding of the similarities and differences between the state BMP guidelines will allow the forested wetland manager to comply with or improve upon existing forestry BMPs for wetlands.

1. Introduction

The Federal Water Pollution Control Act (PL-100-1, Sec. 319) considers a best management practice (BMP) to be any method, measure or practice used to reduce water pollution to include, but not limited to, control of water caused erosion. Forestry activities in wetland areas are exempt from the Section 404 permit process if the activities meet the following conditions:

1. the activity is not a conversion of a wetland to an upland,
2. the activity is part of an on-going operation,
3. the activity has not lain idle so long that hydrological operations are necessary,
4. the activity does not contain any toxic pollutants, and
5. the activity uses normal silvicultural activities that comply with the federal BMPs.

There are fifteen federal wetland BMP requirements; eight of the federal BMPs address road construction activities in wetlands. The remaining seven BMPs provide protection for endangered species, avifauna breeding and nesting areas, shellfish production, public water supplies, wild and scenic rivers, prevention of toxic discharges, and removal of temporary fills (Cubbage *et al.*, 1990; Siegel and Haines, 1990). Landowners and operators failing to comply with these federal BMPs may be fined up to $125,000 per day by the U. S. Environmental Protection Agency for noncompliance, even if state BMPs are voluntary (Virginia Forestry Association, 1993).

Several excellent references are available that document the effects of various forest operations on water quality (Corbett *et al.*, 1978; Dissmeyer, 1980; Lynch and Corbett, 1990; Miller *et al.*, 1988; NCASI, 1980, 1992; Patric, 1976, 1978; Sopper, 1975; Yoho, 1980). Forestry water quality research has tended to concentrate on the impact of forest operations on erosion and sedimentation and most studies have been conducted via paired watershed

Water, Air and Soil Pollution **77**: 457–468, 1994.

studies. These studies are relatively expensive to install and usually have been conducted in upland areas where watersheds are more easily defined. Less research has been conducted on wetland sites where watersheds are less easily defined.

Most states in the Southeast have developed state BMP manuals to aid forest landowners, operators, and managers in the practice of responsible stewardship. As might be expected, the eight states in the southern Appalachian region have developed forestry BMPs that are common in some respects and different in others, particularly with regard to forested wetland BMPs (Alabama Forestry Commission, 1993; Georgia Forestry Association Wetlands Committee, 1993; Georgia Forestry Commission, 1993; Kentucky Silviculture Non-Point Source Task Force, 1992; North Carolina Department of Environment, Health, and Natural Resources, 1990; North Carolina Division of Forest Resources, 1989; South Carolina Forestry Association, 1989; South Carolina Forestry Commission, 1989; Tennessee Division of Forestry, 1993; Virginia Department of Forestry, 1989; West Virginia Forestry Division, 1989a; West Virginia Forestry Division, 1989b). The purpose of this paper is to provide an overview of the similarities and major differences among the state forested wetland BMP recommendations.

2. Results

2.1 STATE FORESTRY BMP MANUALS FOR FORESTED WETLANDS

Four states (Alabama, Georgia, North Carolina, and South Carolina) contain almost 90 percent of the total wetland acreages found within the eight-state Southern Appalachian Region (SAR) (Table I). Three of the states (Georgia, North Carolina, and South Carolina) have specific forested wetland BMP manuals that are currently separate from the standard forestry BMP manual and the wetland recommendations are documents of substantial length. Alabama has a substancial acreage of wetlands, yet the forested wetland BMPs are a relatively minor component of the Alabama BMP document. Virginia has far fewer acres of forested wetlands, yet Virginia has an 18 page section in its forestry BMP manual that is devoted to wetland BMPs (Table I).

2.2 VOLUNTARY vs. MANDATORY

Most of the states in the SAR have voluntary forestry BMP programs that are administered by the respective state forestry or natural resources organizations (Table II). West Virginia is the only state in the SAR that has mandatory forestry BMPs. West Virginia also requires logger certification and a pre-harvest management plan. However, West Virginia only inspects and enforces forestry BMPs after a complaint has been registered. Virginia has a voluntary forestry BMP program, yet all harvesting operations in Virginia are subject to BMP inspections during and after harvest. Virginia also has a state forestry anti-sediment law that can be used to fine or issue stop-work orders to landowners or operators who do not comply with state BMPs, if their activities would likely result in significant stream sedimentation.

Table I. Wetland area, type of forested wetland BMP manual, and manual length for states in the SAR.

State (ha x 1000)	Wetland Area"	Wetland Manual Type	Wetland Manual Length (No. pages)
North Carolina	2300	Separate**	27
Georgia	2144	Separate**	26
South Carolina	1885	Separate**	20
Alabama	1531	Section	5
Virginia	435	Combined	18
Tennessee	318	Section	4
Kentucky	121	Section	3
West Virginia	41	Not considered	0
Total	8775		

* From Dahl (1990).

** Upland and wetland manuals are being combined.

North Carolina has a similar anti-sediment law that strongly encourages forestry BMP compliance. Two states, Alabama and Tennessee, have voluntary forestry BMP programs, but these programs emphasize that compliance with federal BMPs is necessary to maintain exemption status from the permits required for wetlands by the Federal Clean Water Act (Table II).

2.3 FORESTED WETLAND TYPES

Not surprisingly, the states having the preponderance of forested wetland area tend to define more types of forested wetlands. Georgia and North Carolina recognize nine types of wetlands, ranging from almost continually flooded, organic-soil, muck swamps to areas that are barely wet enough to be considered jurisdictional wetlands (wet flats) (Table III). On the other extreme, the West Virginia BMP manual does not recognize any specific type of forested wetland. Also some states recognize locally important types of wetlands such as bay forests (North Carolina) and cypress domes (Georgia). However, none of the states in the region recognize any wetland types that might be found only in the mountain regions, such as mountain bogs or ephemeral springs.

Table II. Primary state agencies involved with forestry BMPs and nature of BMP policy for states in the SAR.

State	Primary Agencies for Forestry BMPs	State Policy
Georgia	Georgia Forestry Commission (for education)	Voluntary
Kentucky	Kentucky Division of Forestry Kentucky Division of Water (in floodplains)	Voluntary
South Carolina	South Carolina Forestry Commission	Voluntary
Alabama	Alabama Dept of Environmental Management (for violations) Alabama Forestry Commission (for education)	Voluntary, but Federal guidelines emphasized
Tennessee	Tennessee Division of Forestry (for education)	Voluntary, but Federal guidelines emphasized
North Carolina	North Carolina Division of Forest Resources	Voluntary, with sediment legislation
Virginia	Virginia Division of Forestry (education, inspections, enforcement)	Voluntary, enforceable criteria & inspections
West Virginia	West Virginia Forestry Division (education) West Virginia Dept. of Natural Resources (enforcement)	Regulatory Inspection & enforcement complaint

2.4 STREAMSIDE MANAGEMENT ZONES

Streamside management zones (SMZs) are vegetative strips along streams that serve to decrease sedimentation into the stream, remove sediment from overbank streamflow, and maintain stream temperatures (Golden *et al.*, 1984; NCASI, 1992). SMZs have also been shown to be important to certain types of wildlife populations and activities and improve the visual aesthetics of timber harvests along streams. SMZs are the largest single BMP costs to landowners, because SMZs usually require retention of a certain portion of the merchantable timber (Cubbage and Lickwar, 1991; Ellefson and Miles, 1985). Therefore, SMZ guidelines are important to landowners, operators, and the general public. However, there is not a consensus among the state BMP manuals concerning SMZ recommendations (Table IV). Alabama and North Carolina recommend that SMZs be maintained on all perennial and intermittent streams. West Virginia does not specify a stream type, rather, they recommend keeping filter strips along streams that have at least a 40 ha (100 acre) watershed. The other states in the region recommend SMZs for all perennial streams (usually interpreted as a solid blue line on a USGS topographic map).

Table III. Types of forested wetlands identified in forestry BMP manuals in the SAR.

Type of Forested Wetland	State							
	AL	GA	KY	NC	SC	TN	VA	WV
Muck Swamp	Yes	Yes	No	Yes	Yes	No	Yes	No
Peat Swamp (pocosin)	Yes	Yes	No	Yes	Yes	No	Yes	No
Wet Flats & Savannahs	Yes	Yes	No	Yes	Yes	No	Yes	No
Red & Black River Bottoms	Yes	Yes	No	Yes	Yes	No	Yes	No
Branch Bottom	Yes	Yes	No	Yes	Yes	No	Yes	No
Piedmont Bottomland	No	Yes	No	Yes	Yes	No	No	No
Gulfs, Coves, & Lower slopes	Yes	Yes	No	Yes	Yes	No	No	No
Bay Forests	No	No	No	Yes	No	No	No	No
Headwater Forests	No	No	No	Yes	No	No	No	No
Cypress Stringers	No	Yes	No	No	No	No	Yes	No
Cypress Domes	No	Yes	No	No	No	No	No	No
Wetlands	-	-	Yes	-	-	Yes	-	No
All Categories	6	9	1	9	7	1	6	0

Table IV. Streamside Management Zone (SMZ) recommendations for states in the SAR.

SMZ Specifications	State							
	AL	GA	KY	NC	SC	TN	VA	WV
Stream type	I*	P**	P	I	P	P	P	40*** ha
Primary Zone Minimum (m)	11	6	9	15	12	8	15	8
Primary Zone Maximum (m)	-	24	50	31	24	44	38	50
Secondary Zone Minimum (m)	-	0	-	-	0	-	-	-
Secondary Zone Maximum (m)	-	24	-	-	24	-	-	-
Determining**** Factors	1 3	2	1	1 4	1	1	1 5	1

* I = intermediate or perennial streams.

** P = perennial streams.

*** West Virginia recommends SMZs for watersheds > 40 ha

**** Determining factor codes

1 = slope restrictions.

2 = mountain, piedmont and coastal plain physiographic provinces guidelines.

3 = soil erodibility restrictions.

4 = stream width (greater or less than 9 m).

5 = trout stream restrictions.

Two states, Georgia and Alabama, recommend use of primary and secondary SMZs (Table IV). The secondary SMZs are generally areas that could be clear-cut, but would not be mechanically site prepared. The primary (or only SMZ in some states) SMZs would be at least partially timbered. Seven of the states use slope as the determining factor for SMZ width, Georgia uses physiographic province (mountain, piedmont, and coastal plain) to determine the recommended SMZ width. The minimum SMZ widths vary considerably - the minimum varying from 6.1 m (20 ft) in Georgia to 15.2 m (50 ft) in Virginia. The maximum SMZ widths range from 50.3 m (165 ft) in Kentucky and West Virginia to 24.4 m (80 ft) in Georgia and South Carolina. Three of the states have additional determining factors for SMZ width: North Carolina considers stream width, Alabama considers soil erodibility, and Virginia considers the presence of trout in the stream (Table IV).

Within the SMZs, all but three states recommend "partial" or selective harvests only, assuming that the landowner desires any harvesting within the SMZs (Table V). The Georgia and Tennessee BMPs suggest that any type of harvest method is appropriate, including clearcuts. West Virginia does not specify a harvest method but does recommend that streams remain shaded. Four states, Alabama, Kentucky, North Carolina, and Virginia, recommend a percentage of the overstory, crown cover, or canopy that should be retained in the SMZs. North Carolina also suggests that a similar species mix be retained and Virginia provides a minimum basal area to aid in the process (Table V). With regard to site preparation in the SMZS, all of the states except Tennessee recommend that mechanical site preparation be eliminated or disturbances should be minimized (Table V)

2.5 REGENERATION SYSTEMS

Forested wetlands may be regenerated by artificial or natural regeneration methods. Four of the states (Georgia, North Carolina, South Carolina, and Virginia) recommend several methods of natural regeneration (Table VI) and three of these states include site criteria for selection of a method. Overall, the recommended regenerations systems follow those outlined by Kellison *et al.* (1988), McKevlin (1992), and Toliver and Jackson (1989).

2.6 FOREST ROAD BMPS FOR FORESTED WETLANDS

All eight BMP manuals have significant sections dealing with forest roads. Forest road construction is the operation that has the largest potential to increase stream sedimentation (Yoho, 1980; Golden *et al.*, 1984), thus states have attempted to reduce the problem by suggesting proper techniques of construction, maintenance, and drainage (Table VII). All of the states made the following recommendations for forest roads:

1. Roads should be located properly prior to installation.
2. Road grades should be minimized. All states providing specific road grade guidelines suggested road slope should be kept below 10 percent.

3. Road stream crossings should be minimized and should be installed carefully so that stream sedimentation is minimized. Bridges, culverts, or fords should be used for stream crossings.
4. All forest roads should be constructed so that control of water is achieved. A variety of mechanisms are suggested (Table VII).
5. Forest roads should not be located within the SMZs except at stream crossings.
6. Permanent forest roads should be maintained regularly (e.g., graded, gravelled, ditch maintenance).
7. Temporary forest roads should be closed when operations are complete.

Table V. Harvest and site preparation guidelines within primary Streamside Management Zones (SMZs) for states in the SAR.

State	Harvest Method	Harvest Level	Site Preparation
Alabama	Partial cut only	50 % crown cover	No mechanical
Georgia	Any method including clear-cut	Any	Minimize traffic No Mechanical
Kentucky	Partial	50 % overstory	No mechanical
North Carolina	Selective	50 % canopy, keep species mix	Mechanical site prep not normally done
South Carolina	Selective	-	No mechanical
Tennessee	Any method, minimize traffic	Shade streams	-
Virginia	Partial	50 % crown cover or leave 12 m^2/ha basal area	Minimize litter disturbance
West Virginia	Minimize traffic	Shade streams	Minimize disturbance

Additional recommendations for road construction in forested wetlands were made by two or more states (Table VII). These recommendations included:

1. Minimizing fill-constructed roads or using specialized materials such as porous materials, mats, or geotextiles to allow water movement through the fill road.
2. Correcting or avoiding excessive rutting.
3. Cutting adjacent trees (daylighting) so that roads would dry more rapidly.
4. Following federal road construction BMPs in wetland areas.

2.7 TIMBER REMOVAL

Timber removal systems also have the potential to increase soil erosion by disturbing soil litter or channeling the flow of water. Therefore, all of the state BMP guidelines make recommendations for skid trails or alternative removal systems (Table IV). Five states recommend the use of specialized machinery for harvesting in wetland areas. The most commonly cited options were use of wide or flotation tires, tracked equipment, and cable or aerial systems. Most of the states (7) recommend that skid trails not be excessively steep, and all states recommend that removals should be done during dry periods so that rutting can be minimized. All states also recommend that stream crossings should be minimized and that trails should be closed by a combination of revegetation, brush, and water bar installation. Five states also recommend that dedicated skid trails should be used when soils are saturated so that disturbance can be contained within a smaller area and mitigation work will be more efficient.

Table VI. Silvicultural regeneration systems recommended by the state forestry BMP manuals for forested wetlands.

State	Natural Systems	Artificial Systems
Alabama	No recommendation	No recommendation
Kentucky	No recommendation	No recommendation
Tennessee	No Recommendation	No Recommendation
West Virginia	No Recommendation	No Recommendation
South Carolina	Clear-cut	Direct Seeding
	Shelterwood	Planting
	Seed Tree	Shearing
	Selection	Chopping
	Group Selection	Discing
		Bedding
		Burning
		Herbicides
Georgia*	Clear-cut	Mechanical Site Prep
	Group Selection	Plant
	Shelterwood	Direct Seed
	Seed Tree	
North Carolina*	Clear-cut	Chemical Site Prep
	Group Selection	Mechanical Site Prep
	Patch Clear-cut	Planting
	Shelterwood	Bedding
	Seed Tree	Shearing
		Burning
		Fertilization
Virginia*	Clear-cut	Plant
	Shelterwood	
	Selection	
	Seed Tree	

* Recommendations made by site type (Table III).

Table VII. Recommended state forest road BMPs for forested wetlands (X = consideration in BMP manual).

Recommended Practices	AL	GA	KY	NC	SC	TN	VA	WV
	-----	-----	-----	State	-----	-----	-----	
Location	X	X	X	X	X	X	X	X
Control Grade	X	X	X	X	X	X	X	X
Stream Crossings	X	X	X	X	X	X	X	X
Minimize Fills	X			X	X		X	
Porous Fill					X		X	
Matting		X		X	X		X	
Geotextiles		X					X	
Control Rutting	X			X		X	X	X
Water Control								
broad-based dips	X	X	X	X	X	X	X	X
rolling dips			X	X			X	
ditches	X	X	X	X	X	X	X	X
culverts	X	X	X	X	X	X	X	X
outslope road	X	X	X	X		X	X	X
water bars	X	X	X	X	X	X	X	X
water turnouts	X	X	X	X	X	X	X	X
Proximity to SMZS	X	X	X	X	X	X	X	X
Daylight Road						X	X	
Maintenance	X	X	X	X	X	X	X	X
Road Closure								
reshape	X	X	X	X		X	X	X
control access	X	X	X	X		X	X	
vegetate	X	X	X	X	X	X	X	X
Federal BMPs	X			X		X		

Table VIII. Recommended Skidding BMPs for forested wetlands (X = mention in BMP manual)

Recommended Practices	AL	GA	KY	NC	SC	TN	VA	WV
	-----	-----	-----	State	-----	-----	-----	
Machinery	X							
wide-tires		X		X*			X*	
cable				X*	X		X*	
tracked		+X		X*	X		X*	
aerial				X*	X		X*	
Grade	X	X	X	X		X	X	X
Season	X	X	X	X	X	X	X	X
Rutting	X	X	X	X	X	X	X	X
Dedicated trails when wet	X	X	X	X		X		
Trail Construction	X						X	X
Stream Crossings	X	X	X	X	X	X	X	X
Trail Closure	X	X	X	X	X	X	X	X

* Recommendations based on site types (Table III).

Discussion

The goal of the states in the SAR is to maintain or improve water quality from forestlands. Solomon (1989) concluded that use of BMPs was an appropriate mechanism for achieving the goal, but stipulated that monitoring was necessary so that BMPs could be modified as feedback was provided. The various agencies responsible for development and educational efforts relating to forestry BMPs recognize the on-going nature of the process; indeed, several of the states are working on third or fourth revisions. Also, the state agencies do cooperate with one another to some degree, and findings from one state commonly are used to modify another state's BMP program. Several states are actively monitoring forest operations in-state, (Alabama, South Carolina, and Virginia), while other states are relying on applied research results from a variety of federal and state research agencies.

The forestry BMP guidelines developed for states in the SAR are most similar with regard to forest road construction and timber removal guidelines. This is because of the required federal road construction BMPs for forested wetlands and because the potential pollution problems associated with forest roads are widely recognized.

With the exception of West Virginia, the state BMPs are voluntary and the state forestry organization is commonly the primary administrator. Almost all of the BMPs are directed toward protection of water quality, primarily as affected by harvesting operations and road construction activities. In most of the manuals, site preparation activities receive less consideration, although site preparation can result in greater erosion rates than harvesting (Golden *et al.*, 1984; Yoho, 1980). Over the next decade, BMP recommendations for site preparation, fertilization, and other activities will likely be developed. Although the BMPs were developed to address water quality concerns, several states have already used the BMP manuals as a mechanism for making recommendations relating to site productivity, wildlife, aesthetics, and forest management.

The wetland portion of the BMP manuals is directed almost exclusively to wetland types located in the coastal plain, primarily because this is where the majority of wetlands are located and would also include areas most likely to be included under the Coastal Zone Management Act. However, the SMZ guidelines are generally considered to be applicable to all portions of the state including mountain riparian areas. Some types of mountainous wetlands, such as ephemeral pools and bogs are not specifically addressed by the manuals. This does not mean that the states will not eventually address BMPs for these areas; it only suggests that the states are first developing BMPs for the most common wetlands. As the knowledge base increases concerning wetlands, the BMP manuals will be adjusted. Several of the states stressed that forestry BMPs will not replace the good judgement of the land manager and that BMPs will be updated and modified as managers and researchers provide additional information.

References

Alabama Forestry Commission: 1993. *Alabama's best management practices for forestry.* Printed by the Alabama Forestry Commission, Montgomery, Alabama. 30 p.

Corbett, E. S., Lynch, J. A, and Sopper, W. E.: 1978. Timber harvesting practices and water quality in the Eastern United States. *J. For.* **76**,484-488.

Cubbage, F. W., Kirkman, L. K, Boring, L. R., Harris, T. G., Jr., DeForest, C. E.: 1990. Federal legislation and wetlands protection in Georgia: legal foundations, classification schemes, and industry implications. *For. Ecol. and Manage.* **33/34**, 271-286.

Cubbage, F. W. and Lickwar, P., 1991: Estimating the costs of water quality protection on private forestlands in Georgia. Georgia Forestry Commission Research Division, Georgia Forest Research Paper 86.. 11 p.

Dahl, T. E: 1990. *Wetland losses in the United States, 1780's to 1980's.* USDI Fish. Wildl. Serv., Washington, D.C. 21 p.

Dissmeyer, G. E: 1980. Predicted erosion rates for forest management activities and conditions in the southeast. p. 42 - 50, Proc: *U. S. Forestry and Water Quality: What Course in The 80's?* Water Pollution Control Federation, Virginia Water Pollution Control Federation, Virginia Forestry Association. June 19-20, 1980, Richmond, VA. 208 p.

Ellefson, P. V. and Miles, P. D., 1985: Protecting water quality in the Midwest: impact on timber harvesting costs. *North. J. Appl. For.* **2**: 57-61.

Georgia Forestry Association Wetlands Committee: 1993. *Best management practices for forested wetlands in Georgia.* Published by Georgia Forestry Commission and Georgia Environmental Protection Division, Macon, Georgia. 26 p.

Georgia Forestry Commission: 1993. *Recommended best management practices for forestry in Georgia.* Georgia Forestry Commission, Macon, Georgia. 24 p.

Golden, M. S., Tuttle, C. L., Kush, J. S., and Bradley, J. M. III: 1984. *Forestry activities and water quality in Alabama: effects, recommended practices, and an erosion-classification system.* Alabama Agricultural Experiment Station Bull. 555, Auburn University, Auburn, AL. 87 p.

Kellison, R. C., Martin, J. P., Hansen, G. D., and Lea, R.: 1988. *Regenerating and Managing Natural Stands of Bottomland Hardwoods.* American Pulpwood Association APA 88-A-6. American Pulpwood Association, Inc., Washington, D.C. 26 p.

Kentucky Silviculture Non-Point Source Task Force: 1992. *Kentucky forest practice guidelines for water quality management.* Kentucky Division of Forestry, Frankfort, Kentucky. 55 p.

Lynch, J. A., and Corbett, E. S.: 1990. Evaluation of best management practices for controlling nonpoint pollution from silvicultural operations. *Water Resour. Bull.* **26(1)**, 41-52.

McKevlin, M. R: 1992. *Guide to regeneration of bottomland hardwoods.* USDA Forest Service Gen. Tech. Rep. SE-76. USDA Forest Service Southeastern Forest Experiment Station, Asheville, North Carolina. 35 p.

Miller, E. L., Beasley, R. S., and Lawson, E. R.: 1988. Forest harvest and site preparation effects on erosion and sedimentation in the Ouachita Mountains. *J. Environ. Qual.* **17**, 219-225.

National Council of the Paper Industry for Air and Stream Improvement (NCASI): 1992. *The effectiveness of buffer strips for ameliorating offsite transport of sediment, nutrients, and pesticides from silvicultural operations.* NCASI Tech. Bull. No. 631. New York, NY. 48 p.

National Council of the Paper Industry for Air and Stream Improvement (NCASI): 1980. *1979 review of the literature on forest management practices and water quality management.* NCASI Tech. Bull. No. 330. New York, NY. 60 p.

North Carolina Department of Environment, Health, and Natural Resources: 1990. *Best management practices for forestry in the wetlands of North Carolina.* North Carolina Division of Forest Resources, Raleigh, North Carolina. 28 p.

North Carolina Division of Forest Resources: 1989. *Forestry best management practices manual.* North Carolina Division of Forest Resources, Raleigh, North Carolina. 67 p.

Patric, J. H: 1976. Soil erosion in the eastern forest. *J For.* **74**, 671-677.

Patric, J. H: 1978. Harvesting effects on soil and water in the eastern hardwood forest. *South. J. Appl. For.* **2**, 66-73.

Siegel, W. C. and Haines, T. K.: 1990. State wetland protection legislation affecting forestry in the northeastern United States. *For. Ecol. and Manage.* **33/34**, 239-252.

Solomon, R. S.: 1989. Implementing nonpoint source control: should BMPs equal standards? p. 155-162. In D. D. Hook, R. Lea, eds., Proc. The Forested Wetlands of the Southern United States. USDA Forest Service Southeastern Forest Experiment Station Gen. Tech. Rep. SE-50. 168 p.

Sopper, W. E: 1975. Effects of timber harvesting and related management practices on water quality in forested watersheds. *J. Env. Qual.* **4,** 24-29

South Carolina Forestry Commission: 1989. *Best management practices for South Carolina's forest wetlands.* South Carolina Forestry Commission, Columbia, South Carolina. 20 p.

South Carolina Forestry Association: 1989. *Voluntary Forest Practice Guidelines for South Carolina.* South Carolina Forestry Association, Columbia, SC. 24 p.

Tennessee Division of Forestry: 1993. *Guide to forestry best management practices.* Tennessee Division of Forestry, Nashville, Tennessee. 41 p.

Toliver, J. R., and Jackson, B. D.: 1989. Recommended silvicultural practice in southern wetland forests. p. 72-80. In D. D. Hook, R. Lea, eds., Proc. The Forested Wetlands of the Southern United States. USDA Forest Service Southeastern Forest Experiment Station Gen. Tech. Rep. SE-50. 168 p.

Virginia Department of Forestry: 1989. *Forestry best management practices for water quality in Virginia.* Virginia Department of Forestry, Charlottesville, Virginia. 76 p.

Virginia Forestry Association. 1993. Forestry News, Newsletter 286, September 8, 1993, Richmond, VA. 4 p.

West Virginia Forestry Division: 1989a. *Clean stream handbook for loggers: keeping mud out of the streams.* West Virginia Forestry Division, Charleston, West Virginia. 34 p.

West Virginia Forestry Division: 1989b. *Guidelines for controlling soil erosion and water siltation from logging operations in West Virginia.* West Virginia Forestry Division, Charleston, West Virginia. 26 p.

Yoho, N. S: 1980. Forest management and sediment production in the South - a review. *South. J. Appl. For.* **4**, 27-35.

RESERVOIR RIPARIAN ZONE CHARACTERISTICS IN THE UPPER TENNESSEE RIVER VALLEY

C.C. AMUNDSEN

Department of Botany and Graduate Program in Ecology
Universityof Tennessee,Knoxville,TN 37996-4015 USA

Abstract. Impoundments in the upper Tennessee River Basin have inundated historic river riparian and mesic valley slope habitats. Fifteen major Tennessee Valley Authority (TVA) reservoirs, upstream of Chattanooga TN, have effected 50,000 ha of modern reservoir riparian zones on formerly mesic terrain. These zones were defined here as winter mudflats between winter pool shores and summer pool shores and summer riparian habitats with diminishing soil saturation from summer shores up to flood zone boundaries. Watts Bar (WB), the largest of these 15 projects in terms of area, was chosen for background and field analyses of the consequential reservoir riparian processes and expressions. WB, at the median age of areal TVA reservoirs, was closed in 1942. Transects were taken in the WB summer riparian forest after consideration of topoedaphics, species composition, and the level of contemporary disturbance. Regional bottomland forests were compared. Coefficients of species similarity showed 70% compositional similarity. The basal area (BA) density of WB summer riparian forests was not similar as was composition in regional bottomland stands. Four (of twelve) subjectively selected WB transect BA densities were compared nonparametrically and were the same (90% confidence) as other bottomland stands of like chronology. Overall however, WB stands averaged 19.6 m^2/ha BA to 30.0 + for regional comparisons. Stocking was low, but measured litterfall was relatively high, resulting in similar biomass productivity corrected on a unit area basis. The winter mudflats showed expanding development of phenologically distinct herbaceous and graminoid communities under a five to six month drawdown exposure regime. Thermal modifications from the winter pool heat sink were determined to alter the mudflat microenvironment. A recent shortening of the drawdown period has interrupted the adapted life history strategy of the counter-seasonal plants, and seed maturity has been restricted. Erosion of the summer shore has been increasing the extent of the mudflats at the expense of the hydrologically influenced summer riparian habitat foreslope. The discernable trend of summer riparian stand succession over mesic expression will be limited by the height and hydric avoidance occasioned by an abrupt slope to mesic, unsaturated profile, conditions. Approximations of biomass production potential pre-and post impoundment have shown a 12:1 diminution from historic carbon detention, on a seasonal basis, within normal pool shore boundaries. Further approximations comparing the WB summer riparian forest carbon retention and detention to historic forest compartments highlighted the importance of a relatively copious, non-lignified, litterfall in the seasonally available carbon budget.

1. Introduction

In 1942, the Watts Bar (WB) Dam was closed at river mile 530 on the Tennessee River. The Tennessee Valley Authority (TVA) impoundment flooded the valley upstream to river mile 602 with a resultant normal or summer pool at 225.9 meters (m) mean sea level (msl) drowning roughly 13,500 hectares (ha) of terrestrially productive field and forest habitat. At river mile 560 ($\sim 84^0$ 35'W, 35^0 50'N) the normal pool obliterated the shallows of the lowest reported water profile surface (211.8m, msl, 1881), obscured the river bank (219.5m, msl) and exceeded the highest known river flood mark (225.4m, msl, 1868; data from personal communications, TVA Reservoir Operations Office).

Water, Air and Soil Pollution **77**: 469–493, 1994.

The catastrophic inundation of the historic riparian habitat and biota has occasioned the appearance of an upslope, neohydric habitat imposed on once mesic terrain. Selected, relatively undisturbed, riparian position forest community expressions and the associated annual drawdown zones of the half-century old WB impoundment have been the target examples for this investigation of upper Tennessee Valley reservoir riparian plant community characteristics.

2. Study Areas and Definitions

There are 20 dams classified as major by the Tennessee Valley Authority (TVA) in the Tennessee River Basin upstream of Chattanooga, Tennessee (TN). Each creates an impoundment which includes an altered reservoir riparian area. The TVA management

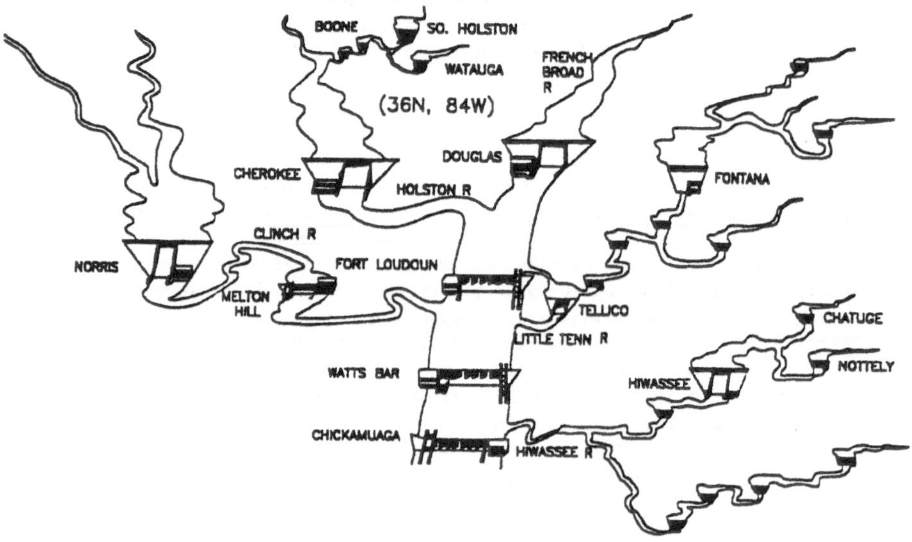

Fig.1 Diagram of TVA impoundments discussed. Watts Bar is 116 km long.

classification (TVA, 1980; other TVA unpublished records) does not include consideration of Aluminum Company of America (ALCOA) dams which are also in the Tennessee Basin watershed. The fifteen larger (Figure 1) of these 20 major TVA projects each have a normal pool with a channel length of more than 20 kilometers (km) along the Sailing Line and a surface in excess of 1500 ha (Table I).

These fifteen impoundments nominally include 50,000 ha of total reservoir riparian habitat, where about one-fifth constitutes terrestrial summer shore reservoir riparian zones and the balance winter, or brumal, drawdown mudflats. Eighty-four hundred km of summer shoreline edges separate these subdivisions horizontally. These TVA impoundments inundated the previous riverine areas and upslope mesic habitats in dam closures dating from 1936 (Norris) to 1979 (Tellico). Most of the others were closed during the World War II decade and are now some 50 years old, the median age of the modern upper Tennessee River reservoir riparian habitats.

Table I

Fifteen major TVA impoundments above Chattanooga, TN in the Tennessee River Basin, managed by the Tennessee Valley Authority. Data from TVA, 1980. For the TVA impoundments considered, the potential reservoir riparian zone total (Mudflats plus Flood Zones) is 50,380 ha. This is ~24% of the reported area of the Tennessee River Basin conserved landscape dominant, The Great Smoky Mountains National Park (210,650 ha).

Dam	Sail Line km	Pool ha	River Bed ha	Mudflat ha	Flood Zone ha	Pool Shore km
Boone	53	1744	291	1118	37	209
Chatuge	21	2853	43	1263	97	213
Cherokee	87	12262	982	7260	308	633
Chickamauga	95	14326	3845	3966	2185	1304
Douglas	69	12303	1283	8013	364	894
Fontana	47	4306	668	2493	93	399
Ft. Loudoun	98	5908	1789	971	364	580
Hiwassee	36	2465	405	1582	57	262
Melton Hill	71	2412	666	453	81	279
Norris	208	13841	1186	7811	2266	1288
Nottely	33	1692	69	1060	52	171
South Holston	38	3068	287	1101	579	270
Tellico	53	6678	863	907	324	499
Watauga	26	2602	127	704	352	171
Watts Bar	154	15783	4186	2549	1902	1241

There were no historic lakes in the upper Tennessee River basin, and although these reservoirs are politically named lakes, they are seldom lentic. The TVA reservoir system was, and is, intended to serve multiple purposes. Navigation and flood control are foremost along with consistent hydroelectric production. The management operations currently take into account societal water supplies (civic, agricultural, and industrial), dilution of anthropogenic additions, insect and aquatic plant pest control, erosion control, habitat enhancement and recreational activities. Management goals and user demands dictate a restricted, but nearly constant flow. Operations result in two substantial riveroid pools at managed levels, specific for each reservoir design, which are out of synchronization with the winter high, summer low, water configurations common to historic and still wild rivers in the moist, mesothermal, Southern Appalachian region. The current WB plan specifies a normal pool (in part, recreation interest driven) of seven plus months, from spring to late summer, with a drawn down pool during the winter months when heavier watershed runoff is expected (Figure 2).

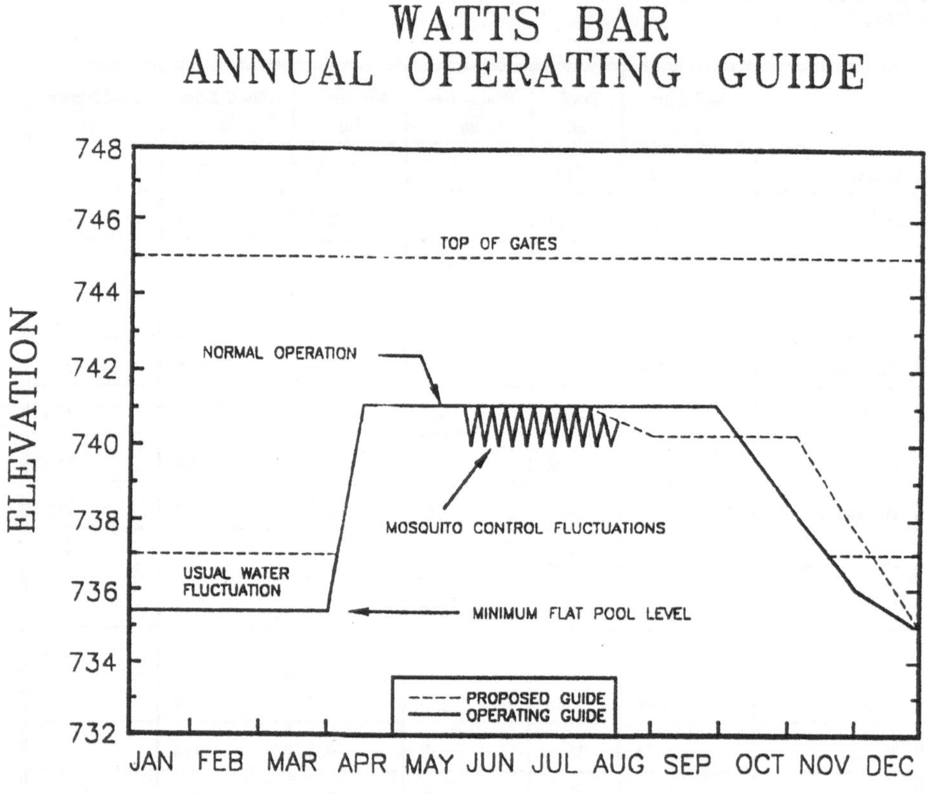

Fig.2 TVA pool levels, Normal 225.9m (741'), Flood 227.1m (745').

The extant riparian zone habitats on otherwise, relatively undisturbed lands along upper Tennessee River Valley reservoirs are transitional bottomland-like, seasonal expressions of variable width. They are positioned between truncated lower to midslope mesic systems and the biannually imposed, impoundment pools. The shore levels of winter and summer pools, once adjusted and barring unforeseen weather events, usually show only short-term, minor (a few decimeters) elevation changes. On WB, these pool shorelines delineate the seasonal upper boundaries of the less productive aquatic habitat. (Figure 2).

The potential flood shoreline is above the summer pool with an upper limit which can be calculated from the height of the hydraulic head upstream, the dynamics of morphometrically influenced excess flow and the elevation of the top of downstream dam gates. This upper border can also be empirically defined by the line of water borne organic debris and anthropogenic waste lodged in the perennial vegetation during infrequent and aperiodic flood pool filling. Two different, periodic, reservoir shore habitats have resulted from the bilevel reservoir impoundment strategy. Where the riparian foreslope allows, the reservoir riparian vegetation or the potential habitat can be defined on WB in the context of the pool levels with (1) a large, counter-seasonal, sparsely vegetated, drawdown zone or mudflat, of 2549 ha, between winter (224m, msl) and summer shores (225.9m, msl) and (2), the growing season, riparian forest habitat from summer shore upland to, or near, the flood zone border (227.1m, msl), often extending up tributary streams. WB maps show 138 first to fourth order tributaries dissecting the 1241 km of summer pool shoreline which separates these seasonal habitats. The summer forest riparian zone may grade into small, hydrogeomorphically and floristically definable wetlands, particularly along gradient interrupted, hanging tributaries, but such inferred distinctions were not investigated here. The WB summer reservoir riparian zone was considered as 1902 ha (Table I), and undisturbed forest communities in the zone were chosen for the study reported here.

The mudflat supports mostly patchy, exposure stimulated seasonal vegetation (with fall to spring life history strategies) as well as expanding stands of summer emergent herbaceous or graminoid perennials. In the absence of other anthropogenic effects, the reservoir riparian summer shore supports closed communities of woody perennials. Under the present conditions which favor riparian species, competitively successful hydric shrubs and trees have migrated from upriver and tributary valleys to establish on the historically mesic, flood zone, terrain.

3. Objectives and Hypotheses

The study design and key targets for the preliminary characterization of WB summer reservoir riparian position forests were conceived in concert with documentary research and field investigations which included ongoing examinations of the ecological conditions and dynamics of the winter mudflats. The undeveloped zones were considered typical of a TVA reservoir within the Ridge and Valley Province of the Southern Appalachians. Four hypotheses were proposed, along with a related objective, for design and commission of the continuing vegetative description and analysis of consequent summer riparian zone communities; as well as consideration of the winter mudflat.

3.1 Null hypotheses

Ho1 Watts Bar (WB) reservoir summer shore forests on riparian topoedaphic terrain show no difference in similarity indices among unmanaged regional, riverine or bottom stands.

Ho2 WB riparian forest stands exhibit no difference in basal area density when compared to regional riverine or bottom stands.

Ho3 Chronologically well developed WB riparian plant community populations have shown no adaptation or morphogenetic response to the altered and dynamic hydric regime which would indicate a shift from mesic to reservoir riparian expression.

3.2 Directional hypothesis

Hd The approximated biomass within impoundments boundaries has changed with inundation and the seasonal detention of carbon is less. Watts Bar is an example of this change.

An additional objective of this study was to continue to characterize the areal and physical dynamics of the drawdown winter mudflat habitat, while summarizing the vegetative ecology of this counterseasonal riparian expression, positioned and enlarging between the aquatic and the forested riparian zones.

4. Design Criteria, Methods and Target Concepts

The design criteria in these reservoir riparian studies included baseline data characterizations and year-round field observations. We employed the use of traditional and documented methods for field measurement in order to arrive at comparable data. Studies have continued since 1988, and intensified in 1992 and 1993. The targets of interest, the affective processes and limitations, are listed in Table II.

4.1 Background and baseline determinations

Studies on Southern Appalachian bottomland natural forest characteristics in the last several decades have apparently been limited by the increased extent of impoundment projects in large and small valleys and the continuing societal development of flood protected bottomlands (Swift, 1984). No reservoir riparian zone ecological reports, in our context, were found. For this study, J. Rosson provided unpublished regional US Forest Service (FS) bottomland inventory data for TN (1992). Information on hydric southern forests was found in Smith *et al.* (1975) for North Carolina (NC), Virginia (VA) and TN. Their red river bottom, wet flat, and bottomland site enumerations, and the FS data, were used to develop baselines for community delimitations.

Mann and Bierner (1975) specify and categorize the vegetation of the Oak Ridge Reservation, (ORNL) which abuts WB on the Clinch River arm, by sites which include the designations swamp, swampy thickets and forested swamps, all interpreted as bottomland habitats. Rheinhardt (1992) includes basal area statistics for mature, fresh water tidal forests in eastern VA which show taxonomic similarities to the bottomland forests of TN, VA, and NC of Smith *et al.*, (1975).

Shanks (1952) published an annotated checklist of woody species of TN. Core and Ammons (1958) noted sylvic habitats for the region. Clebsch (1989) discusses valley vegetation extant in east TN. The long experience of Clebsch (1989, and personal communication) and H.R. DeSelm (personal communication and in Carter and Burbank,

Table II

Targets of interest, affective processes and limitations appropriate to the riparian zones study focused on the Watts Bar impoundment.

Study Design, Key Targets, Watts Bar Riparian Zones	Summer Riparian Forest	Winter Mudflat
History, Cartographics, Operations	Document, Map & Photo, TVA Plan	Sample, Characterize
Reconnaissances	All season, Water relations, Land form	Periodicity, Temperature, pH
Taxonomy and Composition	Student efforts, Verification, Comparison	Life History Strategies
Hydric Alterations, Intrusion	Water measures, Erosion	Dynamics, Aggradation
Stand Type Fate	Mesic displacement, Comparison, Longevity	Colonization, Success
Sample Site Selection	Hydrics, Edaphotopography, Ownership, Stability	Accessibility, Aspect
Community Characteristics	Density, Height, Cover, Pattern, Compartmentalization	Biomass, Distribution, Phenology, Role, Emergents
Physical Characteristics	Dimensions, Substrate dynamics, Microtopography, Storm effects	Thermics, Drainage
Continued Alteration	Surface dynamics, Plant reproduction, Diversification	Periodicity, Tenure
Food, Cover Resource	In progress, Food web relationships, Allochthony	Continuing observations
Regional Importance (Resource)	Biomass/Carbon, Areal status, Aesthetics, Responsibility	Stabilization

1978) allows assignment of pertinent forest species to hydric *vs* mesic categories for this study area. The nationally formalized designations for forest wetland plant indicators as summarized by Tiner (1991), do not take into account local, ecotypical, or latent response of specific taxa within neo-hydric communities of varying compositions, as Tiner pointedly observes. The dichotomy, hydric *vs* mesic, employed in this study was based on autecological observations made in the Tennessee River Valley. Identifications were verified by B.E. Wofford (1989; and personal communications).

McGinnis (1958) provided data on compartmental litter biomass in mesic east TN forests of some 40 years ago and contributed extrapolable information for contemporary comparison. Ploskey (1985) reviewed several contributions concerning nutrient supply losses and reservoir impacted vegetation and litter. Ecological phenomena of riparian zone systems were discussed, including the key value of terrestrial litter, by Fredrickson and Reid, (1986), who alluded to reservoir riparian zones, and by Leidy *et al.* (1992).

The preliminary establishment of upland, ruderal cover species (R, or "weed successional" strategy) on draughty mudflat barrens during prolonged periods of aerial exposure has been noted (Gara and Stuckey, 1992; Amundsen and Walker, 1991; Sain *et al.*, 1984). Resistance to summer shore erosion was occasioned by the presence of well ramified roots and shoots of terrestrial and aquatic-emergent plants ("stable" or K strategy) along scarps, and these defenses have been recorded from our investigations along WB and other valley reservoirs (Amundsen, 1989; Amundsen and Bartlett, 1990).

Seasonal water level changes have affected the subsurface temperatures, the saturation of rhizospheres and the permeability of adjacent substrates (Amundsen and Bartlett, 1990; Amundsen and Walker, 1991; Amundsen, 1992).

As a result of recent ORNL studies, pertinent information on certain aspects of WB sedimentary processes, hydrology and reservoir dynamics have become available (Brenkert *et al.*, 1992). Life history strategies of extant perennial flora, when newly subjected to periodic flood stress, can be shown to be flexible or diminished, by our field observations and by confirmation in the literature (Tiner, 1991).

TVA (1980) lists physical data for each of the 15 impoundments of interest. Surface areas, pool elevations, sailing lines and shoreline measures are included. Flow profiles, pre-impoundment conditions, and intended and realized water level manipulations were obtained from Reservoir Operations (TVA, Knoxville, TN). Pertinent environmental information on utilization, shore-tract development and studied biotal conditions were available from TVA's Natural Resources group (Norris, TN). Meinert (1991) has reported aquatic bioparameters for selected TVA reservoirs. The University of Tennessee, Knoxville (UTK) Library maintains a comprehensive Cartographic Information Center with current and historic cartography of the areas of interest.

4.2 Measurements, reservoir riparian summer shore transects

Riparian position shoreline forest sample sites were chosen for compatible topographic character, absence of ongoing disturbance and taxonomic expression. Gentle slopes, closed stocking without evidence of tree removal and presence of riparian species were considered. No attempt was made to classify the extents of shoreline not susceptible to riparian community development. Wooded habitats predominated along the mostly undeveloped WB summer shores. Twelve preliminary summer shore transects were located and sampled during July and early August, 1993. Three were on fourth order tributary channels, Whites Creek, Clinch River and Emory River. The other nine were between main river miles 541.2 and 588.4. All abutted the shore, along the channel (s), in normal pool coves, or on summer islets. Two of the 12 were marked as permanent, each with four, one m deep PVC slotted pipe (loose capped) wells, at sample centers, for water level monitoring tests.

Transects were based on a line of 20 + meters length, observationally a width of riparian habitat integration, normal to the shore plane. The line was begun at the first, near-shore tree with diameter breast height (dbh) of ~125 mm. To test and determine average stocking of riparian stands, four tally centers were then assumed at evened distances along the line. Stand BA was calculated from plotless samples (Cruz-All, BAF 5) by species record, preceding inland, at each center. The two largest (diameter) trees from each center were segregated (dbh, height) for determination of long-term compositional status and classified as mesic (M) or riparian (R) individuals. Shrubs and saplings (2.5-12.5 cm dbh) were identified and totaled, one meter out from each side of the central line (shore-in) for presence and stem replacement data. Other, untallied tree species, as well as shrubs, encountered in the empirical riparian zone were identified for informational purposes.

Canopy heights were graded in coarse, one meter increments shore-in, using an Abney level and stand canopy closure averages were determined and checked with a Lemmon Densiometer C. Crown pattern was noted. Litter was collected from an 0.25 m^2 quadrat located similarly to center number three on all transects, and dried at 105°C. Litter was considered a year's accumulation, interpreting McGinnis (1958) and Ploskey (1985), and calculated to metric tons per hectare (mt/ha).

Slopes conducive to subterranean water intrusion in the rhizosphere, at normal pool, were a part of the selection criteria, and were determined with an Abney level. The pool level, summer shore scarp height, form and root zone exposure were noted. At each tally center, a five cm bore one meter deep was made if the interception of pool horizon was expected. Depth (to or without) of the saturated profile was recorded. Bores allowed root depth measures, rhizosphere oxidation, stoniness, soil origins and clay lenses to be evaluated. Tree bole shape and surface overwash materials were noted.

4.3 Statistical treatments of forest tree data

Two comparable sets of data, species composition of the WB summer riparian forest stands and basal areas of the sampled stands were, even at the preliminary stage, suitable for elementary statistical testing.

The silvic composition of WB stands is compared to appropriate habitat stands reported by Smith *et al.* (1975) Mann and Bierner (1975) and Rheinhardt (1992). The Sørensen quotient of similarity: CCs = 2c/ s1 + s2; where s1 and s2 are the number of species in stands 1 and 2 respectively and c is the number of species common to both, is used (Brower and Zar, 1984).

The basal area (BA) data collected do not reflect parameters of their entire populations, either in the WB samples to date or in the three named stands of 50 years development reported by Smith *et al.* (1975). Without arbitrary classification, such totals show no equality of variances (Tanner, 1978). Given observational assumptions however, relationships can be sought employing the Mann-Whitney nonparametric test using the statistics: U = (n1) (n2) + [(n1)(n1 + 1)/2] -R1; and U′ = (n1) (n2) -U; where n1 and n2 are the sizes of samples 1 and 2 respectively and R1 is the sum of ranks for sample 1 (side by side columnar ranking from smallest datum, Brower and Zar, 1984).

4.4 Interface shoreline emergent perennials

Although none were found within transect locations, several emergent herbaceous perennials were observed rooted in the mudflat along some summer shore interfaces. Besides their organic contribution, these were in a position to dampen onshore waves, and samples were clipped (to substrate, 0.25 m^2 hinged frame) at their empirical growth peak (July, from observation) for air dry biomass determinations. Examples of such stands were observed throughout the mudflat season for phenological process and expansion.

4.5 Summer shore erosion determinations

Two editions of two USGS quadrats, Bacon Gap 1952 and 1968, and Ten Mile 1952 and 1973, were used to measure mapped shoreline loss of summer riparian habitat at normal pool on WB over those map intervals (Amundsen and Bartlett, 1990). A formulation for wave energy was adapted for the average WB off-shore slope and substrate. Topographic features which showed mapped erosion loss (a 12 location sample) were selected (V. Spicer, 1991, unpublished MS). During this current study, pool scarp dynamics, geometry and bio-physical defense against shore erosion were recorded at each of the 12 transect locations as well.

4.6 Methods, winter mudflat riparian vegetation and microclimate

Selected Watts Bar mudflats have been examined at least once a week in season since January, 1988 (Amundsen, 1992, continuing). Our weekly Watts Bar mudflat investigations included surface water/mud and shallow probe temperatures (each at 5cm) and shore-mud slurry pHs. Developmental phenology of colonizing and persistent, seasonal mudflat flora were noted weekly, once initiated, to seed maturity, if reached before summer pool. Subjective biomass clips in well developed brumal stands were made. The few kinds of perennial mudflat plants including summer emergents, were recorded. Webb and Bates (1989) have also provided biological information on valley impoundment mudflats.

A model was developed to express dewatered mudflat sediment rhizosphere level temperatures as moderated by the sustained subterranean thermal output from the winter pool. The contribution of the pool heat, the depth of plant rooting and the presence of an heat conducting anoxic stratum were checked by probe measurements (Amundsen and Bartlett, 1990).

The pool-continuous, water saturated, anoxic "gray zone" was considered empirically isothermal. The gray zone, as a stabilizing heat sink, provided heat to the unbroken mudflat sediments during cold periods. Atmospheric radiative and convective heat transfer were taken into account, and with conductive heat flow from the gray zone included, the model was designed and developed to allow calculations of effective temperature moderation across the sediment mantled mudflats.

A time dependent, finite difference, heat conduction approach was tested. The model was solved by computer for various ambient conditions. The sediment was considered homogeneous and approximations for small, finite changes in time and distance were inserted and yielded an equation solved at each node in the computer model. The heat equation was modified to include net absorbed surface radiation and surface convective energy transfer. For each solution, constant initial temperatures for both ambient air and the gray zone were assumed.

4.7 Determination of seasonal riparian biomass approximations and altered areal carbon detention

Detailed bottomland forest statistics (Smith *et al.*, 1975), reports from Odum (1993) and Satoo (1970); and selected observations from the WB impoundment area provided information that could be corrected and standardized. We compared "yields of mixed hardwood stands occurring naturally" (sensu Smith *et al.*, 1975) with the reservoir aquatic systems and the with the less productive WB reservoir riparian stands of similar chronology. Data on east TN forest litter accumulations (McGinnis, 1958) and litter collection summations (Ploskey, 1985) provided baselines for WB riparian forest litter comparisons. The bioassay determinations of carbon in TVA reservoirs (Meinert, 1991) and reviews of studies concerning carbon input into impoundments (Ploskey, 1985) gave sufficient information to link the terrestrial - aquatic biomass production/seasonally detained carbon pool approximations, pre and post-reservoirs. We employed conservative projections of historic, net primary productivity for Tennessee River Valley sites.

J.F. Faulkner (UTK, personal communication and unpublished MS, 1992) has adapted a biomass model from Smith *et al.* (1975) which was translated into BASIC and test applied to measures of the frequently encountered riparian forest species Platanus occidentals L. (sycamore). Sycamore was consistently represented in other appropriate bottomland stands (Rosson, 1992; Hall and Smith, 1955; Carter and Burbank, 1978) and was designated an integral indicator of reservoir riparian habitat conditions and response in our WB studies. Faulkner's equation as tested was: $\log_{10}$ CU.BIOMASS = C_{1+} C_2 (1/AGE) + C_3 [($\log_{10}$ HEIGHT)/AGE] +C_4 ($\log_{10}$ BASAL AREA); where C_{1-4} are stand type coefficients from Smith *et al.* (1975); AGE is the stand age; HEIGHT is the average tree height; and BASAL AREA is the m^2/ha.

The equation was developed for our dbh-height scale for WB stand yields against Smith *et al.* (1975) yields for stands of greater density and stature (BA, height). Seasonal below ground biomass increment, which can be substantial in stressed environments (Smith, 1992) was not measured.

McGinnis (1958) recorded annual vegetative litter totals, that once adjusted, showed a contribution one third greater than intact standing tree biomass increments. Adjustments of Ploskey's (1985) data supported the quantifications used here. Comparative calculations, annual increment, were based on fully stocked, sycamore identified stands (all trees 3mt/ha/yr) and mixed hardwood litter fall (4mt/ha/yr). WB riparian stand trees were considered (⅔ BA, $^4/_5$ height) at two thirds, 2mt/ha/yr; litter at 5mt/ha/yr. Total biomass, seasonally adjusted, was 7(3+4) to 7(2+5). Terrestrial carbon was one half biomass (Olson, 1970). Carbon was directly reported by Meinert (1991) for aquatic systems.

Table I. Estimates of the area of wetland loss during the last 200 years in the south-central U.S. (from Dahl, 1990).

State	Wetland Area -1780's X 10^6 ha	Wetland Area -1980's X 10^6 ha	Proportion of Wetlands Destroyed (%)
Alabama	3.0	1.5	50
Georgia	2.7	2.1	23
Mississippi	4.0	1.7	59
North Carolina	4.5	2.3	49
South Carolina	2.6	1.8	27
Tennessee	0.8	0.3	59
Virginia	0.7	0.4	42
West Virginia	0.04	0.04	24

This conference was organized by the Southern Appalachian Man and the Biosphere Program (SAMAB) in two parts. The first was the presentation of invited technical papers. These papers were selected to provide a current assessment of wetland functions, wetland regulation and assessment, management effects, and wetland restoration and creation. The second part of the conference consisted of three working group sessions (1-Research and Information Needs; 2 -Wetland Functional Assessment and Restoration; and 3-Wetland Protection and Conservation) which were designed to summarize the topic and provide recommendations to scientists, resource managers, and regulators.

This paper provides a summary of the major findings and research recommendations presented at the technical session of the conference and the working group sessions.

2. Conference Summary

The following discussion summarizes important findings reported at the technical sessions of the conference.

2.1 WETLAND FUNCTIONS AND VALUES

• Distinctions must be made between wetland functions and values, they are not synonymous terms. Wetland functions are derived from inherent ecosystem processes. Five basic functions can be recognized: hydrology, productivity, biogeochemistry, decomposition, and community dynamics. Each comprise biotic and abiotic processes that affect the structure, composition and dynamics of the wetland ecosystem. While these basic functions are common to all wetlands, they will be expressed differently among wetland types. Value is an anthropocentric interpretation of the quality or importance of an ecosystem function or process. Examples of values ascribed to wetlands include: hunting, fishing, timber production, assimilation of nutrients in waste water or runoff, and flood control, to name a few. (see Richardson, 1994).

• A hydrogeomorphic classification system provides the basis for functional classification of wetlands. Because wetlands require saturated soil to sustain anaerobic conditions, a hydrogeomorphic system provides the means to incorporate geomorphic, hydrologic, and edaphic information into a classification system which reflects properties or processes that affect wetland functions and values (see Brinson, 1993).

• Wetlands are disproportionally important in providing landscape diversity, maintaining biological diversity, and providing refugia for threatened and endangered species. Management and conservation of these wetland ecosystems should involve

19.63 (Table IV). Smith *et al.* (1975) report the three appropriate stand representations (segments of age 50 years) compared here for floristic similarity had an average BA, m^2/ha of 33.7. Rheinhardt, (1992) reported from the tidal swamps of the Chesapeake (maple-sweetgum stands), a BA, m^2/ha of 31.7. Mann and Bierner (1975) do not

Table III

Tree and tree size "shrub" taxa found in the WB summer riparian habitat. Columns list comparable stand taxa of east TN species from reports of Smith et al. (1975), Rheinhardt (1992), and Mann and Bierner (1975). Blanks indicate taxon missing from reported stands.

WB	Smith et al.	Rheinhardt	Mann & Bierner
Acer (negundo)			X
Acer (rubrum)	X	X	X
Acer (saccharinum)			
Alnus		X	X
		Asimina	
Betula	X		
Carpinus	X	X	X
Carya	X		
			Catalpa
Celtis	X		
Cornus			X
Fraxinus	X	X	
	Liriodendron		
Liquidambar	X	X	X
	Magnolia	X	
Nyssa	X	X	
Platanus	X		X
	Populus		
	*Quercus (r)		X
*Quercus (w)	X		X
Salix	X		X
Ulmus	X		

* (w) Bottom white oaks, (r) Bottom red oaks

provide BA values. WB transects No. 3 and No. 8 (Table IV) are close to the cited stockings (31.86 and 31.46). WB No. 3 is on a small islet at river mile 588.4 which showed no sign of recent disturbance to the well developed bottomland forest of birch, silver maple, sycamore and boxelder on river bend, slack, silty sands. No. 8 is on the fourth order tributary Emory River embayment, along the navigable channel, and showed large silver maples along with sycamores and microsite favored, relic, tulip trees.

The Mann-Whitney test (rank-sum) clearly demonstrates that the WB reservoir riparian forest stands BA as determined from the 12 transects to date, did not have the same BA values as the stands reported by Smith *et al.* (1975) which were found overall to be floristically similar. However, when only the values from the most productive four of the twelve WB transect (BA) totals are considered (stand transects 1,3,5 and 8; Table IV) and rank-summed with the selected three Smith *et al.* 50 year stand values (32.6, 36.7, 31.7); Mann-Whitney $U = 11$ and $U' = 1$; these statistics are within the bounds of the critical Mann-Whitney values at the significance level of 10%. Thus, only for the four most productive WB stand's is there 90% confidence that these transects reflect a forest with the same BA yield by the stands selected for comparison. Hypothesis two is rejected. From sampling to date, the WB stand BA's were not the same, in spite of the partial comparison agreement.

5.4 Habitat characteristics and influences

Canopy top heights graded up from the shore, greater than slope rise, inland, on six transects (Table IV) often topping with mesic relics inshore. Mean height for twelve transects was 17.3m. For the three, 50 year old Smith *et al.* (1975) compared stands, the means were 21m, 21m and 21m. Canopy closure varied from 60% to 90%, the median was 80%. Litter weights ranged from 1.1 mt/ha to 8.58 mt/ha, the mean was 5.04 mt/ha (Table IV).

The average slope of the 12 WB transects was 10%. An expected WB areal average slope, winter shore to flood shore, had been calculated, employing the known summer shoreline length, the areas of winter, summer and flood pools, and their respective elevations. For the entire zone of hydrologic influence in the WB impoundment the mean slope was 9%. Two included transects had much steeper slopes (30% and 40%) due to abrupt rises, unseen from offshore, near the end of their 20m transects. No trend is discernable in 12, 20m transects between BA, litter weights and mean slope, or between canopy height, BA and litter weights. There was an inconsistent tendency for canopy closure to relate to litter weights. Table IV shows transects 1,4,6, and 9 had less than 80% closure with an average projected litter of 3.6 mt/ha. Transects 2,5,7,10 and 11 had 80% closure, showed an average 6.3 mt/ha; transects 3, 8 and 12 had 90% closure but litter weights varied from 2.2 (the least but one, collected) to 7.3 mt/ha, the highest, but one. The eight transects with 80% or greater closure had a mean of 5.8 mt/ha.

Water profiles measured below the (regular) surface of the summer transects but not detailed here, showed that the inland projection of pool water elevation was horizontal when scarp was considered, with time lags of saturation and drainage in response to short term, minor, vertical pool fluctuations. Pipe well placements (on Transects 1 and 2) showed pool level controlled substrate water levels were consistent with fresh, unlined bore hole measurements. Riparian woody species were found to be favored along the summer shore where the soil depth to the summer pool level regularly saturated profile was less than 0.50m. A 20m transect, from water level shore inland on a regular five

Table IV

Summer shore riparian zone forest transects, growing season, 1993, WB. Location is by posted rivermile (TVA) designations. All are Tennessee River except E4, Emory; C9.4, Clinch; and W2, Whites Creek, fourth order tributaries which are included in the WB Impoundment. BA, Largest (Lg) BA, M = Mesic, R= Riparian, Litter, Canopy Top, Closure and Slope measurement techniques are detailed in the text.

Transect	River Location	BA m^2/ha	Lg BA Total	Litter mt/ha	Canopy Top	Closure	Slope AV%
1	582.3	25.54	M/0 R/.57	5.07	15-18m	60%	5
2	580.1	15.50	M/.58 R/.50	5.75	12-18m	80%	2
3	588.4	31.86	M/0 R/.88	2.23	12-18m	90%	2
4	575.0	11.50	M/0 R/.23	3.68	6-15m	70%	12
5	546.0	21.24	M/.61 R/0	8.58	18-20m	80%	4
6	556.0	15.21	M/.38 R/.08	1.10	18m	70%	5
7	541.2	16.65	M/.38 R/.20	6.99	12-20m	80%	30
8	E4.0	31.46	M/.23 R/.43	7.30	18-25m	90%	5
9	C9.4	19.52	M/.47 R/0	4.38	18-25m	60%	8
10	556.0	12.92	M/.18 R/.08	6.03	18m	80%	5
11	567.0	14.64	M/.25 R/.30	4.29	15-25m	80%	2
12	W2.0	19.52	M/.16 R/0	5.14	18m	90%	40

percent slope projects a water free profile of one meter at the upper end. This upper slope unsaturated depth ostensibly favored the maintenance of, and continued establishment of, flood intolerant, mesic species. Microedaphic anomalies, particularly undulant scarps, were seen to allow the persistence of some mesic individuals along the shore, but replacement in kind has been limited to higher ground.

Qualified observations showed that stump buttressing and proximal root exposure was common, particularly with Liquidambar, in the near-shore riparian stands. Many trees, of several kinds, appeared to have lost apical dominance and excessively ramified crowns were noticeable, but not evaluated quantitatively. These conditions were attributed to the thinness of the water free soil profile, not to overwash flooding which has been infrequent and short-term according to TVA records. Extant sylvan individuals showed morphological response to the saturated substrate horizon. Growth pattern, replacement process and compositional trends pointed to a slopeward shift of mesic expression.

5.5 Watts bar winter riparian mudflats

The consistent, seasonal flora of the mudflat was seen to germinate within two weeks of continuous winter drawdown exposure. Barring a continuing drought or prolonged, unexpected, pool rise, development continued to seed maturity before summer pool rise. Water rises that inundate mudflat plant communities for less than a week before ebbing have not been observed to preclude this maturity. In the four year period January, 1988, to spring 1991, neither three persistent snow covers nor 48 hours or more of durational air temperatures below 0°C (six of record) have precluded seed set for the common, counter-seasonal mudflat species. Given a sufficient time period, the successful phenological development of the counter-seasonally exposed flora was in large part attributed to the relative absence of frost with the reservoir pool moderated temperatures of the mudflat substrate.

5.6 Mudflat flora

The spatially discontinuous, regular drawdown, brumal plant communities were of two broad life history categories. Near-shore expressions on WB flats were often erratically occurring monotypic stands of summer emergent perennials, many of which exhibited tiller or rhizomatous growth activity throughout the winter. Graminoid genera such as Juncus, Scirpus, Phalaris, Cinna, Cyperus, Leersia, Typha, and Zizaniopsis were locally prevalent. Dicots, such as Justicia, Saururus, and Polygonum occasionally formed large, expanding, clonal stands. On WB, Justicia was found furthest out from the summer shore in up to 0.5m of water, and was found, from photographic record, to have spread widely over the past few years of reconnaissance. Seven clip plots of wave-baffling perennial emergents were made at the time of subjective determination of maximum biomass (July-August). The projected air-dry yields per m^2 were: Zizaniopsis, 580g; Typha, 420g; Phalaris, 440g and 820g; Polygonum, 500g; Justicia, 40g; and Saururus, 460g. There were no submersed aquatic species of consequence in the WB area.

The off shore exposed sediment flats of WB, when vegetated at all, showed patchy expressions of spring maturing Alopecurus, Myosurus, Ludwigia, Sibara and Rorippa among some 25 other kinds of lesser importance. The counter-seasonal, locally dominant Alopecurus carolinianus was harvest-clipped for aerial biomass and yielded as much as 200g air dry material m^2. The consistence criterion for inclusion in the brumal mudflat flora was seed production before summer pool rise. Many volunteers have germinated from mudflat seedbanks, but were not included without development to seed maturation.

5.7 MUDFLAT TEMPERATURES

During the winters of study it was noted that an initial autumn drop in the sediment temperatures did not continue as daily mean air temperatures had lower values, but showed a bottom and a subsequent temperature plateau in the mudflat plant root zone that usually registered well above freezing. Systematic, albeit instantaneous, readings demonstrated that a saturated anoxic zone (the gray zone) beneath the aerobic root zone of the mudflat lithosphere had temperatures approximating those of the on-shore contact pool. Measured pool temperatures during the study period were never lower than 4°C.

The pool-continuous, water saturated, anoxic "gray zone" was isothermal, and, acting as a stabilizing heat sink, provided heat to the unbroken mudflat sediments during cold periods. A time dependent, finite difference, heat conduction modeling approach was successful in predicting the sediment temperatures. Sample run results are shown for four-hour intervals in Figure 3. The model confirmed that as the air temperature declined, the amount of temperature reduction of the ground surface below the air was reduced. As the surface temperature was reduced, there was more heat transferred from the gray zone since heat energy transfer was consequential to temperature difference. During very cold air periods, the sediment surface temperature was much warmer than the air a few centimeters above, and the root zone temperatures were not observed to drop to frost. When the gray zone was closer to the surface, the profile in Figure 3 was shifted to higher temperatures. Field checks verified the model.

5.8 AQUATIC INTERFACE AND MUDFLAT pH

Impoundment channel pH (one m from shore) tended to be basic, in summer; neutral, in winter. Shore-water interface pH tends to be neutral to basic when channel water exchange was occurring (current, wind, boat wave) and slightly acidic when calm and subject to vegetated shoreline run-off or storm flow. Local precipitation made little difference otherwise. On warm, sunny, spring days, in standing micropools, mudflat plants were often subjected to pH which rose from 6.5 to 10.0 as solar warming and algal respiration raised basicity. Such fluctuations have only been noted occasionally. Most mudflat community associated pH (mud slurry) measured 6.5 to 7.5 depending upon the extant weather conditions.

5.9 WATTS BAR MUDFLAT SUBSTRATES AND SUMMER SHORE EROSION

With contemporary reservoir management, the factors of sediment supply, bed morphometry, flood scouring and deposition continue to influence the characteristics of the mudflat substrates. Current drawdown zone alluvial sediment pedogenesis has been precluded by the biannual hydrodynamic cycle and residual historic soil profiles were being degraded by erosion. There were conspicuous residual soils in scoured areas below summer shores which were often soils akin to those found hydrologically modified above the summer shore. Coarse, dynamic alluvial sands have graded to silt and clay textures depending on the local restrictions to riveroid flow and occasional slack flood surges above the summer shore. Alluvial sandy-silts and silty-clays comprise the greater areal

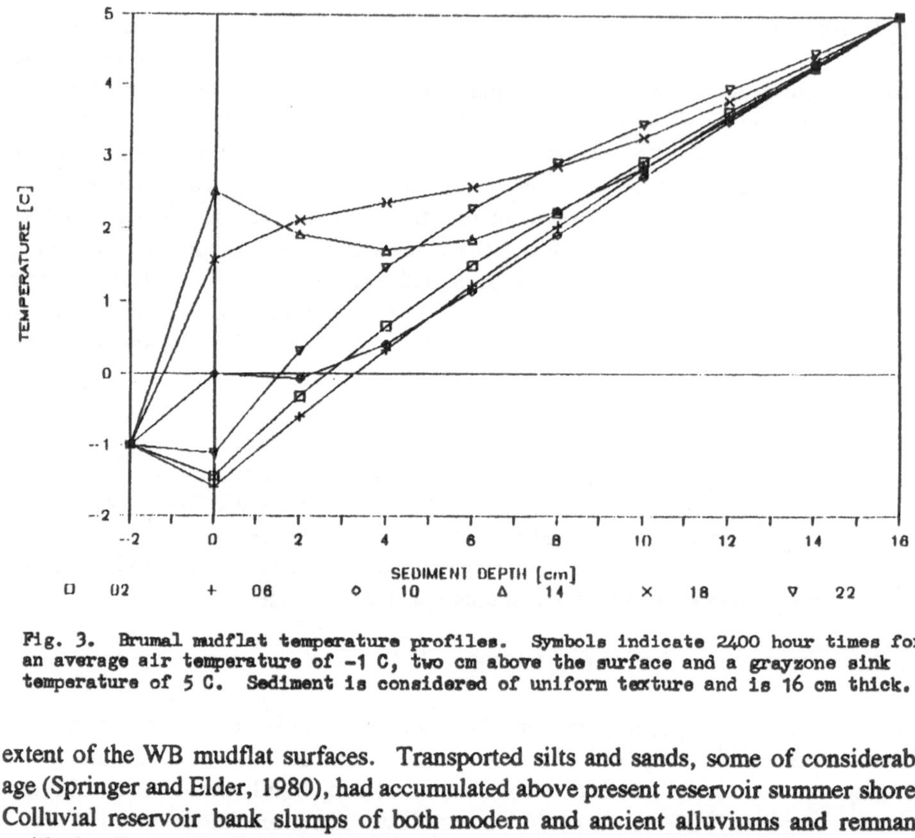

Fig. 3. Brumal mudflat temperature profiles. Symbols indicate 2400 hour times for an average air temperature of -1 C, two cm above the surface and a grayzone sink temperature of 5 C. Sediment is considered of uniform texture and is 16 cm thick.

extent of the WB mudflat surfaces. Transported silts and sands, some of considerable age (Springer and Elder, 1980), had accumulated above present reservoir summer shores. Colluvial reservoir bank slumps of both modern and ancient alluviums and remnant, residual soils usually dispersed rapidly and contributed to the sediment supply.

Notwithstanding constricted current scouring, particularly at the river inflows, boat waves were found to be the major cause of WB summer shore erosion. Stern wakes impact shores with a force of 0.96 kW/m (smooth littoral with a five % slope with a one meter depth at 20m) generated by a 0.5 meter high wave. Consequent topographic maps showed localized, unprotected, summer riparian zone shoreline averaged retreats with loss of riparian habitat of up to two meters annually. An overall one meter summer shoreline retreat would geometrically add 113 ha a year to the WB mudflat. Shore scarp form was generally determined by physical defense; vertical for rock, concave with a rooted overhang limited to the oxygenated summer profile on wooded shores, collapsing convexly when herbaceously vegetated and scoured even beyond the angle of repose where soils were denuded. TVA projections of sediment accumulations for WB (TVA, 1980) were 1.64 x $10^6 m^3$ a year. Bank collapse of extant terrestrial soils plus related sediment accumulations have accounted for both the enlargement of the mudflats of WB, and the retreat of the aquatic interface of the slope constricted summer riparian zone; resulting in a wider, shallower, normal pool.

5.10 Shifts from mesic to riparian expressions

In the summer riparian zone, hydric tree species have become established while mesic species were in decline. In fifty year old stands, BA and height were less than compared bottomland forests. Tree height limits and empirically observed excess branching yielded a canopy of high closure, and litterfall was relatively heavy. The slope of riparian expressions sampled was close to the projected geometric slope for the WB impoundment, and horizontal pool incursion limited mesic species when the unsaturated profile was less than 0.50m. On the mudflats of WB, seasonal stands of grass and herbs reflect a counterseasonal brumal development in a pool temperature moderated microclimate. The pH measurements on the mudflat regularly showed a neutral to slightly basic substrate. Erosion processes showed increased mudflat extent likely with a demonstrated, accompanying retreat of the summer riparian scarp. The riparian stands, defined by their integral depth or width, were being constricted against the elevationally higher, mesic terrain. Hypotheses three is rejected. There is a conspicuous shift from pre-impoundment mesic to post-impoundment hydric or riparian expression.

5.11 Approximations of biomass and alterations in impoundment habitat carbon detention

The imposition of impoundments which inundated the riverine and mesic habitats of the upper Tennessee River basin has drastically altered within pool shores biomass yields and the growth-season detention of carbon. The impoundment areas are dominated by modern aquatic systems which show, for six tested reservoirs, 0.24 mt/ha, corrected to surface exposure, of carbon (Meinert, 1991). The historic mesic habitat above the pre-impoundment river channel seasonally fixed approximately one half the biomass of 7.0 mt/ha or 3.5 mt/ha, surface exposure, of carbon in tree increment and annual litterfall. When the historic river was attributed the same level of carbon detention as the modern pool (0.24 mt/ha), the approximate change, old river plus former mesic habitat, was calculated at an average of a 12:1 diminution from historic to modern, seasonal carbon sub-pool (Table V). The directional hypothesis (Hd) is accepted. There has been a many-fold lessening of biomass and annually sequestered carbon with the imposition of the reservoir systems onto historically terrestrial habitats.

5.12 Watts bar summer riparian biomass and carbon detention

The riparian forest and forest litter sampling along the WB shore allowed a compartmental proportioning of the biomass production and carbon detention. The WB riparian forest was figured to be less productive of standing biomass and therefore of tree retained carbon than the compared bottom stands but, more productive of seasonal litter detained carbon. The predominant biomass/half carbon resource of the riparian is in the more easily available, more rapidly turned over, litter (Ploskey, 1985). The retention of biomass in boles is of long term consequence, but the annually detained supply of the litter as an energy resource, is important (Table VI). It was noted that the overall average production of terrestrial biomass/carbon was approximately the same under several vegetative characterizations.

Table V

Approximations of the categorical differences in reservoir area carbon detention. Calculations include corrected pre-impoundment river and mesic forest valley and modern, normal pool measured carbon.

Reservoir	Normal Pool (ha)	River Bed $(ha)_1$	Mesic C $(mt)_2$	Aquatic C $(mt)_3$	Approximate Change
Cherokee	12262	982	39716	2943	13:1
Chickamauga	14326	3485	37607	3438	11:1
Douglas	12303	1283	41874	2953	14:1
Ft. Loudoun	5908	1789	14846	1419	10:1
Norris	13841	1186	44578	3322	13:1
Watts Bar	15783	4187	41595	3788	11:1
					X=12:1

1. Riverbed area, pre-impoundment
2. Within normal pool shore elevations as mesic habitat minus riverbed hectareage. Multiplier is 3.5 mt/ha biomass considered 50% carbon. The Soil Conservation Service has reported the same biomass for alfalfa on Class 1 soils in East Tennessee.
3. Based on surface and volumes of normal pools; 0.24 mt/ha carbon, corrected from measurements by Meinert (1991).

6. Discussion and Conclusions

6.1 The Summer Shore Reservoir Riparian Zone

The summer shore reservoir riparian forest of WB was floristically similar to other Southern bottomland forests. The overall BA for WB when compared to regional bottom stands was not the same, unless comparisons were selectively limited. Most transects showed BA less than 20 m^2/ha, only two thirds the compared stands. The 50 year riparian condition has allowed successful migration of appropriate species, but not their comparable development in most instances. Six of the eight sampled stands with less than 20 m^2/ha BA had Mesic dominance persisting in the largest, two trees per center, category. The best-stocked stands (represented by transects, 3 and 8, Table IV) showed largest Riparian species, indicative of an earlier onset of riparian expression shift. The large mesic individuals which remain ostensibly viable on the WB transects such as white oak, tulip tree, American elm and black cherry are generally found at the upper end of the transects, on a scarp edge which allows rapid root zone oxygenation at any pool drop, or on raised microtopography along the transect. The saturated profile of

Table VI

Approximated carbon detention and retention* for reservoir riparian forest, Watts Bar. Calculations corrected for biomass comparisons, 50% carbon in metric tons per hectare. Mesic forest sites were pre-impoundment.

Compartment	Mesic Forest Expression	Riparian Forest Expression
Tree	1902 ha x 1.5 mt/ha (2853)	1902 ha x 1.0 mt/ha (1902)
Litter	1902 ha x 2.0 mt/ha (3804)	1902 ha x 2.5 mt/ha (4755)

* In terms of availability, as forage or nutrient substrate, in-tree carbon is retained for an indeterminate period, litter carbon is seasonally available.

gentle, inshore slopes has favored the recent developmental strategy of hydric over mesic species. The presence of hydric shrubs, the appearance of juvenile forms of hydric species and the lack of similar mesic reproduction indicated the hydrologically impacted stands were transitional from mesic to hydric. Based on data collected to date, however, no projection as to a hydric dominance time frame could be made, although ultimate riparian stand width could be linked to horizontal subsurface water intrusion.

The growth season, saturated soil profile was reflected in many of the height contour restrictions of the shore forest canopy tops which rose inland beyond the rise in slope. The water incursion was horizontal; as the slope rose, even at a modest two to five percent, tree heights increased divergently as unsaturated soil profiles increased in depth on one-half the transects.

Closure of canopy and BA did not appear related, but most closures were high. Closure was observationally related to widely branching tree form and limited apical elongation in these instances. Our litter collections were not conclusive evidence, but systematic sampling (same relative point of each transect) showed, with the exception of transect No. 3 (Table IV) that greater closure was linked to more abundant litter on the average. These litter accumulations were some 12% (5.04 mt/ha) higher than those reported by McGinnis (1958), who projected up to 4.5 mt/ha for east TN mesic forest stands. Ploskey (1985) reported more than seven mt/ha for deciduous, monotypic plantation, forest litter for the Southern Appalachians.

The BA was comparatively low, with restricted growth related to saturated substrates, but the litter accumulation was comparatively high for unmanaged TN forests. The detritus component was shown to be a very important part of this riparian expression, a conclusion also emphasized by Fredrickson and Reid (1986). The contribution of detritus inshore, and allochthonously offshore, is a continuing key riparian food and nutrient resource (Ploskey, 1985).

Erosion at the summer aquatic/riparian interface compresses riparian habitat width against mesic elevation rises. The inland ingress of riparian communities is tied to horizontal subsurface saturation which limits mesic tree life history strategies.

6.2 The Winter Drawdown Mudflats

The patchy, brumal mudflat plant communities were rather consistent in their developmental strategy, seed to seed in about five months, with Myosurus and Alopecurus first to germinate, and Myosurus first to mature. Mudflat Alopecurus, typically a robust, bunch habit grass in crowded stands, is likely an ecological segregate of the uncommon, but wide-spread, riverine growing season annual, Alopecurus carolinianus Walt. The development and seed maturation of the Alopecurus ecotype and other mudflat flora was recently interrupted by a one month shorter drawdown regime (Figure 2) at the median elevation for counterseasonal plant development (~223m, msl). The later fall draining of the mudflat has caused the grass development to be set back, with seedheads immature and aborted by subsequent summer pool rise. This has led to an obvious diminution of seedbank replenishment. There have been exceptional individuals which have fully matured, even in the shorter period, and the likelihood of further directional selection toward a briefer term seed maturity strategy for some species seems possible. The noted success of this grass genotype on the counter-seasonal mudflats was proposed to have resulted from a potential for divergence in strategy within the parent riverine population, of decades ago, which was realized with the advent of a new habitat, the mudflat. It is not impossible that all of the maturing taxa on the mudflat are of such divergent or founder populations, taking advantage of the initial (and continuing) dearth of spatial competition as well as the winter pool thermal moderation of the mudflat substrate microclimate.

The known counter-seasonal floral composition had enlarged slowly. New successes were found, sparingly, each winter. There was no basis to project a complete winter cover however, since the explanations for the repeated patchiness of the current brumal vegetation have eluded our mudflat microhabitat investigations, and natural environmental phenomena have scant effect on anthropogenically programmed pool levels. The recent lesser exposure period of the mudflat will probably set back the fruition of such taxa.

Predictions for the perennial, summer shoreline, facultatively clonal, emergents at the shore interface at the upper levels of the mudflat are more straightforward. Justicia, Typha and Zizaniopsus showed evidence of opportunistic colonization along unvegetated shore margins in front of low scarps. Justicia remained most prominent and appeared to withstand the deepest summer water. Those found were mostly clonal populations, but all taxa have been observed in flower. These conspicuous perennial emergents were often in the mudflat slope range which benefits from winter pool heat supplementation, and many continued to show persistent photosynthetic tissue throughout the winter, even though aerially uncovered.

The extensive development of volunteer ruderals, on drawdown zones which have remained partially dewatered throughout the growing season, has been found more often on tributary reservoirs, such as Douglas (Table I), when winter runoff was less than expected, and normal pools did not fill. There was little to distinguish between the resultant flora of such habitats and old field successional flora in the region. Almost without exception, such plants were lost with the next pool rise, although they contribute directly to the aquatic carbon pool as do the mudflat seasonals and shoreline emergent perennials.

Alluvial sandy silts have provided the mudflat habitats for seasonal microbial and faunal predator-prey activity, in studies not reported here. A good portion of the historic residual soils truncated by the normal pool remain steepened and subject to slumping or water and animal traffic (cows) erosion. Where such historic soils were more horizontal, and mantled by contemporary alluvium, they have responded similarly to the deeper alluvium sediments in terms of thermal transfer, depth to anoxic profile, lack of organic incorporations and apparently depauperate decomposer populations. The pH of the pool contact and of the exposed winter substrates as measured was not exceptional. Clay lenses, accumulated in alluvium or exposed by scouring of previous profiles, affected local permeability, but were not found to be extensive.

6.3 Biomass production and carbon detention

Areal bioproductivity and carbon detention (and retention), one half the biomass, were negatively affected by reservoir impoundment. Reservoirs benefit from organic carbon in biomass left behind in impoundment preparation, but such amounts, which depended on the kind of sequestering organic material (as retaining tree trunks vs detaining non-lignified litter) and on the hydrologic retention time of the impounded pool, were noted to diminish rapidly (Ploskey, 1985). Any large losses in reservoir carbon would not be replaced by organic carbon increments at the levels measured by Meinert (1991), Table V. The developing (or successionally directional) reservoir riparian zone forest with litter, of WB, has shown an approximate total equivalent of seasonal carbon detention (one half the biomass) to the mesic forest displaced on an areal basis. The riparian compartments (trees vs litter) were not the same as the mesic, where retained tree carbon was more; while in WB riparian forests the litter compartment was the greater seasonal pool. The arithmetic in Table VI demonstrates the contribution of riparian litter (within the flood zone of the WB impoundment, and not considered in Table V) to the overall organic resource available to the WB areal biota. In any carbon depauperate aquatic-riparian system, the litter of the shoreline vegetation was considered very important (Fredrickson and Reid, 1986; Ploskey, 1985) as it should be for the transitional WB system.

6.4 Summary of hypotheses and judgements

1. The subjectively sampled Watts Bar summer shore reservoir riparian forest has been determined to be similar floristically but not in yield to regional valley bottom forest stands. The WB stands were found in transition and site limited by growing season waterlogging of soil and incomplete establishment of hydric species. There were no precedent habitats matching the characteristics of the ecologically depauperate mudflat or the shoreline forest in lakeless, pre-reservoir, east TN.

2. The mesic forest has been displaced upslope by hydric habitat conditions. Erosion has reduced the summer reservoir riparian habitat width and hydric conditions will not extend up steepened, high foot scarped, slopes. The winter mudflats are expanding.

3. The otherwise unremarkable biogeocoenotic expressions of the relatively new, developing reservoir riparian zones were notable for two important impacts; the thermal moderation of the winter mudflat and the counter-seasonal thinness of the nearshore growing season riparian habitat, unsaturated soil profile.

4. Benefits of the extant reservoir riparian zone were noted in the landscape erosion control offered by emergent perennial and well rooted, woody summer shoreline trees and shrubs, in the resource supply for the conservation of an important faunal habitat, and in the unmeasurable environmental appeal of a naturally forested, stable summer shoreline.

Acknowledgements

Many people have helped in this on-going study without formal credit. They include: (From UTK) C.C. Loy, MS candidate in Ecology, transect and research person; P.L. Walne, E.E.C. Clebsch, H.R. DeSelm and B.E. Wofford, professional colleagues in identification and categorization, J. Minton, curator of the Cartographic Information Center, V. Spicer, who worked up the erosion dynamics, J. Faulkner for biomass computations and the following academic/research assistants - J. Noe, M. Weaver, T. Bailey, S. Lawson-South, B. Hafer, M. Finger, K. Dwyer, R. Kramel, J. Rodgers, J. Duff, A. Rose, E. Gray, H. Rial-Meador, A. White, K. Harter, E. Maclin, K. Black, N. Fraley, and K. Goddard. L. McMillan processed the many revisions. (From Loudon County, TN) S. Wampler, erstwhile mudflat sampler. (From TVA) Wes James, Wetland Ecologist, Natural Resources, the personnel of the Reservoir Operations Office, and many helpful providers of specific information by telecommunications contributed.

TVA contract no. TV 86493V provided a stimulus for the intensive summer riparian zone work of 1992-1993. The UTK Department of Botany provided support, The Graduate Program in Ecology (Director D.L. Bunting) provided field equipment and laboratory facilities and a flexible summer schedule which allowed the intensive field work. Figures 1 and 2 are adapted from drawings supplied by TVA by J. Polson, UTK.

References

Amundsen, C.C.: 1989, Reservoir interpool plant habitat dynamics I, in: Quinones, F. and Balthrop, B.H. (eds), Abstracts of the Second Tenn. Hydrology Symp. p. 92.

Amundsen, C.C.: 1992, Reservoir interpool plant habitat dynamics IV, in: Quinones, F. and Hoadley, K.L. (eds), Abstracts of the Fifth Tenn. Water Resources Symp. pp. 102-105.

Amundsen, C.C.and Bartlett, E.B.: 1990, Reservoir interpool plant habitat dynamics II, in: Sale, M.J. and Presley, P.M. (eds), Extended Abstracts of the Fourth Tenn. Water Resources Symp. pp. 180-183.

Amundsen, C.C.: 1991, Reservoir interpool plant dynamics III, in: Sale, M.J. and Presley, P.M. (eds), Extended Abstracts of the Fourth Tenn. Water Resources Symp. pp. 141-144.

Brenkert, A.L., Brandt, C.C., Rose, K.A., Cook, R.B., Wood, M.A., Beard, L. and Schohl, G.A.: 1992, A comparison of two methods for estimating spatial patterns of sediment accumulation in the Clinch River-Watts Bar reservoir system, in: Quinones, F. and Hoadley, K.L. (eds), Extended Abstracts of the Fifth Tenn. Water Resources Symp. pp. 78-81.

Brower, J.E. and Zar, J.H.: 1984, Field Laboratory Methods for General Ecology. Wm. C. Brown.

Carter, V. and Burbank, J.H.: 1978, Wetland Classification System for the Tennessee Valley Region, Tennessee Valley Authority Tech. Note, No. B24.

Clebsch, E.E.C.: 1989, Vegetation of the Appalachian Mountains of Tennessee east of the Great Valley. J. Tenn. Acad. Sci. 64, 79-83.

Core, E.L. and Ammons, N.P.: 1958, Woody Plants in Winter. Boxwood Press.

Fredrickson, L.H. and Reid, F.A.: 1986, Wetland riparian habitats: a nongame management overview, in: Hale, J.B., Best, L.B. and Clawson, R.L. (eds), Management of Nongame Wildlife in the Midwest: A Developing Art. North Central Section Wildlife Society, Chelsea, MI. pp. 59-96.

Gara, B.D. and Stuckey, R.L.: 1992, Life-form spectra of the mudflat flora on the Scioto River, Delaware County, Ohio. SIDA 15. pp. 289-303.

Hall, T.F. and Smith, G.E.: 1955, Effects of flooding on West Sandy Dewatering Project, Kentucky Reservoir. J. Forestry 53, pp. 281-285.

Leidy, R.A., Fiedler, P.L. and Micheli, E.R.: 1992, Is wetter better? Bioscience 42, pp. 58-65.

McGinnis, J.T.: 1958, Forest litter and humus types of East Tennessee. MS Thesis, University of Tennessee, Knoxville.

Mann, L.K. and Bierner, M.W.: 1975, Oak Ridge, TN, Flora: Habitats of the vascular plants/Revised inventory. E.S.D. Publication No. 775, Oak Ridge National Laboratory. p. 141.

Meinert, D.L.: 1991, Reservoir vital signs monitoring. 1990. Tennessee Valley Authority TVA/WR/WQ-91/10.

Odum, E.P.: 1993, Ecology and Our Endangered Life Systems. Sinauer Associates.

Olson, J.S.: 1970, Carbon cycles and temperate woodlands, in: Reichle, D.E., (ed), Analyses of Temperate Forest Ecosystems. Springer-Verlag. pp. 226-241.

Ploskey, G.R.: 1985, Impacts of terrestrial vegetation and preimpoundment clearing on reservoir ecology and fisheries in the United States and Canada. FAO Fish. Tech. Pap., 258. United Nations.

Rheinhardt, R.: 1992, A multivariate analysis of vegetation patterns in tidal freshwater swamps of lower Chesapeake Bay, U.S.A. Bull. Torrey Botanical Club, 119, pp. 192-207.

Rosson, J.F. Jr.: 1992, Compiled US Forest Service Inventory Data, Tennessee. Starkville, Mississippi.

Sain, J.E., Fonferek, W.J., Simpson, M.S. and Whittingham, K.W.: 1984, First year vegetation following exposure of the Edmonson Lake Bed, Washington County Va. Castanea 49. pp. 158-166.

Satoo, T.: 1970, A synthesis of studies by the harvest method: primary production relations in the temperate deciduous forests of Japan, in: Reichle, D.E. (ed), Analyses of Temperate Forest Ecosystems. Springer-Verlag. pp. 55-72.

Smith, H.D., Hafley, W.L., Holley, D.L. and Kellison, R.C.: 1975, Yields of mixed hardwood stands occurring naturally on a variety of sites in the Southeastern United States. Tech. Rep. 55, School of Forest Resources, North Carolina State Univ.

Smith, R.L.: 1992, Elements of Ecology. Harper and Collins.

Springer, M.E. and Elder, J.A.: 1980, Soils of Tennessee. Univ. Tenn. Ag. Exp. Sta. Bull. 596.

Shanks, R.E.: 1952, Checklist of the woody plants of Tennessee. J. Tenn. Acad. Sci. 27, pp. 223-229.

Swift, B.L.: 1984, Status of riparian systems in the United States. Water Res. Bull. 20, pp. 223-229.

Tanner, J.T.: 1978, Guide to the Study of Animal Populations. Univ. Tenn. Press.

Tilman, D.L.: 1988, Plant Strategies and the Dynamics and Structure of Plant Communities. Princeton Univ. Press.

Tiner, R.W.: 1991, The concept for a hydrophyte for wetland identification. BioScience 41, pp. 236-247.

Tennessee Valley Authority: 1980, Water control projects. Tech. Monog. 55, Volume 1.

Webb, D.H. and Bates, A.L.: 1989, The aquatic vascular flora along rivers and reservoirs of the Tennessee River System. J. Tenn. Acad. Sci. 64, pp. 197-203.

Wofford, B.E.: 1989, Guide to the Vascular Flora of The Blue Ridge. Univ. Georgia Press.

PART VI

WETLAND RESTORATION AND CREATION

DECISION SEQUENCE FOR FUNCTIONAL WETLANDS RESTORATION

M. M. DAVIS

US Army Engineer Waterways Experiment Station, 3909 Halls Ferry Road, Vicksburg, MS 39180 USA

Abstract. As wetland functions are being more clearly evaluated, demand is increasing for the ability to mitigate for specific wetland functions that have been degraded. When wetland restoration project goals specify functions, success of the project depends heavily on proper guidance for project siting, design, implementation, and monitoring. A decision sequence is presented for wetland restoration projects to help achieve functional replacement. This methodology incorporates site selection and design features for specified wetland functions into three phases of a project planning decision sequence. The first phase, site selection, situates a wetland where there is the potential to perform a function. Phases two and three, the incorporation of functional design features into design criteria and project plan development, focus on the optimization of the functional capacity of a site. An example is given of how a wetland restoration project planning team can consider enhancing vegetation diversity during the project plan development phase to achieve a goal of improved wildlife habitat.

1. Introduction

Wetland restoration efforts are become increasingly important as wetlands continue to be degraded throughout the United States. The increase in extent and variety of wetland restoration projects is due to the recognition of the value of wetland functions to society and legislation mandating the protection of wetlands. Wetland restoration, however, is expensive to perform and many projects have not been successful (Kusler and Kentula, 1990). As a result there is a great demand for information about wetland restoration techniques that help achieve project goals and improve restoration success rates as economically as possible. Furthermore, as procedures for assessing wetland functions improve (Brinson, 1993a), demand is increasing for the ability to mitigate for specific wetland functions that have been lost or degraded.

Until recently, wetland restoration project design was based on the assumption that wetland functions followed form. If hydrologic and substrate conditions were established that supported a given type of vegetation, such as emergent herbs or trees, it was assumed that a functioning wetland had been successfully established. However, not all wetlands perform the same functions, nor do all wetlands have the capacity to perform functions to the same level (Adamus *et al.*, 1991, Brinson, 1993b). While limited functions probably were restored, the restored wetland did not necessarily replace the lost functions of the impacted wetland that were being mitigated. The old adage "function follows form" must be more closely examined (Marburger, 1993). Specific functional restoration of future wetland projects will depend on technical guidance for project siting, design, plan development and implementation, and monitoring methods.

Water, Air and Soil Pollution **77**: 497–511, 1994.

Wetland restoration project managers make many decisions at different phases of project development that affect how a wetland will attain functional goals. For example, alternate project sites must be evaluated and selected. Design criteria are specified. A project plan is developed that incorporates the design criteria with the site conditions. The project must be constructed and monitored over time. Information is available for general wetland restoration techniques (e.g., Hammer, 1992, Soil Conservation Service, 1993) and wetland function evaluation techniques (e.g., Adamus *et al.*, 1987, Bartoldus *et al.*, 1993). While elements from the evaluation techniques can be used in determining design criteria for specific functions (Marble, 1990, Bartoldus *et al.*, 1993), there is no guidance to aid wetland project managers in attaining wetland functions in other project development phases. The objective of this paper is to outline a method for attaining functional replacement goals in wetland restoration projects at specific project development phases.

2. Factors Affecting Wetland Functional Replacement

Wetland functions are the result of processes and characteristics occurring in the landscape (Brinson, 1993) and at the site (Adamus *et al.*, 1991). Conditions in the landscape affect the wetland hydrological, geological, chemical, and biological processes, thereby influencing the types and levels of functions performed by a wetland (Table I). Attainment of functional replacement requires that the restored wetland be placed in a landscape setting with the necessary conditions for the performance of the desired functions. For example, if sediment retention is a project goal, the restored wetland must be positioned in a disturbed or denuded watershed where it will receive runoff carrying sediments. This example can be used to illustrate the possibility that wetland projects can be placed where their capacity to perform a function can be overwhelmed. Most wetlands cannot perform the function of sediment retention where enough sediments are received and retained to cover and kill the vegetation. Wetland restoration project planners have relatively little design control over landscape features, and consequently, site selection criteria should be largely determined by the landscape features influencing the desired functions.

Once the landscape setting for the restored wetland has been selected, project planners can manipulate site characteristics (e.g., hydrology, energy, substrate, and vegetation) to determine the levels at which the wetland performs the desired functions. More nutrients will be retained, for instance, in a wetland designed to have no flow or sheet-flow than in a channelized wetland with high energy and turnover rates of water (Brown, 1985, Adamus *et al.*, 1991). Wetland project site characteristics are specified in the design criteria. Wetland functional replacement is not attained, however, until the design criteria are incorporated into a plan and the project successfully constructed.

Table I

Site selection (s) and design (d) features for restoration of wetland functions (based on Marble 1990). It should be noted that the level to which a feature influences the functional capacity of a wetlands differs with the specific function and wetland system (i.e., riparian, depressional, or fringe) (see Adamus *et al.* 1991, Marble 1990). Wetland functions codes: NRT - Nutrient Retention and Transformation; STR - Sediment and Toxin Retention; SS - Shoreline Stabilization; FFA - Floodflow Attenuation; GWR - Groundwater Recharge; PE - Production Export; ADA - Aquatic Diversity and Abundance; and WDA - Wetland Dependent Wildlife Diversity and Abundance.

	WETLAND FUNCTIONS							
	NRT	STR	SS	FFA	GWR	PE	ADA	WDA
LANDSCAPE								
Wetland System		s		s	s	d	s	s
Watershed Land Cover		s		s	s			s
Wetland/Watershed	s	s		s		s	s	
Watershed Size						s	s	s
Water Chemistry					s		s	s
Wetland Acreage							s	s d
Habitat Type and Interspersion							s d	s d
Human Disturbance							s	s
Fetch and Exposure		s	s			s		s
Erosive Conditions			s					
Water Source	s d	s d			s d		s d	
Nutrient Sources	s							
Watershed Soils				s	s			
pH						s		s

SITE HYDROLOGY								
Hydroperiod		d					s d	s d
Flood Extent-Duration		s d				s d	s d	
Artificial Drainage					d			s d
Outlet Characteristic	d	d		d	d	d	d	d
SITE ENERGY								
Sheet Flow			d	d		d		
Channel Gradient and Water Velocity	d	d				d	d	d
Shoreline Geometry			s d				d	d
SITE SUBSTRATE								
Substrate Type		s d				s d	s d	s d
SITE VEGETATION								
Vegetation Class and Form Richness	d	d	d	d		d	d	d
Vegetated Width	d	d	d			d		d
Water/Vegetation and Interspersion		d	d	d		d	d	d
High Plant Product.						d		
Vegetated Canopy							s d	s d
Diversity Enhancement							s d	s d
Special Habitat Feat.							d	d

3. Attaining Wetland Functional Replacement

Functional wetland restoration is the result of the successful implementation of a project plan that has been designed to optimize the performance levels of desired functions for a site in a given location. To accomplish this, project plans are based on a series of decisions that incorporate design criteria and site characteristics with ecological, engineering, and economic considerations (Figure 1). There are three critical phases in the decision sequence in which the features affecting wetland function must be incorporated: site selection, design criteria development, and plan development.

3.1 PHASE 1: SITE SELECTION

Proper site selection is an integral step in attaining desired wetland functions, as is evident from the various relationships of landscape features with different wetland functions (Table I). The ideal wetland restoration site for functional replacement has all of the landscape features required for the desired functions. In cases where the functions to be replaced were degraded by on-site activities, this is on the same site as where the functions were degraded or lost. The relationship between the existing landscape features and the wetland site remains intact and functions can be restored by on-site manipulations. In cases where it is not feasible to restore the wetland on the same site, alternate sites must be evaluated for their relative capacity to perform desired functions. The pertinent landscape features for the desired functions should be evaluated at each site and the sites prioritized on this basis (Figure 1). The site that has the optimum set of landscape features is the best candidate for successful functional wetland restoration. Depending on the availability of suitable sites to provide desired functions, however, site selection may or may not be goal driven in practice.

Practical limitations exist for wetland site selection, such as for mitigation projects resulting from a Section 404 regulatory action. The mitigation memorandum of agreement between the U.S. Army Corps of Engineers and U.S. Environmental Protection Agency specifies a preference for compensatory wetland mitigation to be on-site and in-kind wetland (Ainsley, this volume). Replacement of wetland functions, however, may not be possible in a developed landscape. For example, the wetland hydrology may be drastically altered by changes in runoff quantity and quality due to loss of permeable substrates. Habitat for a protected species may be lost due to lack of access to the wetland through the altered landscape. If site size or alteration by development make on-site restoration impractical, alternate sites must be selected for off-site and in-kind replacement. Limitations on alternate site selection such as availability or ownership may limit the possibilities for locating adequate areas for functional replacement.

Once the wetland project is situated in a viable setting, the next phase in the wetland decision sequence (Figure 1), is to determine design criteria for hydrology, substrate, energy, and vegetation.

FIGURE 1

Decision sequence for wetland restoration project development. Functional wetland restoration can be enhanced during three phases: site selection, identification of design criteria, and project plan development (after Palermo 1992).

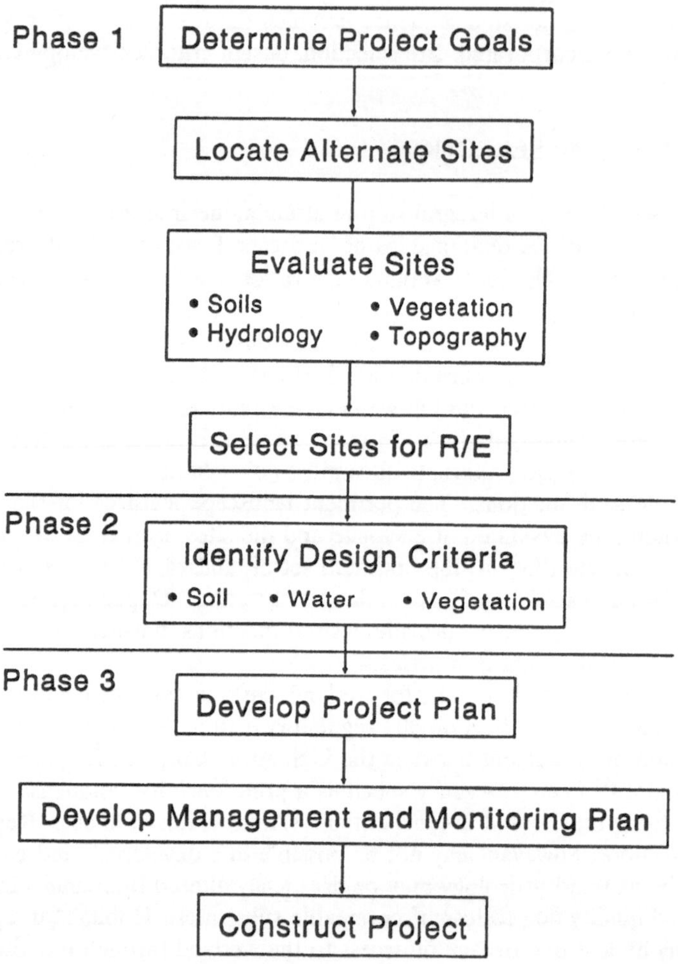

3.2 PHASE 2: DESIGN CRITERIA DEVELOPMENT

Development of design criteria is the objective of the second project planning phase for functional wetland replacement. Design criteria specify the site conditions that must be established for the wetland to perform the desired functions. At this phase, the project planner begins to specify the hydrological conditions, current and wave energy, substrate characteristics, and vegetation composition and distribution that contribute to the wetland functions. To follow through with a previous example, if the functional goal is to improve water quality by retaining sediments received from upstream, the design criteria would specify the site conditions necessary to optimize the sediment retention capacity of the project site. Site hydrologic and energy characteristics could be created to reduce water energy and increase settling time (Table I). The design criteria might include creating a restricted outlet to increase retention time and a gentle gradient to reduce water velocity.

Design criteria serve several purposes. The first objective of design criteria is to specify the basic conditions required to establish a wetland. That is, the substrate must support wetland vegetation in areas that will experience at least saturated conditions during the growing season in most years. Design criteria for protection measures are also necessary to insure that destructive forces (e.g., energy, herbivory, fire) do not cause project failure, particularly during early project developmental phases. Finally, more specific design criteria for hydrology, substrate, energy, and vegetation determine how the functional goals of the wetland project will be attained.

Hydrologic design criteria are integral to the optimization of nearly all wetland functions (Table I). The depth, duration, seasonality, and extent of inundation are primary factors controlling most wetland ecological processes and functions, such as degree of anaerobiosis and plant productivity (Mitsch and Gosselink, 1986). Wetland hydrology is determined by a positive balance between sources and losses of water (Soil Conservation Service, 1992). The capacity of wetlands to perform most functions is improved, therefore, with either a constricting outlet or no outlet to allow for water retention. For example, functional capacities for nutrient retention/transformation and floodflow attenuation are improved with increased duration and extent of inundation resulting from reduced outflow (Adamus *et al.,* 1991).

Hydraulic design criteria determine water energy levels and thus, movement of particulate, nutrients, and toxins by water. Capacities of wetlands to perform many functions are affected by the frictional resistance of water, channelization, water velocity, or direction of impinging wave energy (Adamus *et al.,* 1991). For example, sheet flow has higher frictional resistance of water moving across the wetland floor than channelized water. As a consequence, sheet flow contributes to stabilization of shorelines, attenuation of floodflow, and export of organic matter production in wetlands (Table I).

Substrate design criteria for enhancing functional capacity of wetlands pertain primarily to substrate type (Table I) and underlying soils. Substrate nutrient content, depth, texture, and stability must be sufficient to support wetland vegetation, which in turn is important to the wetlands capacity to perform several functions (e.g., production export, wetland dependent habitat diversity and abundance). Underlying soils determine permeability rates for groundwater recharge (Freeze and Cheery, 1979).

Wetland vegetation contributes to many wetland functions (Table I). In addition to providing food, nesting areas, and cover for fish and wildlife, emergent and aquatic vegetation decrease water energy by increasing flow resistance (Adamus *et al.*, 1991) and modify substrates physically and chemically (Carpenter and Lodge, 1986). Vegetation design criteria for most functions often call for establishing a diversity of vegetation types that are interspersed with areas of open water (Table I).

3.3 PHASE 3: PROJECT PLAN DEVELOPMENT

Incorporation of function-specific design criteria into the project plan is the final phase in which functional restoration can be effectively planned (Figure 1). In order for the project to be built to perform the desired functions, the project planning team must use the design criteria to guide decisions regarding site-specific questions. Design criteria are too general in nature to be incorporated directly into a plan. For example, a certain width of vegetation between the upland and open water may be specified as a design criterion for the shoreline stabilization function in the target wetland. Additional design elements must be specified to insure that shoreline stabilization is achieved. For example, wetland plant species that are effective for stabilization must be selected that will tolerate the site conditions and a plan developed to establish the plants in the target location. Shoreline stabilization will not occur if plant material is obtained in poor health at the wrong time of year, installed incorrectly for the site conditions, or not protected during early development periods.

A wetlands restoration project plan is developed by comparing the site hydrology, energy, substrate, and vegetation conditions with the design criteria. The planning team considers factors such as the site ecology, economic limitations, engineering structures and techniques, and logistics to determine how to best incorporate the design criteria and site conditions into the project plan. Usually the site substrate, hydrology, and energy conditions are planned before the vegetation. The process should be iterative, however, so that all conditions are checked against the others for compatibility and feasibility. The result is a plan that is internally consistent for establishing the hydrology, substrate, and vegetation necessary for the wetland to perform the desired functions. The plans will be further developed into contract specifications to help ensure the design criteria are met as the project is being built.

An example follows to illustrate how a design criterion for diverse vegetation for wetland dependent wildlife habitat can be considered in the plan development phase.

4. Vegetation Establishment Plan Development: An Example

If a wetland is to be established to mitigate for loss of habitat for wetland dependent wildlife, several conditions need to be established. For example, the wetland should be situated in a landscape setting that provides wildlife access to the site (i.e., within migration range of target wildlife species) and environmental conditions that do not threaten the health or perpetuation of these populations. Design criteria may be specified to include a diversity of habitats on site to support a diversity of wildlife (Weins, 1989) at all stages of their life histories, such as feeding, winter cover, and breeding (Heitmeyer *et al.*, 1984, Frazer *et al.*, 1990). Establishment of habitat diversity requires the establishment of diverse vegetation.

Once the substrate and hydrological features have been planned, the vegetation plan can be developed to enhance diversity (Figure 2). The vegetation planning procedure begins by determining whether desirable vegetation exists on or near the project that is capable of colonizing the site. If desirable plant species will not naturally colonize, a plan to establish vegetation must be developed. The following decision sequence illustrates considerations for attaining diverse vegetation at all points in the vegetation establishment plan development phase.

4.1 SPECIES SELECTION

The standard guidance for species selections is to use locally occurring species that tolerate planned site conditions and meet project objectives (Hammer, 1992, Soil Conservation Service, 1992). If the project objective is to maximize species diversity then species within each moisture zone should be selected with consideration of plant form, mode of reproduction, and stratum. In marshes, herbaceous plant forms need to be compatible. For example, a tight sod forming grass species may inhibit growth and reproduction of slower growing or single-stemmed species. If maximizing structural diversity is the priority, then herbaceous plants and midstory tree species should comprise the majority of the selected species. These species are often shade tolerant species, and attention may be required to the provision of nurse plants to ameliorate initially harsh site conditions (see below).

4.2 PLANT SOURCE AND ACQUISITION

Wetland plant material is acquired from natural sites and from commercial sources. Seeds and vegetative propagules collected from local natural wetlands are likely to be tolerant of regional conditions and have the genetic diversity necessary to adjust to changing site conditions. Collection of plant materials from natural areas is, however, limited by the degree of site disturbance. Seeds can be collected with little impact, but digging of vegetative propagules is not advisable unless the site is going to be developed. It is desirable, therefore, to have as great a diversity of commercially available plant material as possible.

The most important factors that restrict supply of diverse plant materials from commercial nurseries are a lack of demand and limited knowledge about

FIGURE 2

Detailed decision sequence for wetland vegetation establishment in the project plan development phase (after Palermo 1992).

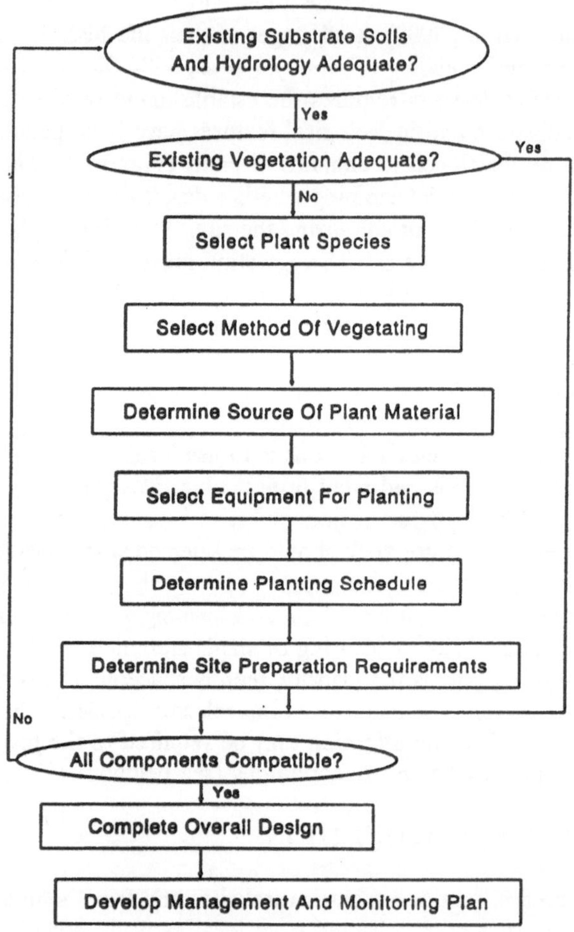

species handling requirements. These factors are interrelated, and so must be addressed together. Demand for diverse plant material needs to be increased by wetland project managers who are aware of the advantages of using such material. Wetland planting guides which consolidate important information regarding plant selection and characteristics are scarce (e.g., Thunhorst, 1993), and additional sources are needed. Much information on common species simply does not exist. Species-specific experience is required on techniques for plant propagation, handling, establishment, and management.

4.3 PLANTING METHODS

An effective means of actively obtaining natural species diversity in a wetlands project is to move topsoil containing seeds and vegetative propagules from a natural wetland to the wetland project site. Topsoil can be moved with the plants intact, either as sod (Figure 3) or in smaller plugs. Successful marshes have been created by using the topsoil as a mulch and spreading it over a contoured ground surface (Figure 4). Caution needs to taken, however, that stockpiling time is minimized and that stockpiles are not placed in wetlands (Garbisch, 1986). In addition, careful matching of hydrclogical conditions between topsoil donor areas and recipient areas facilitates the formation of vegetation zones in the project area. Use of topsoil from natural wetlands is limited to cases where the donor wetland will be developed.

4.4 PLANTING SCHEDULE

Certain plants, primarily shade-tolerant species, grow best in the low-light and cooler interiors of swamps. These species survive and grow best when protected from harsh conditions that are commonly found in wetland projects. These plants require the presence of hardier plants that may provide shade, protection from wind, or improved soil fertility. These cover or nurse crops can be planted at the same time or prior to the establishment of less tolerant species (Clewell and Lea, 1990). Interplanting species after the establishment of an initial complement of plants prolongs the involvement period with the project, but greatly increases the potential for using species not commonly used in wetland projects.

4.5 SITE PREPARATION

Species diversity in natural wetlands is greater where the surface is uneven. Hummocks, fallen logs, and depressions provide a variety of hydrological conditions which are exploited by many species. For example, a plant growing on a hummock can escape long periods of inundation it would experience directly on the wetland floor. Creation of a rough surface in a wetland project does not look "neat", but it adds a feature found in diverse natural areas.

(a) (b)

FIGURE 3

Movement of intact pieces of sod salvaged from wetlands that will be developed and moved to restored wetlands is an innovative technique for establishing diverse wetland vegetation. A) Sod mat on modified frontend loader being moved to a flatbed truck for transport. B) Sod reconstructed in center of restored wetland.

(a) (b)

FIGURE 4

Spreading topsoil from a donor wetland that will be developed on a created or restored wetland is a technique that has been successfully used for rapidly establishing wetland vegetation over relatively large areas. A) The floor of the created wetland soon after the wetland mulch was applied. B) The dense wetland vegetation that became established within two years. Caution should be used not to transfer undesirable or aggressive species to the restored wetland with the topsoil seedbank.

5. Conclusions

Success of functional restoration can be enhanced with a multi-phased approach to wetland project planning. Proper site selection insures that a wetland is situated where there is the potential to perform a function. Incorporation of functional design features into design criteria and project plan development further optimizes the potential for a site to perform desired functions. All three phases are interrelated and are integral to the replacement of wetland functions in restoration projects.

While decision sequences for wetland project planning such as presented here will help attain functional wetland restoration and establishment, it should not be overlooked that we are a long way from fully understanding wetland ecological processes and functions. We may be even further from being able to restore wetland functions similar to natural systems. It is clear that there is no simple solution. The effort will require continuing research and experience. Functional wetland replacement is a goal worthy of this commitment.

Acknowledgements

This work was sponsored by the U.S. Army Corps of Engineers Wetlands Research Program. Helpful comments on earlier drafts were received from Candy Bartoldus, Carl Trettin, and Dan Smith. Ann Marble and Mike Palermo were generous with their consent to incorporate their work into this paper.

References

Adamus, P. R., Clairain, E. J., Smith, R. D., and Young, R. E.: 1987: *Wetland evaluation technique (WET) Volume II: Methodology (Operational Draft Report).* Environmental Laboratory, U.S. Army Engineer Waterways Experiment Station, Vicksburg, MS.

Adamus, P. R., Stockwell, L. T., Clairain, E. J., Morrow, M. E., Rozas, L. P., and Smith, R. D.: 1991. *Wetland Evaluation Technique Volume 1: Literature Review and Evaluation Rationale*, Wetlands Research Program Technical Report WRP-DE-2, U.S. Army Engineer Waterways Experiment Station, Vicksburg, MS, p. 287.

Ainsley, W. B.: 1995. Rapid wetland functional assessment: its role and utility in the regulatory arena, *Water, Air, and Water Pollution* this volume.

Bartoldus, C. C., Garbisch, E. W., and Kraus, M. L.: 1994. *Evaluation for planned wetlands.* Environmental Concern. St. Michaels, MD. p. 179.

Brinson, M. M.: 1993a. *A Hydrogeomorphic Classification for Wetlands*. Wetlands Research Program Technical Report WRP-DE-4, U.S. Army Engineer Waterways Experiment Station, Vicksburg, MS, p. 79.

Brinson, M. M.: 1993b. Changes in the functions of wetlands along environmental gradients, *Wetlands* 13:65-74.

Brown, R. G.: 1985. Effects of wetland channelization on runoff and loading. *Wetlands* 8:123-133.

Clewell, A. F. and Lea, R.: 1990. Creation and restoration of forested wetland vegetation in the southeastern United States, p. 195-232 In: Kusler, J. A. and Kentula, M. E. (eds), *Wetland Creation and Restoration: the Status of the Science*, Island Press, Washington, D. C.

Crabtree, A., Day, E., Garlo, A., Stevens, G.: 1992. Evaluation of wetland mitigation measures Volume 1: Final report. Federal Highway Administration Report No. FHWA-RD-90-083, U.S. Department of Transportation, Washington, D. C., p. 353.

Carpenter, S. R. and Lodge, D. M.: 1986. Effects of submersed macrophytes on ecosystem processes. *Aquatic Botany* 20:341-370.

Frazer, C., Longcore, J. R., McAuley, D. G.: 1990. Habitat use by post-fledgling American Black ducks in Maine USA and New Brunswick Canada. *Journal of Wildlife Management* 54: 451-459.

Freeze, R. A. and Cherry J. A.: 1979. Groundwater. Prentice-Hall, Inc., Englewood Cliffs, NJ. p. 604.

Garbish, E. W.: 1986. Highways and wetlands: Compensating wetlands losses. Federal Highway Administration Report No. FHWA-IP-86-22, U.S. Department of Transportation, Washington, D. C., p. 60.

Hammer, D. A.: 1992. *Creating freshwater wetlands.* Lewis Publishers, Boca Raton, FL. p. 298.

Heitmeyer, M. E. Fredrickson, L. H., and Krause, G. F.: 1991. Water relationships among wetland habitat types in the Mingo Swamp, Missouri. *Wetlands* 11:55-66.

Kusler, J. A. and Kentula, M. E. (eds): 1990. *Wetland Creation and Restoration: the Status of the Science*, Island Press, Washington, D. C., p. 594.

Marble, A. D.: 1990. *A Guide to Functional Wetland Design*, Federal Highway Administration Report No. FHWA-WP-90-010, U. S. Department of Transportation, Washington, D. C., p. 222.

Marburger, J. E.: 1993. *Wetland plants: materials, technology, and development.* U.S. Department of Agriculture, Soil Conservation Service. p. 54.

Mitsch, W. J. and Gosselink, J. G.: 1986. *Wetlands*, Van Nostrand Reinhold Company, Inc., New York, N.Y. p. 539.

Palermo, M. R.: 1992. Wetlands engineering: Design sequence for wetlands restoration and enhancement. Wetlands Research Program Technical Note WG-RS-3.1, U.S. Army Engineer Waterways Experiment Station, Vicksburg, MS, p. 4.

Soil Conservation Service: 1992: *Engineering field handbook Chapter 13: Wetland restoration, enhancement, or creation.* U.S. Department of Agriculture 210-EFH,1/92, Washington, D. C. p. 79.

Thunhorst, G. A.: 1993. *Wetland planting guide for the northeastern United States: Plants for wetland creation, restoration, and enhancement.* Environmental Concern, Inc. St. Michaels, MD p. 179.

Wiens, J. A. 1989. The ecology of bird communities Volume 1: Foundations and patterns. Great Britain, Cambridge University Press. p. 372.

DESIGN AND IMPLEMENTATION OF FUNCTIONAL WETLAND MITIGATION: CASE STUDIES IN OHIO AND SOUTH CAROLINA

SUE ANN MCCUSKEY[1], ALLEN W. CONGER[2], AND HILBURN O. HILLESTAD[1]
[1]*Law Environmental, Inc., 114 TownPark Drive, Kennesaw, Georgia 30144*
[2]*Law Engineering, Inc., 3820 Faber Place Drive , Charleston, S.C. 29405*

Abstract. Wetland development offers the opportunity to replace and enhance ecological functions lost through permitted wetland impacts. Components necessary for the restoration and creation of wetlands are presented and examples of wetland construction are described to illustrate the application of wetland design. Land contours, top soil, hydrology and vegetation were manipulated to develop wooded wetlands at sites in Ohio and South Carolina. In Ohio, approximately 30 ha of former crop land/sod farm were modified to bring water from the adjacent creek onto the site and hold it to saturate soils for wetland development. A 2.8 ha ponding area and channels were constructed, berms were built to slow the exit of stormwater runoff, and trees were planted in spring 1994. The mitigation site lies adjacent to a park and high school, thereby also providing community benefits and wetland education opportunities. In South Carolina, 9.5 ha of an abandoned soil borrow pit were converted into wooded wetlands, hydrologically connected to an adjacent swamp. Native plants were removed from the 4 ha of isolated wetlands to be impacted, and were augmented with nursery stock to create the mitigation wetland. Monitoring of vegetation, hydrology and wildlife usage of the constructed system continues to document wetland development and success.

1. Introduction

Wetland creation and restoration is a new science, driven by governmental rules and policies. The Clean Water Act requires that wetlands that are permitted to be filled must be replaced; former President George Bush in 1991 directed that this replacement result in "no net loss" of wetland function and value. Creation and restoration may be initiated as part of a Section 404 permit, in reparation for a wetland fill violation, or for a major federal action proposed in an environmental impact statement (see Hammer, 1992). Newly created or restored wetlands add ecological functions that compensate for those that were lost by development. Wetlands are also created to reduce environmental pollution by the acidic runoff of abandoned mines or the nutrient-rich effluents of sewage treatment plants and feedlots (see Reed, 1990).

Because wetlands are defined by their three basic components: wetland hydrology, hydric soils and hydrophytic vegetation, these are the three characteristics that must be given careful consideration in the design of restored and/or created wetlands. In wetland permitting, the term "restoration" refers to a site that was once a wetland but has been altered to lose one or more of the three components listed above. Frequently a restoration site is a farmed field with hydric soils and a stream that has been channelized to prevent it from flooding the crops periodically. Wetland "creation" refers to construction of a wetland where only one or none of the three components were present originally.

In the early days of wetland construction for mitigation, once the site was graded and planted, both the permit holder and regulatory agencies were no longer involved, that is, no follow-up monitoring was required. When some of these small mitigation sites were revisited years later, many had failed (National Wetland Newsletter, 1992). We have become more sophisticated in wetland design since those earlier efforts and agencies have also frequently built in a monitoring requirement as part of the wetland permit conditions. Additionally, a contingency plan may be required to specify proposed actions if the created or restored wetland site is not meeting one or more of the defined success criteria.

This paper has two objectives. The first is to describe criteria by which wetland functions are evaluated by the U.S. Army Corps of Engineers for Section 404 determinations, and the second is to present two case studies in which these criteria were

Water, Air and Soil Pollution **77**: 513–532, 1994.

applied. Preliminary monitoring results from these case studies indicate that the functions of created or restored wetlands can replace the functions of permitted wetland fills.

2. Wetland Construction Methodology

Wetland restoration and creation involves (1) evaluation of wetland function, (2) design, and (3) monitoring. Each of these three critical aspects are discussed below.

2.1 WETLAND EVALUATION

Documentation of wetland functions is necessary before developing the wetland restoration or creation design. The following list of functions and values that may be served by wetlands is provided in the U.S. federal regulations at 33 CFR, Part 320.4(b)(2). According to this Federal Register citation, wetlands considered to perform functions important to the public interest include:

(i) Wetlands which serve significant natural biological functions, including food chain production, general habitat and nesting, spawning, rearing and resting sites for aquatic or land species;

(ii) Wetlands set aside for study of the aquatic environment or as sanctuaries or refuges;

(iii) Wetlands the destruction or alteration of which would affect detrimentally natural drainage characteristics, sedimentation patterns, salinity distribution, flushing characteristics, current patterns, or other environmental characteristics;

(iv) Wetlands which are significant in shielding other areas from wave action, erosion, or storm damage. Such wetlands are often associated with barrier beaches, islands, reefs and bars;

(v) Wetlands which serve as valuable storage areas for storm and flood waters;

(vi) Wetlands which are ground water discharge areas that maintain minimum baseflows important to aquatic resources and those which are prime natural recharge areas;

(vii) Wetlands which serve significant water purification functions; and

(viii) Wetlands which are unique in nature or scarce in quantity to the region or local area.

Not all of these functions are assumed to be characteristic of all wetlands. The spatial distribution of wetlands of the same type also can vary in function across areas (Brinson, 1993). For example, bottomland hardwoods in a broad lowland area of the Coastal Plain will serve different ecological functions than a narrow band of bottomland hardwood wetland located at the base of steep slopes in the Piedmont. Additionally, all wetlands are influenced by seasonal changes.

The evaluation methodology may be one published in the literature (such as the Wetland Evaluation Technique; Adamus et al. 1987) or it may be an original investigation to determine specific functions. Regardless of the evaluation approach, the same criteria should be used to evaluate both the functions of the destroyed wetland and the mitigation wetland.

The purpose of the evaluation is to determine the wetland functions at the impact site. This information is then used as part of the decision process in designing the mitigation wetland with the goal of balancing functions lost against functions gained. Table I presents a conceptual example of an evaluation of wetland functions at an isolated cattail (*Typha sp.*) wetland to be impacted relative to a proposed mitigation wetland that would be forested and adjacent to a watercourse.

In this example, the total wetland function units (wfu) for the cattail wetland are 3.9 and for the proposed forested wetland, 7.8. The product of the area of impact (4 ha in this example) and the wfu determine the goal for wetland mitigation (e.g., impact total = 4 ha x 3.9 = 15.6 wfu goal). With creation/restoration of a forested wetland like the system evaluated in Table I, the same wfu goal can be reached with half the mitigation area (e.g., mitigation total = 15.6 wfu ÷ 7.8 = 2 ha).

Consequently, the loss of 4 ha of monoculture cattails in an isolated wetland pocket could be mitigated by development of 2 ha of planted bottomland hardwood trees to create a wetland near a waterway, because the mitigation wetland in this example could provide a higher effectiveness in performing the wetland functions desired.

TABLE I

Selected wetland functions assessed in the Wetland Evaluation Technique (Adamus *et al.*, 1987) where 0.3 = low effectiveness, 0.6 = moderate effectiveness, and 0.9 = high effectiveness. Functional values are summed for the impact and mitigation wetlands. The impact wetland represents an isolated cattail monoculture wetland; the mitigation wetland represents a seasonally flooded forested wetland adjacent to a stream.

Functions	Impact Wetland	Mitigation Wetland
Groundwater recharge/discharge	0.3	0.3
Floodflow alteration	0.3	0.9
Sediment stabilization	0.6	0.6
Sediment/toxicant retention	0.9	0.9
Nutrient removal/transformation	0.3	0.9
Production export	0.3	0.6
Wildlife diversity/abundance:breeding	0.3	0.9
Wildlife diversity/abundance:migration	0.3	0.9
Wildlife diversity/abundance:wintering	0.3	0.9
Aquatic diversity/abundance	0.3	0.9
Total wetland function units (wfu)	3.9	7.8

2.2 WETLAND MITIGATION GOALS AND CONSTRUCTION DESIGN

2.2.1 Mitigation Goals

The mitigation goals identify those functions we wish to replace at the created or restored wetland. The wetland design is determined by these objectives. For example, if flood flow attenuation is a goal, then removing old stream levees, recontouring adjacent to the waterway, and/or planting trees and shrubs may be methods used in the wetland design to meet this objective. Only when the objectives are defined will we be able to design the construction components and then measure wetland mitigation success against the pre-determined set of objectives.

In most wetland mitigation plans, site modifications such as pond construction and planting are specified but the wetland function goals remain unstated for the following reasons: (1) insufficient data are available to document baseline functions for comparison after wetland creation or restoration, (2) wetland functions such as sediment/toxicant retention are influenced by inputs from the entire watershed while the mitigation wetland area may represent less than 1 % of the total watershed, and (3) for regulatory purposes, mitigation success criteria must be measurable within the time frame (generally 3 to 5 yrs) of the permit conditions. Nevertheless we concur with Clewell and Lea (1990) that a close correlation exists between forest composition and physiognomy and most functional attributes of a natural forested wetland.

2.2.2 Wetland Hydrology

Water to sustain the mitigation wetland may be derived from precipitation, ground water, and surface water overflow. Generally, each of these sources will be important at different seasons. For a mitigation site to meet regulatory requirements, the wetland must have saturated soils (to the root zone) during approximately 15 % or more of the growing season during most years (Environmental Laboratory, 1987, pg. 36). Methodologies typically used to manage water include construction of berms, water control structures, and grading to ground water level.

Berms may be constructed to hold rainfall or surface water on the site so that it does not drain off rapidly. This may be appropriate for a depressional wetland that receives surface water from off-site portions of the local watershed.

To cause overbank flooding during storm events, water control structures may be used on a site where the stream has been channelized. Depending on site specific conditions, structures may be designed to allow minimum flows in the channelized stream during dry portions of the year. Structures such as an adjustable weir may be lowered or raised if site monitoring indicates that either too much or insufficient water is being supplied to the restored or created wetland (Figure 1). For ground water supplied wetlands, site grading may be appropriate to provide this hydrology source within the root zone of wetland vegetation. This source of hydrology for wetland construction is most effective when a slough can be dug at the water elevation of an adjacent undisturbed system, thereby tying the two together. Baseline evaluation of the water table and annual stream flow will be necessary to adequately design water control features. Hydrologic modeling may also be required through use of existing data on the subject waterway or through collection of baseline data. For both of the hydrological modifications described, potential impacts to adjacent property need to be given full consideration.

2.2.3 Hydric Soils

When wetlands are constructed in upland soils as mitigation for permitted wetland impacts, the hydric soils from wetlands to be lost may be stockpiled and spread over the wetland creation site (Willard et al., 1990). When it is not possible to move wetland soils to the creation site, plant development will still be enhanced if rich topsoil can be spread at a depth of 15 to 30 cm over the constructed wetland prior to seeding. This will facilitate root development by providing nutrients and favorable soil texture for new vegetation.

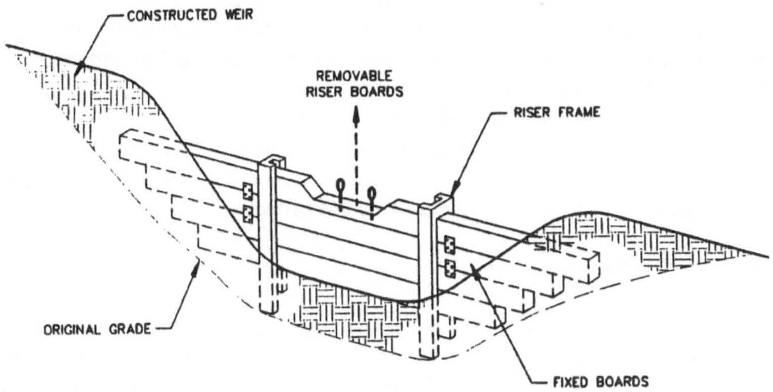

Fig. 1. Conceptual design of adjustable weir.

After a created wetland site has been contoured, topsoil or hydric soil applied, and the site stabilized with herbaceous and possibly wooded vegetation, the site would be ready for implementation of hydrological changes. Depending on the soil type on the site, it may take months or years for the created wetland to exhibit hydric soil field characteristics. Nevertheless, with the hydrology and vegetation in place, the created wetland will perform most of the wetland functions and values typical of undisturbed wetlands while the hydric soil is still forming.

2.2.4 Hydrophytic Vegetation

At a restored or created wetland site, hydrophytic vegetation can be allowed to invade naturally or trees and herbaceous species can be planted. Many aggressively invading weeds are classified as facultative or facultative-wet species (USFWS, 1988). In fact, some of these species such as cattails (*Typha* spp.) can choke out other herbaceous plants and thereby limit species diversity. Trees such as red maple (*Acer rubrum*), and sweetgum (*Liquidambar styraciflua*) also can invade at such high densities that it may take 20 to 30 years before species more useful as a wildlife food source can become established (Clewell, 1986).

The extent of vegetation management in a created wetland system depends on the wetland type desired. For herbaceous marshes, natural invasion from a nearby marsh or topsoil from another marsh are good potential sources of seeds. If undesirable monoculture begins to dominate the system, the nuisance species should be eliminated or kept to a low density (Hammer, 1992). On the other hand, if rapid establishment of diverse species is the mitigation goal, herbaceous wetland species are available from wetland plant nurseries. For best survival, plant stock should come from local sources. Plant selection significantly affects the wetland functions created. Species can be selected to provide high food value for waterfowl, nesting and cover for birds and other animals, endangered plant and animal habitat, and for site water retention and nutrient uptake.

Tree species can be selected and planted to provide community diversity and wildlife food value thereby minimizing the early successional stage of an overabundance of wind seeded species such as box elder (*Acer negundo*) and optimizing the diversity of oaks

(*Quercus* spp.). Trees can be planted on 3- to 4-m centers in hydrological zones appropriate to the species (Hook, 1984, Taylor, et al, 1990). To determine the appropriate species for planting at various hydrological zones or elevations, and to find the tree density expected in a mature system, a reference wetland may be evaluated and these data applied to the design of the created wetland (Hammer, 1992). Where necessary to protect nursery stock from wildlife browsing, tree shelters can be installed over bare root seedlings. The shelters are designed to photo-biodegrade in the field after 3 to 4 yrs, at which time trees should be large enough to be able to survive some browsing.

2.3 WETLAND MONITORING

Monitoring functions of the created or restored wetland requires on-site data collection once or twice each year for up to 5 years. Necessary monitoring typically includes documentation of site hydrologic conditions, vegetation species composition, and soil inspection. It is important to select measurable wetland function criteria to document mitigation progress. Examples include well readings at specific stations to document ground water levels at a particular depth, and for vegetation, a pre-determined percent survival of specific tree species.

3.0 Case Studies

Two case studies are presented as examples of mitigation plans which have been accepted by regulatory agencies and implemented. The first example describes a wetland mitigation site that is in the early stages of development in Ohio and the second describes an established site in South Carolina.

3.1 WETLAND CREATION/RESTORATION ADJACENT TO A CREEK IN OHIO

The central Ohio wetland impacted by development in 1990 was an isolated herbaceous system comprised of approximately 4 ha of cattails and 11 ha of diverse grasses and flowering weed species surrounded by 11.7 ha of upland. The site had been in row crop and hay production from about 1948 to 1988. The 26.7-ha site was located in an urban setting bordered by four-lane roads, commercial, and residential development (Figure 2).

Mitigation activities (1) evaluated the wetland functions lost at the urban wetland, (2) identified wetland function goals for the mitigation site, (3) developed a wetland creation/restoration plan, (4) constructed the mitigation design at a separate site, and (5) monitored measurable criteria by which to evaluate mitigation success.

3.1.1 Functions of Original Wetlands

To evaluate the impact wetland and as part of wetland permitting, vegetation, hydrology, soil type, wildlife use, and the potential for protected species were investigated. Wetland areas supported cattail, barnyard grass (*Echinochloa crusgalii*), sedges (*Carex sp.*), rushes (*Juncus sp.*), and tickseed (*Bidens sp.*). Upland areas on site were dominated by weedy successional species such as frost aster (*Aster pilosus*), goldenrod (*Solidago sp.*), Canada thistle (*Circium arvense*), and foxtail (*Setaria faberi*). Precipitation was the source of hydrology; the nearest stream was one mile away from the site. Two soil types were identified in about equal proportions on the 26.7-ha site: one was a hydric soil and one was non-hydric.

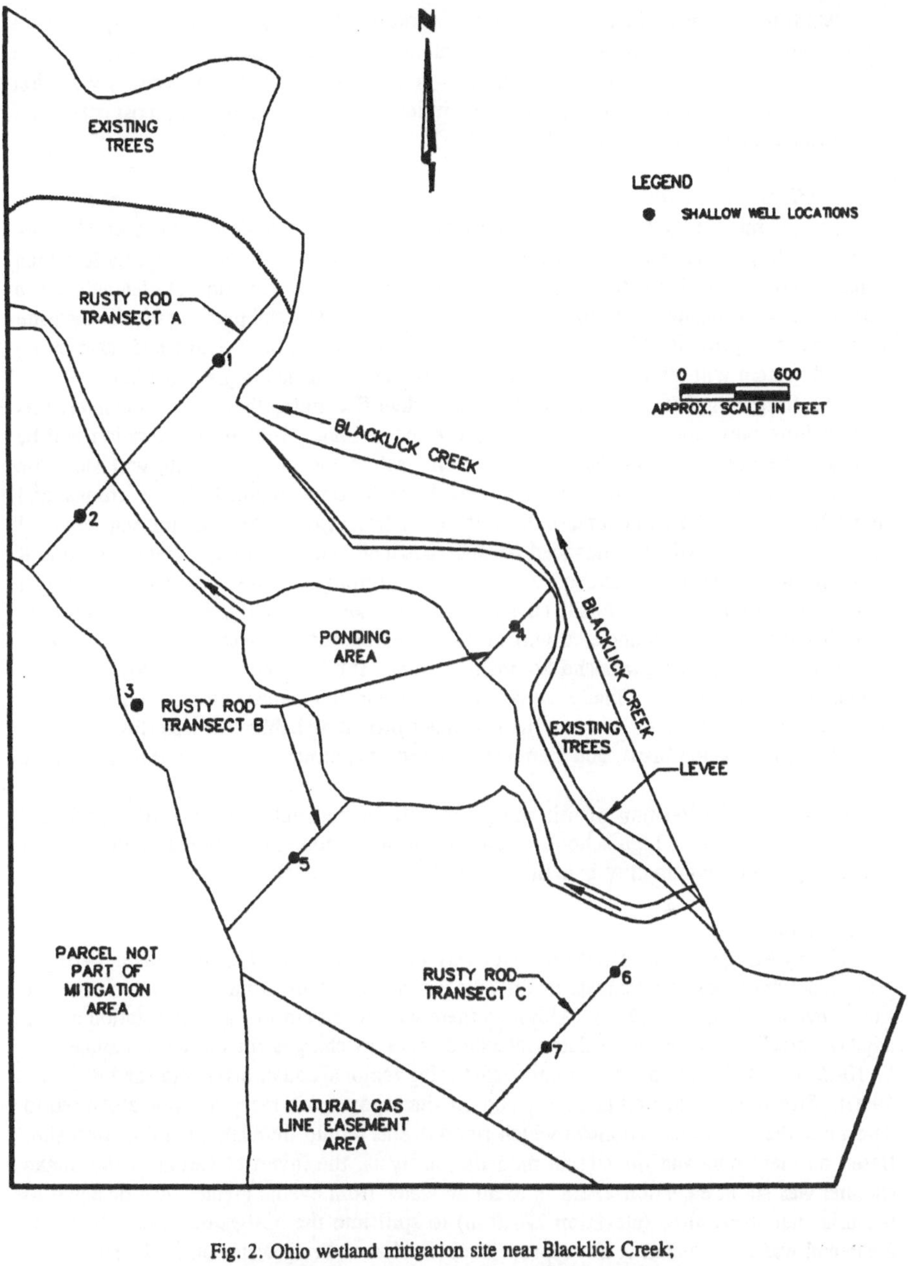

Fig. 2. Ohio wetland mitigation site near Blacklick Creek; rusty rod transect locations and shallow well sites are shown

Wildlife use was limited primarily to birds and small rodents. The lack of a permanent water source precluded development of a community of aquatic fauna. No species of state or federal protected plants or animals was found on site. The Wetland Evaluation Technique analysis was applied and resulted in low or moderate ratings in most categories (see results for the impact wetland in Table I).

3.1.2 Mitigation Goals

To off-set wetland losses, the mitigation plan was designed to create/restore approximately 30 ha of forested off-site wetland at a mitigation ratio of slightly less than 2 ha of created wetland for every 1 ha of impact. The construction of channels and a ponding area to enhance hydrologic input to the mitigation site and planting of mast-bearing tree species to provide wildlife food and habitat will convert the existing sod farm into a wetland system with diverse vegetation structure and permanent aquatic habitat.

The following four wetland functions are identified mitigation goals to convert this monoculture grass species sod farm into a forested wetland. Floodflow alteration will be enhanced because weirs in the constructed channel of the mitigation site will slow the runoff of flood waters and once the trees have become established, the trunks and understory vegetation will decrease the velocity of flood flows. Second, nutrient removal and transformation will be enhanced because nutrient input from upgradient agricultural fields in the 158-sq km watershed will be filtered through the mitigation wetland forest instead of running off the fertilized sod farm as under pre-mitigation site conditions. Wildlife diversity and abundance will be increased because vegetation structure will be modified from mowed grassland to include trees, shrubs, and wet meadow areas of potential wildlife habitat. Finally, aquatic diversity and abundance will be enhanced by the construction of a 2.8 ha ponding area and channels providing habitat for waterfowl, benthic organisms, fish, amphibians, and mammals that require permanent water within their home range.

Because of the location of this mitigation site immediately upstream of a park and across the creek from a high school, an additional mitigation value offered by this project is that it provides community benefits and wetland educational opportunities.

3.1.3 Methodology

The hydrology of the mitigation site was investigated in 1992 by first preparing a HEC-2 stream modeling analysis of the adjacent stream to determine the annual average flow elevation (USACE, 1988). Although there was no streamflow gauging station on the adjacent creek or in the immediate watershed, peak discharges for the mean annual, 2-, 5-,10-,50-, and 100-yr floods were calculated using regional equations (Kolton and Roberts, 1990). Stream cross section geometry and on-site topographic mapping were also used to determine the stream elevations at which flood events would inundate the mitigation site. Based on these data and the stream modeling analysis, the invert elevation of the intake channel was set at elevation 222.8 m to allow water from events greater than or equal to mean annual streamflow (elevation 222.9 m) to spill into the mitigation area. In 1993, a channel was constructed at the appropriate elevation to bring water on to the site, fill a shallow 2.8 ha pond, and continue in a channel across and off the property, back into the creek. Weirs were constructed in the channel to hold water on the site. Soil berms were also constructed to prevent storm water runoff from exiting the site as quickly as it did prior to the site modifications.

Topsoil that was removed to excavate the ponding area and channels was stockpiled and spread over the final graded water conveyance features. A temporary grass cover was seeded to stabilize the sides of the channel and ponding area until native wetland species invade. The site was constructed in the summer of 1993 and allowed to adjust to the new hydrologic conditions over the 1993-1994 winter.

Bare root tree seedlings were planted in early spring 1994 at a density of approximately 746 trees per ha. Because most of the mitigation area varied in surface elevation by less than 1 m, vegetation was planted according to only 3 zones: aquatic edge herbaceous species such as pickerel week (*Pondederia cordata*), bulrushes (*Scirpus sp.*) smartweed (*Polygonum sp.*), and lizard's tail (*Saurus cemuus*); bottomland hardwood wetland species such as oaks (*Quercus bicolor, Q. michauxii, Q. palustris,* and *Q. shumerdii*), red-osier dogwood (*Cornus stolonifera*), river birch (*Betula nigra*) and sycamore (*Plantanus occidentalis*); and upland tree species such as pecan (*Carya illinoensis*) and bur oak (*Q. macrocarpa*) planted on berms.

3.1.4 Monitoring

Monitoring of the creation/restoration site following construction and planting is essential in order to measure the presence of the three wetland criteria of hydric soils, appropriate hydrology, and hydrophytic vegetation. The first two criteria were documented by the "rusty rod" technique (Gibb and Coffman, 1993; McKee, 1978; Carnell and Anderson, 1986) and by automatic-recording shallow wells.

The rusty rod technique documents the depth of soil aeration over a period of several weeks. Mild steel welding rods 91 cm long and 0.24 cm in diameter are driven 61 cm into the soil and examined after several weeks. In soils which are not saturated, the presence of oxygen in soil pores causes the mild steel to rust; where soils are saturated, the rods will not rust. The point on the rod at which rusting ceases reflects the depth to saturated (anaerobic) soil conditions.

At the Ohio site, 29 rusty rod stations were established along three transects (Figure 3); the rods were read every 4 wks during the growing season and replaced with new rods. Results from April through June 1994 showed that the depth to soil saturation during the first period (April 6 to May 6, 1994) averaged 12.2 cm (range = 2.5 to 20.3 cm), and for the second period (May 6 to June 3, 1994) averaged 24.9 cm (range = 7.6 to 53.3 cm). These data indicate that saturated soil conditions were present within the top 30 cm (the root zone) during the early part of the growing season.

To collect shallow water table data, 7 slotted PVC pipe wells, model WDS WL-40, were installed at points throughout the wetland creation/restoration site (Figure 3). These wells are self-contained data loggers set to measure and store water table levels at 12-hr frequencies. Data were downloaded every 4 wks and showed a range of groundwater levels (from April 5 through June 3, 1994) from surface inundation to 109 cm below surface. These results closely tracked precipitation events in the study area.

Vegetation monitoring will be conducted at the end of each growing season. Permanent sampling plots are established to document the survival rate of planted seedlings and record species composition and density of natural regeneration. Dead seedlings will be replaced as necessary to meet the regulatory requirement of 70 % survival of planted tree species.

3.1.5 Case Summary: Ohio Wetland Example

The functions of the impact wetland in the Ohio case study were severely limited by its isolated location in an urban setting while the mitigation site offered the opportunity to create a high value wetland. The vegetation communities affected by fill, the original

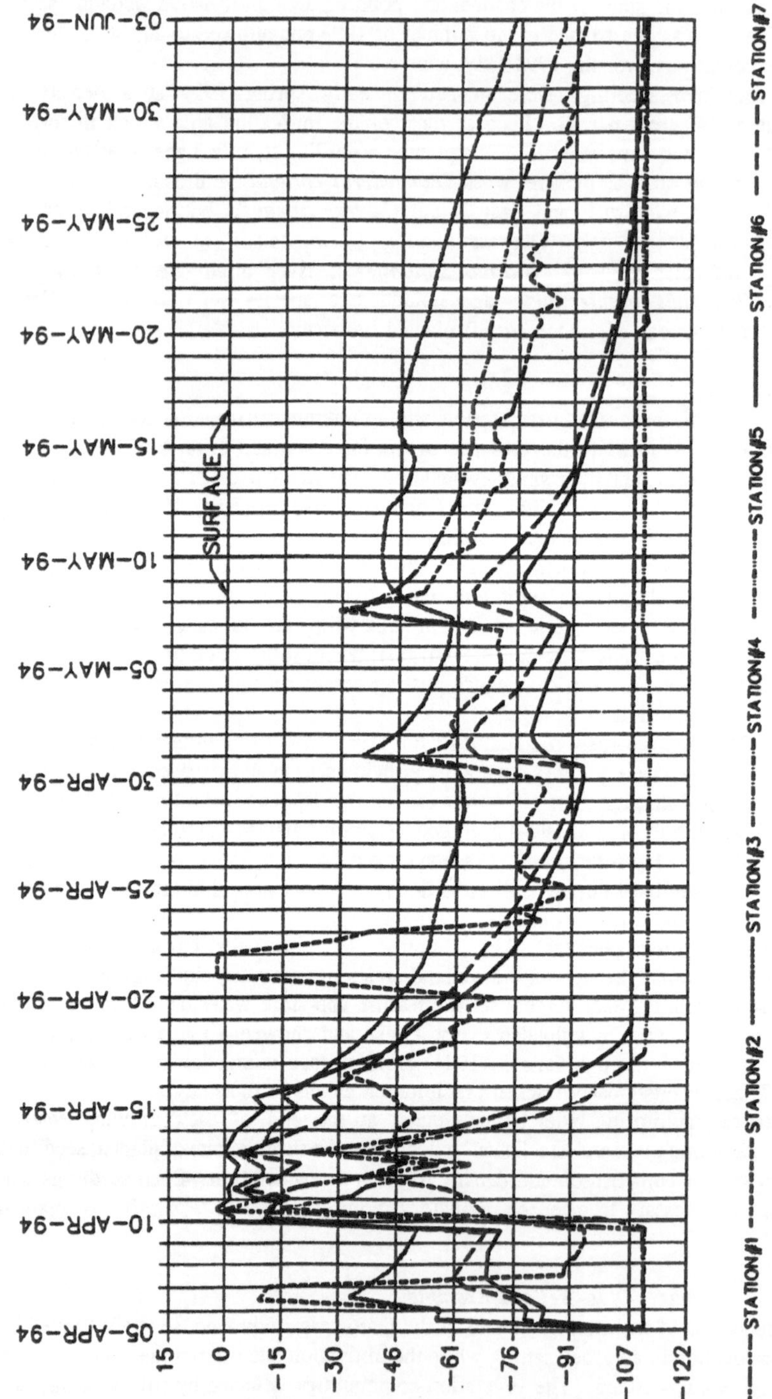

Fig. 3. Graph showing depth to groundwater at 7 well locations at the Ohio mitigation site during April - May 1994.

mitigation site conditions, and conditions following wetland creation/restoration are presented in Table II.

TABLE II

For the Ohio case study, the hectares of upland and wetland vegetation communities are shown below for (1) the urban site where wetlands were filled, (2) the sod farm mitigation site prior to site modifications, and (3) the mitigation site after construction of the ponding area, channels, and planting of the wetland vegetation.

Vegetation Community	Impact Site	Pre-Mitigation Sod Farm	Post-Mitigation Sod Farm
UPLAND:			
mixed weeds	11.7	-	-
mono-species grass	-	28.4	-
existing woods	-	6.1	6.1
WETLAND:			
cattails/mixed weeds	15.0	-	-
emergent grasses/water	-	-	3.2
bottomland hardwoods	-	-	25.2
Total area (ha)	26.7	34.5	34.5

Through modifications of on-site hydrology, soil saturation was increased, waters were retained in the ponding area and channels, and new habitat was created for waterfowl and aquatic fauna. Mast producing tree species were planted and will grow to provide a diversity of vegetation structure, wildlife food, shelter, and potential nest sites. The increase in woodland in this portion of the county is especially valuable because most lands have been cleared for farming. Therefore, this mitigation wetland will be able to provide hydrologic functions such as flood flow attenuation and provide high quality diverse habitat for waterfowl, aquatic, and terrestrial wildlife.

3.2 WETLAND CREATION AT A SOIL BORROW AREA IN SOUTH CAROLINA

The second case study describes wetland mitigation at a solid waste landfill in South Carolina, where expansion of the existing facility required placement of fill in a total of 4 ha scattered throughout six pockets of isolated wetlands (Figure 3). The vegetation types, hydrology, wildlife use, and the potential presence of protected species were investigated as part of the permitting process. Wetland impacts to the site were avoided to the extent practicable, impacts were minimized and limited to those areas of low to moderate wetland value.

3.2.1 Functions of Original Wetlands

The isolated wetlands impacted by expansion of the landfill footprint and soil borrow areas were pockets of jurisdictional areas impacted by silvicultural activities conducted by previous landowners. The mixed hardwood forest in the proposed 49 ha landfill expansion area had been clear cut or selectively logged and left to regenerate naturally. At the initiation of this study in 1990, the forest was an uneven-aged stand consisting of a few large trees (approximately 20 per ha), 40 to 50 cm dbh but dominated by regenerating trees averaging 10 to 15 cm dbh. Dominant species included swamp blackgum (*Nyssa biflora*), blackgum (*N. sylvatica*), loblolly pine (*Pinus taeda*), sweetgum (*Liquidambar styraciflua*), red maple (*Acer rubrum*), willow oak (*Quercus phellos*), and water oak (*Q. nigra*).

The wetlands to be filled were characterized by species composition, canopy closure, ground cover, average dbh, stand age, hydrology, and wetland size. Based on these field observations, a relative value (low, medium, high) was assigned (Table III). Value was considered high in swamp forest wetlands where hydrology had not been altered by ditching or draining, in older mixed hardwood forest stands, and in larger wetlands ($\geq$ 2 ha) relative to those evaluated in this case study. A low relative value was attributed to those small isolated wetland pockets that had been ditched and bermed, where the majority of trees were $\leq$ 20 yr old, and where planted pine replaced the original hardwood species.

TABLE III

The size, existing vegetative structure, approximate stand age, and hydrologic characteristics were determined for each of the six wetlands to be filled in the South Carolina case study in order to evaluate relative wetland value. Wetland locations are shown in Figure 4.

Wetland/ Area(Ha)	Species Comp.	% Cover			Mean dbh(cm)	Stand Age(yr)	Hydrology	Relative Value
		Canopy	Subcanopy	Ground				
A 1.00	MH[1]	40	80	80	48	20	SS[3], with ditches	med.
B 0.79	MH	30	90	90	38	20	SS, no diches	med.
C 0.08	MH	30	90	90	30	20	SS, no ditches	low
D 1.91	MH	90	70	30	46	50	SS with ditches and berms	med.
E 0.21	MH	90	70	30	46	50	SS with ditches and berms	med.
F 0.04	LP[2]	100	50	5	20	15	SS limited hydrology	low
Total 4.03								

[1]MH = mixed hardwood
[2]LP = loblolly pine
[3]SS = seasonally saturated

3.2.2 *Mitigation Goals*

Wetland mitigation activities resulted in a mitigation to impact ratio of more than 2:1 through creation of approximately 9.5 ha of forested, scrub-shrub, and herbaceous wetlands in an abandoned soil borrow pit.

The objective of the wetland creation was to build a system that would: (1) provide natural biological functions including general habitat and nesting, spawning, and rearing sites as described in the list of wetland functions according to the U.S. Federal Register [33 CFR, Part 320.4(b)(2)], (2) serve as a storage area for storm and flood waters, (3) provide ground water discharge/recharge, and (4) create wetland habitat scarce in the local area. These objectives were met by incorporating hydrology, elevation and plant communities of an adjacent alluvial swamp forest into the design of the mitigation wetland.

3.2.3 Methodology

As shown in Figure 4, an existing swamp forest lies immediately southeast of the landfill and adjacent to the soil borrow area where wetlands were created. Prior to modifications, the soil borrow site was sterile sandy to clayey subsoil with no trees, little herbaceous cover, and no hydrologic connection to the adjacent swamp.

As a first step in mitigation wetland design, contour elevations and vegetation species were studied in the adjacent wetland to determine the elevations necessary to enable the created wetland to support hydrophytic vegetation and act as part of one large alluvial system. In order to determine the elevations at which existing wetland plant communities were found, data were collected along a series of 5 line transects surveyed through the reference wetland. These elevations along with the associated plant community were the basis for the grading plan implemented in the 9.5 ha wetland creation area.

The grading plan was designed to link the created wetland to the adjacent reference wetland by using swales that provided hydrologic connection between the two areas. The created wetland has a large meandering aquatic swale through the 9.5 ha area with small sloughs and cypress-gum ponds extending off the main swale. Grading activities in the mitigation area began in the fall of 1991 after the completion of the vegetation and elevation study of the reference wetland.

A planting plan was created from this grading plan that reflected the plant communities and associated elevations found in the reference wetland. Species ranged from those that are adapted to almost permanent inundation to those adapted to seasonally saturated soil conditions (Table IV). Wetland herbaceous and sapling species were planted according to their hydrologic tolerances (Hood, 1984, Taylor, et al., 1990).

TABLE IV

List of scientific and common names of vegetation planted in the South Carolina created wetland. The wetland indicator status (USFWS, 1988) of each species, obligate (OBL), facultative wetland (FACW) and facultative (FAC), indicates the tolerance of the selected species to each hydrologic zone. Planting zones are shown in Figure 5.

Species	Common Name	Indicator Status
AQUATIC WETLAND: ZONE 1		
Peltandra virginica	arrow acrum	OBL
Pontidaria cordata	pickerelweed	OBL
Sagittaria lancifolia	bull-tongue arrowhead	OBL
Sagittaria latifolia	broad leaf arrowhead	OBL
Saururus cernuus	lizard's tail	OBL
Scirpus americanus	three-square bulrush	OBL
Scirpus validus	soft-stem bulrush	OBL
SWAMP FOREST: ZONE 2		
Cephelanthus occidentalis	buttonbush	OBL
Clethera alnifolia	sweet pepperbrush	FACW
Fraxinus pennsylvanica	green ash	FACW
Nyssa aquatica	water tupelo	OBL
Quercus lyrata	overcupoak	OBL
Taxodium distichum	bald cypress	OBL

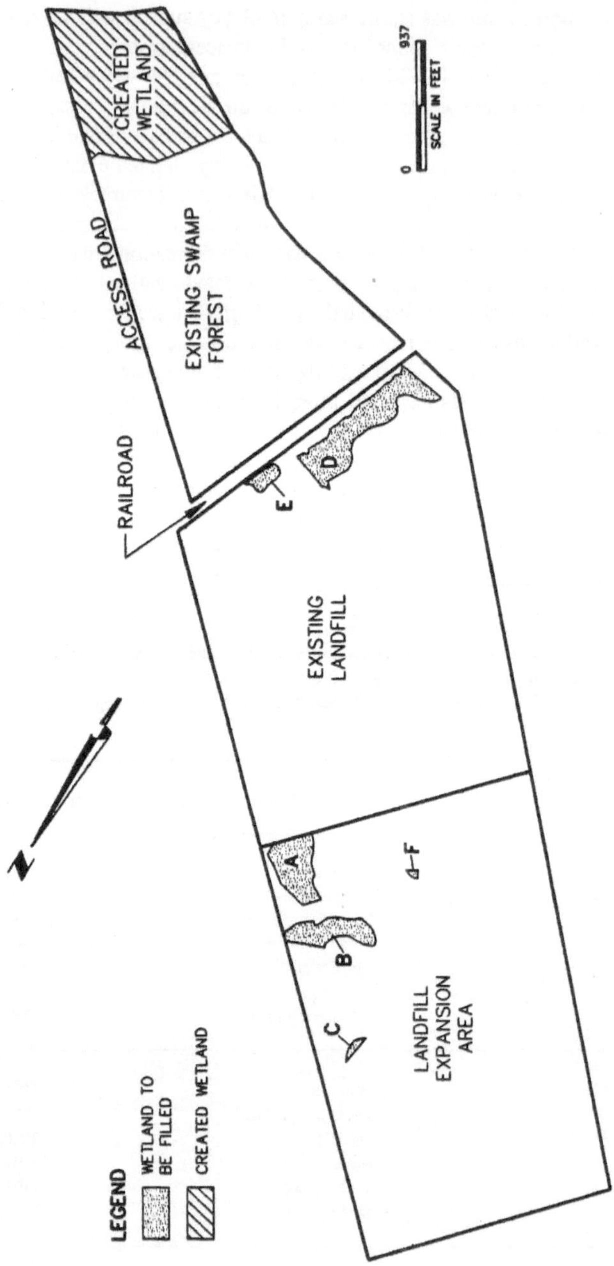

Fig. 4. Location of South Carolina landfill expansion area, adjacent reference wetland, and soil borrow pit where wetlands were created.

Table IV - continued

LOW HARDWOOD FOREST: ZONE 3		
Acer rubrum	red maple	FAC
Cyrilla racemiflora	titi	FACW
Diospyros virginiana	persimmon	FAC
Fraxinus pennsylvania	green ash	FACW
Lyonia lucida	fetterfush	FACW
Quercus michauxii	chestnut oak	FACW
Quercus phellos	willow oak	FACW
MEDIUM HARDWOOD FOREST: ZONE 4		
Acer rubrum	red maple	FAC
Arundinaria gigantea	giant cane	FACW
Cliftonia monophylla	buckwheat-tree	OBL
Diospyros virginiana	persimmon	FAC
Gordonia lasianthus	loblolly bay	FACW
Ilex verticillata	winterberry	FACW
Itea virginica	Virginia willow	FACW
Liquidamber styraciflua	sweetgum	FAC
Myrica cerifera	southern bayberry	FAC
Pinus taeda	loblolly pine	FAC
Quercus nigra	water oak	FAC
Quercus virginiana	live oak	FACU
Rhododendron serrulatum	toothed azalea	OBL
Vaccinium corymbosum	highbush blueberry	FACW
Viburnum dentatum	arrowwood	FAC

Zone 1 is the aquatic bed habitat type characterized by predominantly open water with aquatic emergent vegetation along the water's edge. Planted species included bulrushes, arrowhead, and pickerelweed. Zone 2, the swamp forest hydrologic zone, is represented by cypress/gum ponds and the forested area immediately adjacent to the meandering slough within the 9.5 ha mitigation area. These areas are frequently flooded for long durations and support species that typically occur in inundated areas. Dominant planted species represented in the Zone 2 hydrologic zone are bald cypress, water tupelo, green ash, and overcup oak. Zones 3 and 4, low hardwood wetlands and medium hardwood wetlands, make up the remainder of the created wetland. These areas are seasonally inundated with soils frequently saturated to the surface. This area provides an abundance of plant diversity with the dominant sampling species including swamp chestnut oak, willow oak, red maple, and persimmon. Both bare-root and containerized stock were planted in the mitigation area.

Once the correct grade was constructed, with some portions at low elevation to support aquatic bed habitat and some areas graded to higher elevations, wetland topsoil was spread over the area to an average depth of 15 cm. Topsoil from the wetlands that would be filled had been stockpiled for this use. The wetland creation site was then planted in accordance with the planting plan using a variety of tree, shrub and herbaceous species including both native plant material and nursery grown stock. Hardwood saplings were planted in early spring 1992. The species described above were planted on 3.1 m center

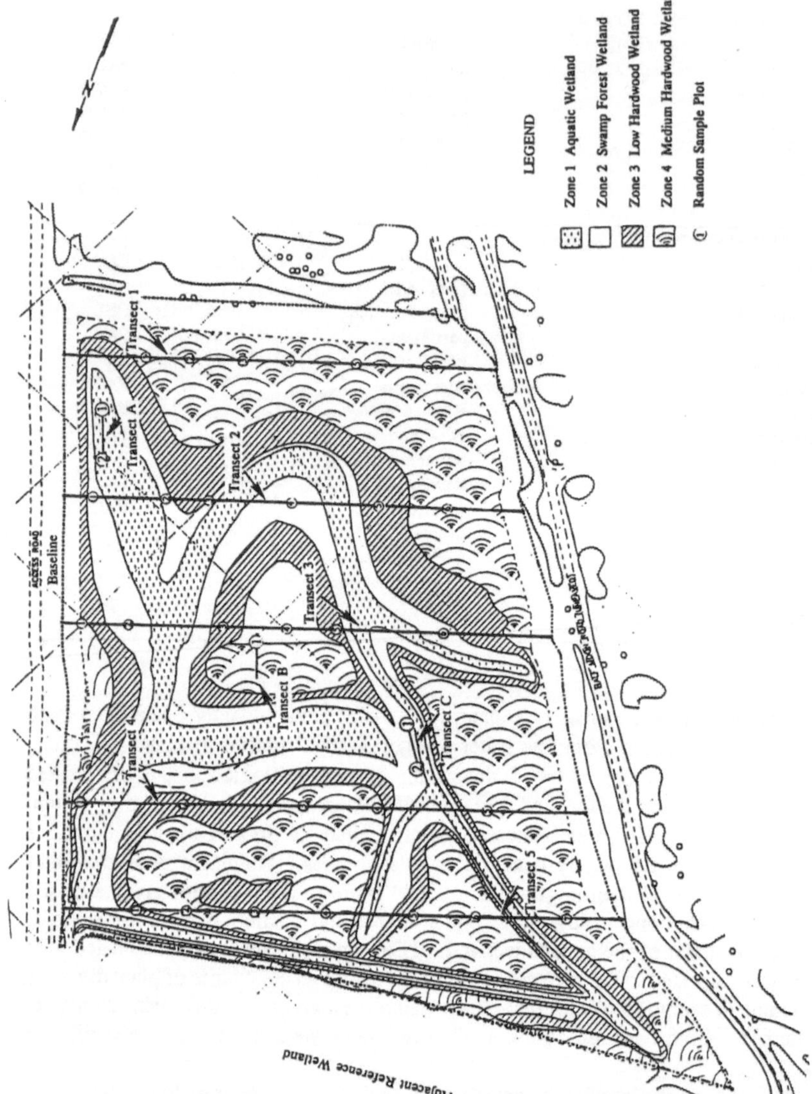

Fig. 5. Configuration of planting zones in South Carolina wetland creation area.

spacings. In addition to wetland creation, wood duck boxes were installed throughout the mitigation area and in the adjacent reference wetland in order to provide nesting structures in a maturing forested wetland.

3.2.4 Monitoring

Wetland monitoring will be carried out for a period of 3 yrs to assess vegetation development and abundance, species composition, survival, and growth of planted saplings. Vegetation monitoring was conducted to evaluate the regeneration, survivability, and success of plant species occurring in conjunction with wetland creation activities.

Five line transects were established perpendicular to topographic rise in the monitored mitigation area (Figure 5). Sample plots were permanently established within every 46 m segment of the transect; plot location was selected randomly by the use of a random numbers table. This methodology was chosen to eliminate bias in identification of data collection points while collecting a representative sample of vegetation located throughout each transect. A minimum of 5 sample plots and a maximum of 7 were positioned along each transect. Due to differences in length of each transect, the number of sample plots along each transect varied. All transects and plots were permanently marked to facilitate identification and future monitoring.

Vegetative species composition and dominance were measured within each sample plot. Trees, saplings, and shrubs were monitored within 9.2 m radius circular plots. All species were identified and the number of stems and percent areal cover recorded for each. Tree density was calculated from the data obtained. Within each plot, a minimum of one representative specimen of each of the dominant species was selected for yearly growth measurements. Total sapling height was measured for a record of annual growth.

Herbaceous vegetation and vines were monitored within 1.5 m radius circular plots along the 5 transects. All herbaceous species were identified and the percent areal cover recorded for each. The mean percent cover was calculated from the data obtained. Plant species present were identified to the lowest taxonomic unit and the National Wetland Inventory indicator status recorded for each.

Because the constructed swale and associated herbaceous aquatic and wetland vegetation meanders across several of the transects, herbaceous vegetation was measured wherever it was encountered on the transect sampling plots. In order to ensure adequate monitoring of this vegetative strata, 3 additional transects with a total of 5 plots were installed along the littoral shelf to determine the success of the planted and naturally regenerating herbaceous plant material along these areas.

The results of measurements recorded at the end of the growing season (September) in 1992 and 1993 are summarized in Table V. Overall, the number of tree and shrub species on each transect stayed the same or increased from 1992 to 1993; the tree density generally increased, tree height increased, and for herbaceous vegetation, both the number of species and % cover increased from 1992 to 1993. In terms of vegetation composition and hydrology, the created wetland is developing into a typical alluvial swamp forest (Nelson, 1985) with the relatively unique feature of having more open water and herbaceous vegetation than might be characteristic of a more mature system.

Based on wildlife observations made in 1992 and 1993, amphibians and reptiles including treefrogs (*Hyla sp.*), bullfrogs (*Rana catesbeiana*), and the American alligator (*Alligator mississippiensis*) use the South Carolina created wetland. White-tailed deer (*Odocoileus virginianus*), raccoon (*Procyon lotor*), and gray squirrel (*Sciurus carolinensis*) have been recorded on site. Additionally, the following 17 species of birds have been observed: American anhinga (*Anhinga anhinga*), mallard (*Anas platyrhyncos*), wood duck

TABLE V

At the end of the growing season in 1992 and 1993 at the South Carolina created wetland, the number of sapling, shrub, and herbaceous species are shown for each transect and the littoral shelf. The mean tree density (range in parenthesis) averaged over each sampling plot in each transect is shown for both years. The mean tree height (and range) and % cover of herbaceous vegetation also are presented. Sampling transect locations are shown in Figure 5.

Transect	1992				1993			
	Species Number	Trees per ha	Tree Height(m)	Herbaceous Cover %	Species Number	Trees per ha	Tree Height m)	Herbaceous Cover %
1	11 sap. 6 sh. 20 herb.	1037 (568-1823)	0.86 (0.18-2.77)	78	12 sap. 7 sh. 52 herb.	1015 (531-1638)	1.02 (0.28-2.77)	104
2	9 sap. 4 sh. 20 herb.	551 (153-911)	0.74 (0.15-2.06)	72	9 sap. 4 sh. 41 herb.	741 (153-1329)	0.83 (0.18-1.26)	95
3	8 sap. 4 sh. 18 herb.	1324 (267-2660)	1.11 (0.15-1.85)	61	8 sap. 5 sh. 33 herb.	1672 (608-3117)	1.23 (0.15-5.08)	85
4	6 sap. 2 sh. 16 herb.	729 (76-1215)	0.65 (0.09-1.23)	42	8 sap. 1 sh. 28 herb.	845 (114-1899)	0.71 (0.28-1.29)	79
5	8 sap. 5 sh. 18 herb.	1786 (647-3648)	0.77 (0.28-2.31)	47	9 sap. 5 sh. 20 herb.	2075 (608-4863)	0.89 (0.40-2.49)	84
Littoral shelf	17 herb.	-	-	63	35 herb.	-	-	87

(*Aix sponsa*), blue-winged teal (*Anas discors*), red-tailed hawk (*Buteo jamaicensis*), eastern wild turkey (*Meleagris gallopavo*), common snipe (*Gallinago gallinago*), great egret (*Casmerodius albus*), bobwhite quail (*Colinus virginianus*), osprey (*Pandion haliaetus*), bald eagle (*Haliaeetus leucocephalus*), wood stork (*Mycteria americana*), killdeer (*Charadrius vociferous*), American coot (*Fulica americana*), Carolina wren (*Thryothorus ludovicianus*), prothonotary warbler (*Protonotaria citrea*), and great crested flycatcher (*Myiarchus crinitus*).

Results from 2 yrs of monitoring indicate that the vegetation is surviving and growing and that a variety of wildlife typically found in wetlands are using the site. The mitigation goals are being met in that this created wetland provides wildlife habitat, serves as a storage area for storm and flood events, is connected to the groundwater table and, therefore, serves seasonal groundwater discharge and recharge functions, and provides a wetland habitat relatively scarce in the local area.

3.2.5 Case Summary: South Carolina Wetland Example

The majority of wetlands impacted by this project were isolated depressional forested wetlands previously impacted by silvicultural activities. Mitigation in this case offered the opportunity to not only replace functions lost from the small impacted wetlands but also to create valuable alluvial swamp forest habitat from the soil borrow area.

Through studies of the hydrology, land elevation, and vegetation at the adjacent reference wetland, design criteria were developed to link the created wetland to the existing system. Wetland hydrology was established on the site through excavation of a meandering slough that has provided habitat for aquatic plants and animals. Wood storks, bald eagles, osprey, and alligators have been documented in the created wetland. Because the existing substrate was originally sterile sands, nutrient and seed-rich hydric soil was spread across the site and now supports lush vegetation. Trees were planted to match the appropriate degree of soil saturation. This created wetland system now provides aquatic and terrestrial habitat for wildlife including protected species. Local school groups frequently visit the site to learn about the value of wetlands, offering a long-term benefit to the community.

4.0 Conclusion

The science of wetland mitigation and particularly of wetland creation is rapidly growing with our increased knowledge of the dynamic nature of these systems. Hydrologic modeling, grading of contours to exact elevation specifications, stockpiling and spreading of suitable topsoil, and the careful selection of appropriate vegetation are important components of wetland construction. With these features in place, it remains necessary to monitor the site over a 3-to 5-yr period to ensure success of the hydrologic modifications and planted vegetation survival.

When the hydrology and vegetation are functioning, other wetland functions will naturally evolve. Wildlife will use the habitat, hydric soils will form, trees and shrubs will slow the velocity of flood waters, water will be absorbed into wetland soils when the site is not fully saturated. The construction of well-designed wetland systems offers a unique opportunity to enhance the environment through mitigation for unavoidable losses.

5.0 References

Adamus, P.R., Clairain, E.J., Smith, R.D., and Young, R.E.: 1987, *Wetland Evaluation Technique*. Volumes I and II. U.S. Army Corps of Engineers Waterways Experiment Station, Vicksburg, MS, 206 pp.

Brinson, M.M.: 1993, Changes in the functions of wetlands along environmental gradients. Wetlands 13:65-74.

Carnell, R. and Anderson, M.A.: 1986, A technique for extensive field measurement of soil anaerobism by rusting steel rods. Forestry 59(2):129-140.

Clewell, A.F., and Lea, R.: 1990, Creation and restoration of forested wetland vegetation in the Southeastern United States, in: Kusler, J.A., and Kentula, M.E. (eds), *Wetland creation and Restoration: The Status of the Science*. Island Press, Washington, D.C. pp. 195-231.

Clewell, A.F.: 1986, Assessment of 5.75 acres north of Old Dike at McMullen Branch Restoration Area, Brewster Phosphates, Bradley, Florida.

Environmental Laboratory: 1987, *Corps of Engineers Wetlands Delineation Manual*. U.S. Army Engineer Waterways Experiment Station, Technical Report Y-87-1, 75 pp.

Gibb, D.M., and Coffman, G.C.: 1993, Soil saturation determination for wetland mitigation monitoring using "rusty rods" - a case study, in: Hatcher, K.J. (ed), Proceedings of the 1993 Georgia Water Resources Conference, The University of Georgia, Athens, GA (in press).

Hammer, Donald A.: 1992, *Creating Freshwater Wetlands*. Lewis Publishers, Inc., Chelsea, MI, 266 pp.

Hook, D.D.: 1984, Waterlogging tolerance of lowland tree species of the south. Southern Journal of Applied Forestry. vol. 8, no. 3, 136-149.

McKee, W.H.: 1978, Rust on iron rods indicates depth to soil water tables, in: Balmer, W.E. (ed), Proc. Soil Moisture Site Productivity Symposium. U.S. Department of Agriculture, pp. 286-291.

National Wetlands Newsletter: 1992, Focus Issue: Wetland Mitigation Banking. Vol. 14:1

Nelson, J.B.: 1985, The natural communities of South Carolina, South Carolina Wildl. and Marine Reserve Dep., Columbia, S.C. 55 pp.

Reed, Sherwood C., chair: 1990, *Natural Systems for Wastewater Treatment*. Water Pollution Control Federation, Alexandria, VA, 260 pp.

Taylor, J.R., Cardamore, M.A., Hitsch, W.J.: 1990, Bottomland hardwood forests: their functions and values, in : Gooselink, J.G. Lee, L.C., and Huir, T.A. (eds), *Ecological Processes and Cumulative Impacts: Illustrated by Bottomland Hardwood Wetland Ecosystems*. Lewis Publishers, Inc., Chelsea, MI, 13-86.

U.S. Army Corps of Engineers, Hydraulic Engineering Center: 1988, Computer Program, HEC-2 Water Surface Profiles, Davis, CA.

U.S. Fish and Wildlife Service: 1988, *National List of Plant Species that Occur in Wetlands: Northeast (Region 1)*. Biological Report 88(26.1), 111 pp.

U.S. Fish and Wildlife Service: 1988, National list of plant species that occur in wetlands: Southeast (Region 2). U.S. Fish Wildl. Serv. Biol. Rep. 88(26.2), 124 pp.

Willard, D.E., Finn, V.T., Levine, D.A., and Klarquist, J.E.: 1990, Creation and restoration of riparian wetlands in the agricultural midwest, in: Kusler, J.A., and Kentula, H.E. (eds), *Wetland Creation and Restoration: The Status of the Science*. Island Press, Washington, D.C. pp. 327-350.

EPW: A PROCEDURE FOR THE FUNCTIONAL ASSESSMENT OF PLANNED WETLANDS

C.C. BARTOLDUS
Environmental Concern Inc., P.O. Box P, St. Michaels, MD, USA.

(Received April 11, 1994; revised June 12, 1994)

Abstract. The practice of compensating wetland losses through wetland construction, restoration, or enhancement has become more commonplace; however, an appropriate method for assessing replacement of wetland function has been lacking. The Evaluation for Planned Wetlands (EPW) was developed to meet this need. It is a rapid assessment procedure which documents and highlights differences between a wetland assessment area and planned wetland based on their capacity to provide six functions: shoreline bank erosion control, sediment stabilization, water quality, wildlife, fish (tidal, non-tidal stream/river, and non-tidal pond/lake), and uniqueness/heritage. The differences between wetlands are expressed in terms of individual elements, Functional Capacity Indices, and Functional Capacity Units. The results provide information on individual design elements and measures of functional capacity which are a necessity under current regulatory programs that require tangible goals and a method for calculating planned wetland size. EPW includes functional assessment models, a procedure for using these models during the planning/mitigation process, and guidelines for functional design.

1. Introduction

Wetland assessment techniques are used for a variety of purposes including impact analysis, watershed management, and priority rating for wetland acquisition and protection. Decisions regarding constructed, restored, and enhanced wetlands are generally based upon policy and professional judgement, without the benefit of an assessment technique because they are time and labor intensive (e.g., Habitat Evaluation Procedure (USFWS, 1980)) or have been deemed inappropriate (Bartoldus *et al.*, 1994). There is a need for a rapid assessment technique which can be used to evaluate planned wetlands, i.e., a design or an implemented design for constructed, restored, or enhanced wetlands. The assessment technique must be suitable for (a) evaluating and comparing functional capacity of the planned wetland and a wetland assessment area (e.g., impact area) and (b) serve as a guide to planned wetland design. The purpose of this paper is to introduce a new technique, the Evaluation for Planned Wetlands (EPW), and to describe how it has been designed to meet these needs.

2. Using an Assessment Technique to Evaluate Planned Wetlands

2.1 BACKGROUND

A planned wetland is designed to increase the capacity of a site to perform one or more functions. It may or may not be associated with a project which requires a permit. In either case, EPW can be used to compare the planned wetland to a wetland assessment area. Since most planned wetlands are a product of permit actions in the United States, EPW will be used most within the framework of the U.S. Army Corps of Engineers and state wetland permit programs. The Corps reviews proposed activities in wetlands pursuant to Section 404 of the Clean Water Act and assesses the impact these activities might have on the capacity of a wetland to perform specific functions. If necessary, an assessment technique may be employed during the alternatives analysis stage using one of several applicable techniques [e.g., HEP (USFWS, 1980); Hollands and McGee, 1986; WET 2.0 (Adamus *et al.*, 1987)]. After measures to avoid or minimize impacts are addressed, then measures to mitigate (compensate) for unavoidable project impacts must be considered. EPW can be used during this stage to set mitigation goals and to determine whether a planned wetland can compensate for functions lost in a wetland assessment area.

2.2 PROPERTIES OF ASSESSMENT TECHNIQUES NECESSARY FOR THE EVALUATION OF PLANNED WETLANDS

The application of several assessment techniques to a variety of planned wetlands revealed basic properties which limit the application of these techniques to the planning/mitigation process (Bartoldus, 1992). An assessment technique must have the following properties to be applicable for the evaluation of a planned wetland:

- Document and display procedure and results to facilitate the design and review of the planned wetland
- Provide validated threshold values for design elements
- Include elements applicable to planned wetland design
- Not use opportunity elements to describe functional capacity
- Be sensitive enough to detect differences between wetlands

The format must allow the designer and decision maker to readily identify elements which are important to each function. Information must be easily extracted so that changes can be made to improve functional capacity in the planned wetland. Threshold values should not be used, unless they can be literature-validated or validated through consultation with experts. Threshold values are cutoff values used in the evaluation, above or below which it is believed that a wetland's capacity to perform a function substantially changes. For example, the assessment technique may operate under the assumption that a ≥ 2 ha wetland will provide the best habitat for the wildlife function; anything less would be considered to have minimal habitat value. Examples of wildlife function thresholds using different assessment techniques include 5 ha for U.S. wetlands (Adamus *et al.*, 1987), 4 ha for Connecticut wetlands (Ammann *et al.*, 1986), 4 ha for North Central U.S. grasslands (USCOE and MEQB, 1988), and 203 ha for North Central forested wetlands (USCOE and MEQB, 1988). If the assessment technique is used as a design guide, then substantially different design criteria could be obtained depending upon the technique employed. Based upon the example given, the recommended minimum wetland size could vary from 2 to 203 ha.

Rapid wetland assessment techniques often use a minimum number (e.g., 3-5) of elements to assess each function. Although it lengthens the assessment time, it is important that elements critical to wetland design also be included in the technique.

Elements describing opportunity should only be used to note conditions which could reduce the planned wetland's functional capacity. Opportunity elements are those characteristics of a wetland or its surroundings which determine if the opportunity is available for that wetland to perform a function. In most procedures, opportunity elements are used with other structural elements to describe functional capacity. The rationale is that the wetland is more valuable when the opportunity for performing the function is present. Many of the opportunity elements describe conditions which, if excessive, could change a wetland's functional capacity. For example, it is often assumed that greater pollutant input makes a wetland more valuable for the water quality function. This assumption may be invalid. Studies on the use of wetlands for wastewater treatment have demonstrated that, after several years, some wetlands which initially served as nutrient sinks reached their assimilatory capacity for certain chemical constituents (e.g., Kadlec, 1985; Girts and Knight, 1989). Unfortunately, past assessment techniques have not set an upper limit on opportunity elements. Without an upper limit, the assessment technique may erroneously assign a high rating when the capacity of the wetland to perform a function may be minimal or exceeded due to excessive pollutant input.

Finally, the assessment technique must be sensitive enough to detect planned improvements in function. Any differences should be expressed in terms of a measure of functional capacity and also detected at the level of individual design element changes.

3. Description of EPW

3.1 FUNCTIONS

EPW provides a process by which a wetland assessment area and a planned wetland can be compared based on their capacity to provide six functions (see Table I). Although Uniqueness/heritage is referred to as a function, it identifies characteristics that are often labelled as red flags or wetland "values". All of the functions listed, whether considered a function or value, represent the major categories most often addressed during the planning process.

3.2 UNITS OF COMPARISON

EPW documents and highlights differences between the wetland assessment area and planned wetland. These differences are expressed in terms of the element score, Functional Capacity Index (FCI), and Functional Capacity Units (FCUs).

The element score is the most basic unit of comparison. EPW uses 7 to 20 elements to evaluate each function, for a total of 81 elements. An element is a physical, chemical, or biological characteristic of the wetland or landscape that dominates the wetland's capacity to perform a function. The element score is a unitless number from 0.0 to 1.0

TABLE I
Definitions of EPW functions

Function	Definition
Shoreline bank erosion control (SB)	capacity to provide erosion control and to dissipate erosive forces at the shoreline bank
Sediment stabilization (SS)	capacity to stabilize and retain previously deposited sediments
Water quality (WQ)	capacity to retain and process dissolved or particulate materials to the benefit of downstream surface water quality
Wildlife (WL)	degree to which a wetland functions as habitat for wildlife as described by habitat complexity
Fish: tidal (FT) non-tidal stream/river (FS) non-tidal pond/lake (FP)	degree to which a wetland habitat meets the food/cover, reproductive, and water quality requirements of fish
Uniqueness/heritage (UH)	presence of characteristics that distinguish a wetland as unique, rare, or valuable

where 1.0 represents the optimal condition for maximizing functional capacity and 0.0 represents the worst condition (see Figure 1). When data sheets are completed, differences between the wetlands are calculated (*Difference = Score for planned wetland - Score for wetland assessment area*). The difference for each element indicates whether the planned wetland improves upon (positive change), replaces (0 change), or does not replace (negative change) these function conditions.

The Functional Capacity Index (FCI) is a measure of functional capacity expressed as an index, where 0.0 represents no functional capacity and 1.0 represents optimal functional capacity. The FCI for a function is based on an assessment model that combines elements scores based on the relationship between elements and the function (Figure 2). The assessment models are founded on a thorough review of the literature and the knowledge and experience of the authors (refer to Bartoldus *et al.*, 1994).

FCIs are used to derive Functional Capacity Units (FCUs), a measure of functional capacity expressed in terms of quantity per unit area, which accounts for difference in space and time. FCUs are calculated as

$$FCI_Y \times AREA_Y = FCUs_Y$$

where FCI_Y = Functional Capacity Index for function "Y", AREA = Size of wetland area with specified FCI for wetland function "Y", and $FCUs_Y$ = Functional Capacity Units for function "Y".

FIGURE 1
Example of element and conditions as presented in EPW data sheets

ELEMENT	SELECTION OF SCORES FOR ELEMENT CONDITIONS	SELECTED SCORES FOR ELEMENTS		DIFFERENCE IN SCORES (Planned-WAA)
		WAA	Planned Wetland	
12b. Ratio of cover types (Consider canopy cover of each cover type in each layer).				
a. Approximately equal proportions.	1.0	0.5	1.0	(+)
b. Intermediate condition.	0.5			
c. Predominantly 1 cover type.	0.1			

FIGURE 2
Relationships of elements and components used in the Wildlife FCI model.

Elements	Components		
Disturbance of wildlife habitat Gross contamination Wetland size		Features which reduce habitat value	Wildlife FCI
Layers Condition of layers Spatial pattern of shrubs/trees	Vegetation strata	Habitat complexity	
Cover types Ratio of cover types Cover type interspersion Disturbance species	Vegetation cover types		
Percent open water Vegetation/water interspersion	Vegetation/water proportions		
Shape of upland/wetland edge Wildlife attractors Islands	Physical features		

TABLE II

Comparison of the Charles County wetland assessment area and planned wetland: NA = Not applicable, FCUs = FCI x AREA, Target FCI = FCI goal established by decision makers, *R = 1.5 (the multiplying factor established by decision makers), and Target FCUs = FCU_{WAA} x R. All goals for the planned wetland were met.

Function	WAA			Goals for Planned Wetland		Planned Wetland			Check if goals met
	FCI	AREA (ha)	FCUs	Target FCI	Target FCUs*	FCI	AREA (ha)	FCUs	
SB	NA					0.71	4.86	3.45	
SS	0.99	2.43	2.41	> 0.80	3.62	0.88	4.86	4.28	✓
WQ	0.82	2.43	1.99	> 0.80	2.99	0.89	4.86	4.33	✓
WL	0.44	2.43	1.07	> 0.40	1.61	0.63	4.86	3.06	✓
FS	0.25	2.43	0.61	> 0.30	0.92	0.60	4.86	2.92	✓
UH	NA					NA			

3.3 CONDUCTING EPW

The process of evaluating and comparing a wetland assessment area and planned wetland involves 7 steps, each of which are described below. A detailed description of EPW is provided in Bartoldus *et al.* (1994). The Charles County project is provided to illustrate the basics operation of EPW. This project involved the loss of 2.4 ha of forested wetland in Maryland, USA due to the expansion of a wastewater treatment plant and a proposal to compensate for this loss with a planned wetland.

3.3.1 Define scope of evaluation (Step 1)

The objectives of the evaluation are defined in sufficient detail to avoid any misunderstanding and unnecessary work. For example, the wetlands to be compared and the number of comparisons are stipulated, in addition to the assumed "stage" of development for the planned wetland. Normally, all applicable functions are evaluated, although there may be a decision to evaluate only select functions. The Charles County project involved the simple evaluation of a single continuous wetland assessment area. The intention was to compare this to the predicted conditions for a single planned wetland at the end of the first growing season.

3.3.2 Characterize Wetland to be Assessed (Step 2)

The project area and specific Wetland Assessment Areas (WAAs) are identified and illustrated on a map. Criteria for distinguishing different WAA include wetland class and spatial separation. The WAA is important to the evaluation process because it provides measures of functional capacity which become the basis for establishing the goals and making comparison to the planned wetland. For the Charles County project, the WAA was the 2.4 ha wetland impact area.

3.3.3 Evaluate wetland assessment area (WAA) (Step 3)
The WAA is evaluated by completing the data sheets, and calculating the FCI and FCUs. Results are then summarized (Table II).

3.3.4 Set goals (Step 4)
Decision makers define the goals of the planned wetland based upon the results of the evaluation on the WAA (Step 3) and federal, state, or local agency recommendations. These goals are expressed in terms of FCI and FCUs. For example, the goal may be to provide the same functions at or above the functional capacity levels recorded for the WAA ($\geq$ FCI and $\geq$ FCUs for the WAA). If the goal is to provide greater compensation (e.g., 2:1 mitigation ratio), then Target FCUs are derived by multiplying the FCUs by the appropriate multiplying factor (e.g., R = 2). In the Charles County project, the goal was to provide greater number of FCUs due to (a) the loss of wildlife functional capacity during the time taken for the planned wetland to achieve a mature forest, and (b) anticipated failure of a portion of the planned wetland from initial plant die off. A multiplying factor (R= 1.5) was used to establish the Target FCUs (Table II). Decision makers also determined that the planned wetland be located in the same watershed and have a minimum size of 4.86 ha.

3.3.5 Select planned wetland site (Step 5)
Potential sites are identified and screened to eliminate unacceptable ones, and final selection is made based upon a more detailed examination. Other factors, besides wetland function, are considered including institutional constraints, economic feasibility, and construction constraints. During this step, EPW is used as a reference document to determine which site(s) can provide or can be modified to provide the conditions necessary to attain the Target FCI and FCUs. The Charles County project involved a review of 18 potential sites. Reasons for rejecting all but one site included the decision not to destroy upland forest wildlife habitat, recreation impacts, exorbitant excavation costs, and uncertainty regarding sufficiency of water source.

3.3.6 Design planned wetland (Step 6)
The conceptual plan is prepared which provides a brief description of the planned wetland through drawings and text. Detailed construction plans and specifications are not required, but the plans should contain sufficient information to perform the assessment. The designer refers to EPW to identify the best conditions for maximizing functional capacity. The Charles County planned wetland is designed to include 4.86 ha adjacent to an existing stream with plantings of a variety of trees saplings and emergents.

3.3.7 Evaluate planned wetland (Step 7)
The planned wetland is evaluated by completing the data sheets, and calculating the FCI and FCUs. Results are recorded in the summary tables (Tables II and III). The evaluation

TABLE III
Charles County project: summary of wildlife function evaluation results

FCI		Elements with different scores		
WAA	Planned Wetland	No.	Difference	Explanation
0.44	0.63	11a	(-)	Fewer layers in planned wetland
		12a	(-)	Fewer cover types in planned wetland
		12b	(+)	Higher percentage of cover with near equal proportions of each over type in planned wetland, compared to dominance of one cover type
		13a	(+)	% open water nearer to 50% optimum in planned wetland
		13b	(+)	greater vegetation/water interspersion in planned wetland
		23	(+)	Islands present in planned wetland

results of the WAA and the planned wetland are compared to determine if the goals are met. For the Charles County planned wetland, the FCI and FCUs goals were exceeded. The comparison of individual elements revealed several differences which contributed to changes in FCIs for each function. For example, the improvement in the Wildlife FCI was primarily attributed to a positive change in four elements (Table III). If the comparison reveals that the goals are not met, the plan should be revised and re-evaluated.

4. Summary

The need for a technique which can serve as a guide to wetland functional design has become more apparent with the increase in mitigation and restoration activities. Environmental Concern Inc. initiated the development of EPW as it became apparent that existing techniques contained characteristics that not only limited their application in the planning/mitigation process, but also lead to unfounded design criteria.

EPW provides an approach and assessment technique which can be used as a tool to support professional judgement in the development of planned wetlands. Major benefits of EPW include: (a) it provides a standardized, reproducible assessment approach for the evaluation of planned wetlands, (b) it is compatible with the time and resources available during the wetland permit process, and (c) the format of the results (e.g., functional capacity indices and units) satisfies the need to establish tangible goals for planned wetlands and wetland banks during the wetland permit process.

A criticism of the EPW, and other assessment techniques, is that the simplified assessment models compromise the need for accuracy and precision to achieve a rapid assessment. Since this compromise is necessary, an effort was made to include guidelines for functional design that go beyond the assessment technique. For example, the Wildlife function assessment model simply looks for the presence or absence of islands. The manual includes a discussion elaborating on recommended island size, shape, distance from the mainland, method of construction, and location based upon the reviewed literature. These guidelines, which are provided for each function and address EPW elements and additional factors, are beneficial for functional design.

Acknowledgements

EPW was prepared by myself, Edgar W. Garbisch, and Mark L. Kraus of Environmental Concern Inc. Partial funding for EPW was provided by the National Cooperative Highway Research Program. I would like to thank Dan Smith, Alan Ammann, and Paul Garrett who provided comments on the draft EPW manual and helped refine the terminology and approach. Thanks are also extended to the anonymous reviewers for their constructive comments.

5. References

Adamus, P.R., Clairain, Jr. E.J., Smith, R.D., and Young, R.E.: 1987, *Wetland Evaluation Techniques (WET). Volume II. Technical Report Y-87.* U.S. Army Corps of Engineers, Waterways Experiment Station, Vicksburg, Mississippi. 206 pp. + Appendices.

Ammann, A.P., Franzen, P.W., and Johnson, J.L.: 1986, *Method for the Evaluation of Inland Wetlands in Connecticut. DEP Bulletin No. 9.* Connecticut Department of Environmental Protection, Hartford, Connecticut. 68 pp. + Appendices.

Bartoldus, C.C.: 1992, EPW: A wetland evaluation procedure for the mitigation process, in: Kusler, J.A., and Lassonde, C. (eds), *Effective Mitigation: Mitigation Banks and Joint Projects in the Context of Wetland Management Plans*, Association of State Wetland Managers, Berne, New York, pp. 144-147.

Bartoldus, C.C., Garbisch, E.W., and Kraus, M.L.: 1994, *Evaluation for Planned Wetlands.* Environmental Concern Inc., St. Michaels, Maryland, 310 pp.

Girts, M.A., and Knight, R.L.: 1989, Operations optimization, in: Hammer, D.A. (ed), *Constructed Wetlands for Wastewater Treatment: Municipal, Industrial, and Agricultural*, Lewis Publishers, Chelsea, Michigan, pp. 417-429.

Hollands, G.H., and McGee, D.W.: 1986, A method for assessing the functions of wetlands, in: Kusler, J.A. (ed), *Proceedings of the National Wetlands Assessment Symposium.* Association of State Wetland Managers, Berne, New York, pp. 108-118.

Kadlec, J.A.: 1985, Aging phenomena in a wastewater wetland, in: Godfrey, P.J., Kaynor, E.R., Pelczarski, S., and Benforado, J. (eds), *Ecological Considerations in Wetlands Treatment of Municipal Wastewaters*, Van Nostrand Reinhold, New York, pp. 338-347.

U.S. Army Corps of Engineers and Minnesota Environmental Quality Board: 1988, *The Minnesota Wetland Evaluation Methodology for the North Central United States.* U.S. Army Corps of Engineers, St. Paul District, Minnesota, 97 pp. + Appendices.

U.S. Fish and Wildlife Service: 1980, *Habitat Evaluation Procedure (HEP) (ESM 102)*, Washington, D.C., 76 pp. + Appendices.

CONFERENCE PARTICIPANTS

SAMAB CONFERENCE ON WETLAND ECOLOGY, MANAGEMENT AND CONSERVATION

September 28-30, 1993
Knoxville, Tennessee

D. Briane Adams
U.S. Geological Survey
3850 Holcomb Bridge Road
Suite 160
Norcross, GA 30092

Bill Ainslie
U.S. Environmental Protection Agency
Region 4
345 Courtland Street
Atlanta, GA 30365

Clif Amundsen
Department of Ecology
University of Tennessee
108 Hoskins Library
Knoxville, TN 37996

Mike Aust
Department of Forestry
Virginia Polytechnic Institute and State University
Blacksburg, VA 24061-0324

Sam Austin
Virginia Department of Forestry
P.O. Box 3758
Charlottesville, VA 22903

Jane Bacchieri
School of the Environment
Duke University
Box 90328
Durham, NC 27708-0328

Hal Bain
North Carolina Department of Transportation
1 S. Wilmington Street
Raleigh, NC 27611

Dave Baker
U.S. Army Corps of Engineers
Regulatory Field Office-Asheville
37 Battery Park
Asheville, NC 28801

Mary Ball
Biology Dept.
Carson-Newman College
P.O. Box 557
Jefferson City, TN 37760

Candy Baroldus
Environmental Concern, Inc.
8710 Margaret Lane
Annandale, VA 22003

Joe Beeler
Biology Dept.
Lincoln Memorial University
Cumberland Gap Parkway
Harrogate, TN 37752

Cyndi Bell
North Carolina Department of Transportation
1 S. Wilmington Street
Raleigh, NC 27611

Joe Bergman
Weyerhaeuser Forestry Research
P.O. Box 1391
New Bern, NC 28562

Gene Berry
University of North Carolina-Asheville
7385 S. Mountain Institute Road
Nebo, NC 28761

Brad Bishop
U.S. Army Corps of Engineers
P.O. Box 1070
Nashville, TN 37202-9752

Wayne Bowman
Martin Marietta Energy Systems, Inc.
104 Union Valley Road
FEDC, MS-8218
Oak Ridge, TN 37831

Allan Boynton
North Carolina Wildlife Resources Committee
209 Ervin Road
Morgantown, NC 28655

W.E. Brode
T.D.O.T.
1190 Bald Eagle Drive
Kingston Springs, TN 37082

Richard Brooks
Department of Health & Natural Resources
Division of Soil and Water
3800 Barrett Drive, Suite 101
Raleigh, NC 27609

Sally Browning
U.S.D.A. Forest Service
8 Sloan Road
Franklin, NC 28734

Bradley Bryan
U.S. Geological Survey
1013 N. Broadway
Knoxville, TN 37917

Marianne Burke
U.S.D.A. Forest Service
Southeastern Forest Experiment Station
2730 Savannah Highway
Charleston, SC 29414

Don Byerly
University of Tennessee
Department of Geological Sciences
Knoxville, TN 37226-1410

Steve Chaplin
U.S. Army Corps of Engineers
Regularoey Field Office-Asheville
37 Battery Park
Asheville, NC 28801

Leo Collins
Tennessee Valley Authority
17 Ridgeway Road
Norris, TN 37828

Cary Coppock
Oak Ridge National Laboratory
P.O. Box 2008, MS-6038
Oak Ridge, TN 37831

Steve Cottrell
Tennessee Valley Authority
17 Ridgeway Road
Norris, TN 37828

David Daugherty
Global Management Services
4779 Highway 58, Suite 7
Chattanooga, TN 37416

Mary Davis
U.S. Army Engineering Waterways Experiment Station
CEWES-ER-W
3909 Halls Ferry Road
Vicksburg, MS 39180-6199

Stan Davis
Tennessee Valley Authority
17 Ridgeway Road
Norris, TN 37828

Ron DeLaune
Louisiana State University
Wetland Biogeochemistry Institute
Baton Rouge, LA 70803

Joyce Dickerman
Oak Ridge National Laboratory
P.O. Box 2008, MS-6038
Oak Ridge, TN 37831

H. Lewis Dorn
Western Carolina Alliance
110 Raoul Road
Highlands, NC 28741

Jeffery R. Duncan
University of Tennessee
1642 Hillwood Drive #G-2
Knoxville, TN 37920

Craig Earnest
Bowater, Inc.
Southern Division Woodlands
5020 Highway 11 South
Calhoun, TN 37903-0188

Gerry Edwards
Tennessee Valley Authority
T.V.A. Forestry Building
Ridgeway Road
Norris, TN 37828

Rhonda Evans
U.S. Environmental Protection Agency
345 Coutland Street
Atlanta, GA 30365

Mark H. Eisenbies
Oak Ridge National Laboratory
P.O. Box 2008, MS-6038
Oak Ridge, TN 37831

Lee B. Ficks, Jr.
U.S. Environmental Protection Agency
Wetlands Division
401 M. Street S.W. (A-104F0)
Washington, DC 20460

Steve Fritts
Barge, Waggoner, Sumner and Cannon
1093 Commerce Park Drive
Oak Ridge, TN 37830

Wes Fuemmeler
Soil Conservation Service
38 Old Hickory Crove, Suite A-100
Jackson, TN 38302-1664

Cindy Gabrielsen
Oak Ridge Nation Laboratory
P.O. Box 2008
Oak Ridge, TN 37831

Tim Gangaware
Tennessee Water Resources Research
Center
422 South Stadium Hall, UT-K
Knoxville, TN 37996

Jennifer Graf
Michael Baker, Jr., Inc.
615 Tennessee Avenue
Charleston, WV 25302

Gary Hahn
Department of Surface Mining
Technical Analyses Station
Frankfort, KY 40601

William R. Harms
U.S.D.A. Forest Service
Southeastern Forest Experiment
Station
2730 Savannah Highway
Charleston, SC 29414

Charles Harris
Chattanooga Stormwater Management
City Hall, Room 225
Chattanooga, TN 37402

Edward J. P. Hauser
Environmental Consulting Services
31 Nichols Hill Drive
Asheville, NC 28804

Michael Hayes
Oak Ridge National Laboratory
P.O. Box 2008, MS-6038
Oak Ridge, TN 37831-6038

John Hefner
U.S. Fish and Wildlife Service
75 Spring Street, S.W.
Atlanta, GA 30303

Bob Hickman
National Park Service
75 Spring Street, S.W.
Atlanta, GA 30303

Stephen G. Hildebrand
Oak Ridge National Laboratory
P.O. Box 2008-MS-6035
Oak Ridge, TN 37831

Hubert Hinote
Southern Appalachian Man and the
Biosphere Program
1314 Cherokee Orchard Road
Gatlinburg, TN 37738

Donal Hook
Clemson University
Department of Forest Resources
Clemson, SC 29634-0362

Robert Hubbard
U.S.D.A
Agricultural Research Service
P.O. Box 946
Tifton, GA 31794

Chuck Hunter
U.S. Fish and Wildlife Service
75 Spring Street, S.W.
Suite 1276
Atlanta, GA 30303

Wesley K. James
Tennessee Valley Authority
17 Ridgeway Road
Norris, TN 37828

Bob Johnson
U.S. Army Corps of Engineers
Regulatory Field Office-Asheville
37 Battery Park
Asheville, NC 28801

Ralph Jordan
Tennessee Valley Authority
17 Ridgeway Road
Norris, TN 37828

Steve Karr
Biology Dept.
Carson-Newman College
P.O. Box 557
Jefferson city, TN 37760

Keith Langdon
Great Smoky Mountains National Park
107 Park Headquarters Road
Gatlinburg, TN 37738

R.J. Lewis
University of Tennessee
Department of Plant and Soil Sciences
Knoxville, TN 37901-1071

Zhijun Liu
Louisiana State University
School of Forestry
Wildlife and Fisheries
Baton rouge, LA 70803

Eva Long
Environmental Protection Agency
345 Courtland Street, NE
Atlanta, GA 30365

Steve Lund
U.S. Army Corps of Engineers
Regulatory Field Office-Asheville
37 Battery Park
Asheville, NC 28801

Linda Mann
Oak Ridge National Laboratory
P.O. Box 2008 MS-6038
Oak Ridge, TN 37831-6038

Howard Marotto
Oak Ridge National Laboratory
P.O. Box 2008 MS-6038
Oak Ridge, TN 37831-6038

George Martin
U.S.D.A. Forest Service
P.O. Box 2010
Cleveland, TN 37320

Alan Mays
Tennessee Valley Authority
17 Ridgeway Road
Norris, TN 37828

Sue McCuskey
Law Environmental, Inc.
112 Townpark Drive
Kennesaw, GA 30144-5599

Forrest McDaniel
U.S. Army Corps of Engineers
P.O. Box 465
Lenoir City, TN 37771

William H. McKee, Jr.
U.S.D.A. Forest Service
Southeastern Forest Experiment Station
2730 Savannah Highway
Charleston, SC 29414

Timothy B. Merritt
U.S. Fish & Wildlife Service
446 Neal Street
Cookeville, TN 38501

Charlie Miller
Miller-McCoy, Inc.
911 Creekside Road
Chattanooga, TN 37406

Tina Mohr
Kemron Environmental Services, Inc.
2300 Wall Street
Suite 6
Cincinnati, OH 45212

Kevin K. Moorehead
University of North Carolina-Asheville
Environmental Studies
One University Heights
Asheville, NC 28804

Patrick Mullholland
Oak Ridge National Laboratory
P.O. Box 2008 MS-6038
Oak Ridge, TN 37831-6038

Nora Murdock
U.S. Fish and Wildlife Service
330 Ridgefield Court
Asheville, NC 28806

Eric A. Nelson
Savannah River Technology Center
WSRC
Buildling 773-42-A
Aiken, SC 29808-0001

Janice Nicholls
U.S. Fish and Wildlife Service
330 Ridgefield Court
Asheville, NC 28806

Wade L. Nutter
University of Georgia
School of Forest Resources
Athens, GA 30602

John Ogden
U.S. Department of Commerce
Economic Development
Administration
401 Peachtree Street NW
Suite 1820
Atlanta, GA 30308--3510

Stacy R. Patton
Olgethorpe Power Corporation
2100 E. Exchange Place
P.O. Box 1349
Tucker, GA 30085-1349

Scott Pearson
Biology Department
Mars College
Mars Hill, NC 28754

S.R. Pezeshki
Louisiana State University
Wetland Biogeochemistry Institute
Baton Rouge, LA 70803

Tom Phillips
B.U.R.N.T.
Box 40041
Nashville, TN 37204

Dan Pittillo
Western Carolina University
Biology Department
Colluwhee, NC 28723

Pat Presley
Oak Ridge National Laboratory
P.O. Box 2008 MS-6038
Oak Ridge, TN 37831-6038

Burline Pullin
Tennessee Valley Authority
17 Ridgeway Road
Norris, TN 37828

Wendy Hudson Ramsey
Oak Ridge National Laboratory
P.O. Box 2008 MS-6038
Oak Ridge, TN 37831-6038

Steve Reaves
U.S. Army Corps of Engineers
Federal Building
400 Wells Street, NE, Room 234
Decatur, AL 35601-9990

Barbara Howell Rector
Tennessee Department of Environment
and Conservation
761 Emory Valley Road
Oak Ridge, TN 37831

Jim Renner
Golder Associates
3730 Chamblee Tucker Rd.
Atlanta, GA 30341

Curtis J. Richardson
Duke Wetland Center
Duke University
Durham, NC 27706

Tom Roberts
Tennessee Technological University
P.O. Box 5063
Cookeville, TN 38505

Janet Rock
Great Smoky Mountains National Park
107 Park Headquarters Road
Gatlinburg, TN

Barbara Rosensteel
Oak Ridge National Laboratory
P.O. Box 2008 MS-6038
Oak Ridge, TN 37831-6038

Marti Salk
Oak Ridge National Laboratory
P.O. Box 2008 MS-6038
Oak Ridge, TN 37831-6038

Stephen Schoenholtz
Department of Forestry
Mississippi State University
P.O. Box 9681
Mississippi State, MS 39762

Tom Scott
Chattanooga Stormwater Management
City Hall, Room 225
Chattanooga, TN 37402

Ted Shear
Department of Forestry
North Carolina State University
Box 8008
Raleigh, NC 27695-8008

Jim Shepard
NCASI
P.O. Box 141020
Gainesville, FL 32614-1020

Cherri Lee Smith
DEHNR
Division of Environmental
Management
P.O. Box 29535
Raleigh, NC 27626-0535

Dan Smith
U.S. Army Engineering Waterways
Experiment Station
CEWES-ER-W
3909 Halls Ferry Road
Vicksburg, MS 39180-6199

Judy Stout
Dauphin Island Sea Laboratory
P.O. Box 369
Dauphin Island, AL 36528

Lori Sutter
North Carolina Div. of Coastal
Mangement
P.O. Box 27687
Raleigh, NC 27615-7225

David Szymanski
School of the Environment
Duke University
Box 90328
Durham, NC 27706

Bob Trentham
Biology Department
Carson-Newman College
P.O. Box 557
Jefferson City, TN 37760

Carl Trettin
U.S.D.A. Forest Service
Southeastern Forest Experiment
Station
2730 Savannah Highway
Charleston, SC 29414

Liz Upchurch
Center for Geography &
Environmental Education
212 Claxon Education Buildling
Knoxville, TN 37996-3400

Charles Van Sickle
U.S.D.A. Forest Service
Southeastern Forest Experiment
Station
P.O. Box 2680
Asheville, NC 28802

James Wakeley
U.S. Army Eingineering Waterways
Experiment Station
CEWES-ER-W
3909 Halls Ferry Road
Vicksburg, MS 39180-6199

Mark Walbridge
Department of Biology
George Mason University
440 University Drive
Fairfax, VA 22030-4444

Alan Weakley
North Carolina Natural Heritage
Program
P.O. Box 27687
Raleigh, NC 27611

Mike Wefer
Tennessee Valley Authority
17 Ridgeway Road
Norris, TN 37828

Joe Wisniewski
The Journal of Water, Air and Soil
Pollution
6862 McLean Province Circle
Falls Church, VA 22043

Garnet Wood
U.S.D.A. Forest Service
100 Vaught Rd.
Winchester, KY 40391

Joseph B. Yavitt
Carnell University
Department of Natural Resources
Fernow Hall
Ithaca, NY 14853

Alan Yeakley
Virginia Polytechnic Institute and State
University
Department of Biology
Blacksburg, VA 24061

AUTHOR INDEX

SUBJECT INDEX